Seemingly Unrelated Regression Equations Models

STATISTICS: Textbooks and Monographs

A SERIES EDITED BY

D. B. OWEN, Coordinating Editor

Department of Statistics
Southern Methodist University
Dallas, Texas

Vol. 1: The Generalized Jackknife Statistic, *H. L. Gray and W. R. Schucany*
Vol. 2: Multivariate Analysis, *Anant M. Kshirsagar*
Vol. 3: Statistics and Society, *Walter T. Federer*
Vol. 4: Multivariate Analysis: A Selected and Abstracted Bibliography, 1957-1972, *Kocherlakota Subrahmaniam and Kathleen Subrahmaniam* (out of print)
Vol. 5: Design of Experiments: A Realistic Approach, *Virgil L. Anderson and Robert A. McLean*
Vol. 6: Statistical and Mathematical Aspects of Pollution Problems, *John W. Pratt*
Vol. 7: Introduction to Probability and Statistics (in two parts), Part I: Probability; Part II: Statistics, *Narayan C. Giri*
Vol. 8: Statistical Theory of the Analysis of Experimental Designs, *J. Ogawa*
Vol. 9: Statistical Techniques in Simulation (in two parts), *Jack P. C. Kleijnen*
Vol. 10: Data Quality Control and Editing, *Joseph I. Naus* (out of print)
Vol. 11: Cost of Living Index Numbers: Practice, Precision, and Theory, *Kali S. Banerjee*
Vol. 12: Weighing Designs: For Chemistry, Medicine, Economics, Operations Research, Statistics, *Kali S. Banerjee*
Vol. 13: The Search for Oil: Some Statistical Methods and Techniques, *edited by D. B. Owen*
Vol. 14: Sample Size Choice: Charts for Experiments with Linear Models, *Robert E. Odeh and Martin Fox*
Vol. 15: Statistical Methods for Engineers and Scientists, *Robert M. Bethea, Benjamin S. Duran, and Thomas L. Boullion*
Vol. 16: Statistical Quality Control Methods, *Irving W. Burr*
Vol. 17: On the History of Statistics and Probability, *edited by D. B. Owen*
Vol. 18: Econometrics, *Peter Schmidt*
Vol. 19: Sufficient Statistics: Selected Contributions, *Vasant S. Huzurbazar (edited by Anant M. Kshirsagar)*
Vol. 20: Handbook of Statistical Distributions, *Jagdish K. Patel, C. H. Kapadia, and D. B. Owen*
Vol. 21: Case Studies in Sample Design, *A. C. Rosander*
Vol. 22: Pocket Book of Statistical Tables, *compiled by R. E. Odeh, D. B. Owen, Z. W. Birnbaum, and L. Fisher*
Vol. 23: The Information in Contingency Tables, *D. V. Gokhale and Solomon Kullback*
Vol. 24: Statistical Analysis of Reliability and Life-Testing Models: Theory and Methods, *Lee J. Bain*
Vol. 25: Elementary Statistical Quality Control, *Irving W. Burr*
Vol. 26: An Introduction to Probability and Statistics Using BASIC, *Richard A. Groeneveld*
Vol. 27: Basic Applied Statistics, *B. L. Raktoe and J. J. Hubert*
Vol. 28: A Primer in Probability, *Kathleen Subrahmaniam*
Vol. 29: Random Processes: A First Look, *R. Syski*

Vol. 30: Regression Methods: A Tool for Data Analysis, *Rudolf J. Freund and Paul D. Minton*
Vol. 31: Randomization Tests, *Eugene S. Edgington*
Vol. 32: Tables for Normal Tolerance Limits, Sampling Plans, and Screening, *Robert E. Odeh and D. B. Owen*
Vol. 33: Statistical Computing, *William J. Kennedy, Jr. and James E. Gentle*
Vol. 34: Regression Analysis and Its Application: A Data-Oriented Approach, *Richard F. Gunst and Robert L. Mason*
Vol. 35: Scientific Strategies to Save Your Life, *I. D. J. Bross*
Vol. 36: Statistics in the Pharmaceutical Industry, *edited by C. Ralph Buncher and Jia-Yeong Tsay*
Vol. 37: Sampling from a Finite Population, *J. Hajek*
Vol. 38: Statistical Modeling Techniques, *S. S. Shapiro*
Vol. 39: Statistical Theory and Inference in Research, *T. A. Bancroft and C.-P. Han*
Vol. 40: Handbook of the Normal Distribution, *Jagdish K. Patel and Campbell B. Read*
Vol. 41: Recent Advances in Regression Methods, *Hrishikesh D. Vinod and Aman Ullah*
Vol. 42: Acceptance Sampling in Quality Control, *Edward G. Schilling*
Vol. 43: The Randomized Clinical Trial and Therapeutic Decisions, *edited by Niels Tygstrup, John M. Lachin, and Erik Juhl*
Vol. 44: Regression Analysis of Survival Data in Cancer Chemotherapy, *Walter H. Carter, Jr., Galen L. Wampler, and Donald M. Stablein*
Vol. 45: A Course in Linear Models, *Anant M. Kshirsagar*
Vol. 46: Clinical Trials: Issues and Approaches, *edited by Stanley H. Shapiro and Thomas H. Louis*
Vol. 47: Statistical Analysis of DNA Sequence Data, *edited by B. S. Weir*
Vol. 48: Nonlinear Regression Modeling: A Unified Practical Approach, *David A. Ratkowsky*
Vol. 49: Attribute Sampling Plans, Tables of Tests and Confidence Limits for Proportions, *Robert E. Odeh and D. B. Owen*
Vol. 50: Experimental Design, Statistical Models, and Genetic Statistics, *edited by Klaus Hinkelmann*
Vol. 51: Statistical Methods for Cancer Studies, *edited by Richard G. Cornell*
Vol. 52: Practical Statistical Sampling for Auditors, *Arthur J. Wilburn*
Vol. 53: Statistical Signal Processing, *edited by Edward J. Wegman and James G. Smith*
Vol. 54: Self-Organizing Methods in Modeling: GMDH Type Algorithms, *edited by Stanley J. Farlow*
Vol. 55: Applied Factorial and Fractional Designs, *Robert A. McLean and Virgil L. Anderson*
Vol. 56: Design of Experiments: Ranking and Selection, *edited by Thomas J. Santner and Ajit C. Tamhane*
Vol. 57: Statistical Methods for Engineers and Scientists. Second Edition, Revised and Expanded, *Robert M. Bethea, Benjamin S. Duran, and Thomas L. Boullion*
Vol. 58: Ensemble Modeling: Inference from Small-Scale Properties to Large-Scale Systems, *Alan E. Gelfand and Crayton C. Walker*
Vol. 59: Computer Modeling for Business and Industry, *Bruce L. Bowerman and Richard T. O'Connell*
Vol. 60: Bayesian Analysis of Linear Models, *Lyle D. Broemeling*
Vol. 61: Methodological Issues for Health Care Surveys, *Brenda Cox and Steven Cohen*
Vol. 62: Applied Regression Analysis and Experimental Design, *Richard J. Brook and Gregory C. Arnold*

Vol. 63: Statpal: A Statistical Package for Microcomputers – PC-DOS Version for the IBM PC and Compatibles, *Bruce J. Chalmer and David G. Whitmore*
Vol. 64: Statpal: A Statistical Package for Microcomputers – Apple Version for the II, II+, and IIe, *David G. Whitmore and Bruce J. Chalmer*
Vol. 65: Nonparametric Statistical Inference, Second Edition, Revised and Expanded, *Jean Dickinson Gibbons*
Vol. 66: Design and Analysis of Experiments, *Roger G. Petersen*
Vol. 67: Statistical Methods for Pharmaceutical Research Planning, *Sten W. Bergman and John C. Gittins*
Vol. 68: Goodness-of-Fit Techniques, *edited by Ralph B. D'Agostino and Michael A. Stephens*
Vol. 69: Statistical Methods in Discrimination Litigation, *edited by D.H. Kaye and Mikel Aickin*
Vol. 70: Truncated and Censored Samples from Normal Populations, *Helmut Schneider*
Vol. 71: Robust Inference, *M. L. Tiku, W. Y. Tan, and N. Balakrishnan*
Vol. 72: Statistical Image Processing and Graphics, *edited by Edward J. Wegman and Douglas J. DePriest*
Vol. 73: Assignment Methods in Combinatorial Data Analysis, *Lawrence J. Hubert*
Vol. 74: Econometrics and Structural Change, *Lyle D. Broemeling and Hiroki Tsurumi*
Vol. 75: Multivariate Interpretation of Clinical Laboratory Data, *Adelin Albert and Eugene K. Harris*
Vol. 76: Statistical Tools for Simulation Practitioners, *Jack P. C. Kleijnen*
Vol. 77: Randomization Tests, Second Edition, *Eugene S. Edgington*
Vol. 78: A Folio of Distributions: A Collection of Theoretical Quantile-Quantile Plots, *Edward B. Fowlkes*
Vol. 79: Applied Categorical Data Analysis, *Daniel H. Freeman, Jr.*
Vol. 80: Seemingly Unrelated Regression Equations Models : Estimation and Inference, *Virendra K. Srivastava and David E. A. Giles*

ADDITIONAL VOLUMES IN PREPARATION

Seemingly Unrelated Regression Equations Models

Estimation and Inference

Virendra K. Srivastava

Department of Statistics
Lucknow University
Lucknow, India

David E. A. Giles

Department of Economics
and Operations Research
University of Canterbury
Christchurch, New Zealand

CRC Press
Taylor & Francis Group
Boca Raton London New York

CRC Press is an imprint of the
Taylor & Francis Group, an **informa** business

First published 1987 by Marcel Dekker, Inc.

Published 2019 by CRC Press
Taylor & Francis Group
6000 Broken Sound Parkway NW, Suite 300
Boca Raton, FL 33487-2742

First issued in paperback 2019

ISBN 13: 978-0-367-45148-6 (pbk)
ISBN 13: 978-0-8247-7610-7 (hbk)

Visit the Taylor & Francis Web site at
http://www.taylorandfrancis.com

and the CRC Press Web site at
http://www.crcpress.com

Library of Congress Cataloging-in-Publication Data

Srivastava, Virendra K.
Seemingly unrelated regression equations models.

Includes bibliographies and index.
1. Estimation theory. 2. Least squares. 3. Regression analysis. I. Giles, David E. A.
II. Title.
QA276.8.S65 1987 511.4 87-6672
ISBN 0-8247-7610-0

To the memory of my parents (V. K. S.)

To my sister, Kathryn (D. E. A. G.)

Preface

The purpose of this book is to bring together, for the first time, the rather scattered literature associated with the seemingly unrelated regression equations (SURE) model used by econometricians and others. Since it was first discussed by Arnold Zellner in 1962, this statistical model and various associated estimators, test statistics, and generalizations have generated a substantial body of literature. Although much of this written work has appeared in the econometrics journals, much of it has been less accessible to applied researchers as many important results have been confined to the various mathematical statistics outlets. In providing what we believe to be a relatively comprehensive survey we hope that applied economists and others whose work leads them to use the SURE model will now have good access to this important body of literature.

We have attempted to discuss this literature in some detail in this book. Clearly, there will be omissions, and our emphasis will not satisfy every reader, but we believe that we have given a fair representation of the development of research associated with the SURE model and its current state. We have focused our attention on the theoretical statistical results associated with the SURE model. In particular we have emphasized finite-sample results wherever these are available. Although the SURE model has sound practical motivations in economic analysis and in the social sciences generally, we have not devoted space to the numerous empirical applications of the model to be found in the literature.

We recognize that a concise and unified précis of estimation and inference in the context of the SURE model may be given by viewing the latter as a multivariate regression model with an unknown disturbance covariance matrix and particular linear cross-equation restrictions on the structural parameters. This being so, standard frequentist (e.g., likelihood-based) or Bayesian estimators and tests may be constructed. However, the fact remains that the estimation of the SURE model's parameters has spawned a variety of procedures which differ in terms of their finite-sample properties, and so a summary and assessment of these procedures is in order.

Our purpose in writing this book, then, is essentially twofold. First, we have attempted to summarize and interpret the large body of literature associated with this important statistical model, and in doing so we have attempted to provide a relatively exhaustive bibliography. Second, we have tried to point out those aspects of the analysis of the SURE model that deserve further attention. We hope that this will serve as a warning to applied workers of the limitations of our knowledge in this area, and also provide opportunities for those who wish to extend the theoretical results associated with the SURE model. We hope that the outcome is a book that will assist graduate students in coming to grips with the literature associated with the SURE model, and benefit researchers in theoretical econometrics and various applied areas by providing them with a comprehensive source of reference material.

The book's layout is straightforward. After introducing the SURE model and some of its most common estimators we move to a discussion of the latter's asymptotic and finite-sample properties. A broad class of iterative estimation procedures is then discussed before we turn to more specialized extensions of the model and its estimators. Each chapter contains a set of exercises and a comprehensive set of references, and equations are numbered sequentially within chapters, prefixed by the chapter number. The reader should note the distinction between the asymptotic distribution (and its moments) of an estimator and the large sample asymptotic approximation to the exact distribution (and its moments) of that estimator.

This book was prepared while the second author was Professor of Econometrics at Monash University. The Faculty of Economics and Politics at Monash University provided financial support for a visiting appointment for Srivastava while the manuscript was being finalized, and for research assistance. We are grateful to Tony Hall, Grant Hillier, Chan Low, Peter Phillips, Jim Richmond, Stephen Satchell, Dudley Wallace, and others for giving us permission to refer to their unpublished work. Jan Kmenta, Chan Low, Aman Ullah, and Arnold Zellner read various parts of the manuscript and provided us with constructive criticism, questions, and valuable comments which helped us to clarify a number of issues. Chris Skeels read the entire draft manuscript with great care and patience, and saved us from numerous errors. In thanking all of these people we take full responsibility for the remaining shortcomings of the finished product.

We would like to acknowledge the assistance of Vickie Kearn of Marcel Dekker, Inc., and the dedicated work of several typists at Monash University. Hellen Collens and Lesley Eamsophana typed some early draft material and Mary Englefield provided secretarial assistance throughout the preparation of the entire book. Our special thanks go to Carol Clark for the skill, speed, and patience with which she converted a very difficult manuscript into final copy. Our task would have been impossible without her outstanding efforts. Finally, we thank our families, and especially Kumud and Pat, for their consistent support and encouragement of this project and for the various sacrifices which they have made to enable us to complete the task of writing this book.

Virendra K. Srivastava

David E. A. Giles

Contents

Seemingly Unrelated Regression Equations Models

1
The Seemingly Unrelated Regression Equations Model

1.1 INTRODUCTION

This book deals with a variety of estimation and inferential issues associated with a particular type of statistical model that has broad applicability in the analysis of behaviour in the social sciences and elsewhere. This model takes account of the fact that subtle interactions often may be present between individual statistical relationships when each of these relationships is being used to model some aspect of behaviour. To make the discussion of this point more specific we shall consider a set of individual linear multiple regression equations, each "explaining" some economic phenomenon. This set of regression equations is said to comprise a simultaneous equations model if one or more of the regressors (explanatory variables) in one or more of the equations is itself the dependent (endogenous) variable associated with another equation in the full system. On the other hand, suppose that none of the variables in the system are simultaneously both explanatory and dependent in nature. There may still be interactions between the individual equations if the random disturbances associated with at least some of the different equations are correlated with each other. That is, the equations may be linked statistically, even though not structurally, through the jointness of the error terms' distribution and through the non-diagonality of the associated variance covariance matrix. This possibility was discussed in a seminal paper by Zellner (1962), who coined the expression "seemingly unrelated regression equations" (SURE) to reflect the fact that the individual equations are in fact related to one another, even though superficially they may not seem to be.

The jointness of the equations which comprise the structure of the SURE model, and the form of the associated disturbances' variance covariance matrix, introduce additional information over and above that available when the individual equations are considered separately. This suggests that treating the model as a collection of separate relationships will be suboptimal when drawing inferences about the model's parameters. Indeed, as we shall see, in general the sharpness of these inferences may be improved by taking account of the jointness inherent in the SURE model, rather than ignoring it. It was the recognition of this point, especially with regard to the asymptotic efficiency of estimators of the parameters of the SURE model, that motivated Zellner's original study.

The precise mathematical details of the SURE model, and the underlying assumptions which will form at least the initial basis for our discussion, are given in the next section. However, before examining these it is of interest to consider some specific examples of economic phenomena and models which may give rise to a SURE specification. A first example is that used by Zellner (1962) to illustrate his proposed SURE estimator, and subsequently discussed by such authors as Kmenta (1971; pp. 527-528) and Theil (1971; pp. 295-302). Separate regression equations were specified to explain investment on the part of two large corporations. In each case, real gross investment by the firm is supposed to be determined by the value of its outstanding shares at the beginning of that period, and by the opening value of the firm's real capital stock. The firms in question were General Electric and Westinghouse. It seems reasonable to suppose that the error terms associated with these two investment equations may be contemporaneously correlated, given the presence of common market forces influencing each of these firms and the likelihood that similar factors may be responsible for the random effects necessitating the inclusion of the error terms in the regressions. For instance, if the error term in the first equation reflects (at least in part) the omission of some unobservable variables then these same variables, or others which are highly correlated with them, may be important determinants of the variability of the error term in the other equation. Thus, the two equations are apparently or "seemingly" unrelated regressions, in the sense described earlier, rather than independent relationships. Zellner (1971; pp. 244-246) provided a Bayesian analysis of an extension of this problem involving ten corporations.

As a second example, suppose that we wish to estimate simple demand relationships, for a particular commodity, for each of several household types. For instance, either cross-section or time-series data may be available in a form enabling us to specify and estimate separate demand equations for households of different ethnic background, or which differ in terms of the length of time that the household head has been resident in the country in question. Price and income data may be viewed as exogenous at the household level, but again one might anticipate an inherent jointness of the set of demand equations through their error variance covariance structure and the equations may be viewed as a SURE model. A similar situation can arise when considering a set of demand equations for different commodities, although in this case Engel aggregation implies certain singular restrictions on the variance covariance matrix which may complicate the estimation of the model (see Powell, 1969).

Finally, a common situation which may suggest a SURE specification is where regression equations explaining a certain economic activity in different geographical locations are to be estimated. For instance, Giles and Hampton (1984) considered Cobb-Douglas production functions for five different regions of New Zealand during the period of that country's industrial development, and used a SURE framework to allow for the inter-regional correlations likely to exist between the regressions' error terms. Similarly, Donnelly (1982) used the SURE model as the basis for estimating petrol demand equations for six different Australian states; White and Hewings (1982) used the SURE model to estimate (for each of several industrial categories) employment equations for five multi-county regions within the State of Illinois; and Giles and Hampton (1986) used an extended SURE model to estimate demand systems for four expenditure groups across six regions of New Zealand.

Other studies involving the SURE model abound, but these examples should illustrate the wide range of empirical applications for which this model is appropriate. In several of the studies the authors concerned compared conventional (e.g., ordinary least squares (OLS)) parameter estimates with those obtained by estimating the equations as a SURE system. Frequently, the differences in the estimates obtained are significant, not only in the usual statistical sense, but also in terms of their more general (e.g., economic) implications or interpretations. This empirical evidence reinforces the theoretical results, discussed in detail in this book, in support of adopting the SURE model framework in a variety of situations. The SURE model is continuing to receive widespread attention, in terms of both theoretical developments and empirical applications. In our view this attention is amply justified, and the principal objective of this book is to provide a comprehensive survey and discussion of the theoretical results and literature that this model has generated, rather than present many new results.

1.2 THE MODEL

The basic model that we are concerned with comprises M multiple regression equations:

$$y_{ti} = \sum_{j=1}^{K_i} x_{tij}\beta_{ij} + u_{ti} \qquad (1.1)$$

$$(t = 1, \ldots, T;\ i = 1, \ldots, M)$$

where y_{ti} is the t'th observation on the i'th dependent variable (the variable to be "explained" by the i'th regression equation); x_{tij} is the t'th observation on the j'th regressor or explanatory variable appearing in the i'th equation; β_{ij} is the coefficient associated with x_{tij} at each observation; and u_{ti} is the t'th value of the random disturbance term associated with the i'th equation of the model.

In matrix notation, this M-equation model may be expressed more compactly as

$$y_i = X_i\beta_i + u_i \quad (i = 1, \ldots, M), \qquad (1.2)$$

where y_i is a $(T \times 1)$ vector with typical element y_{ti}; X_i is a $(T \times K_i)$ matrix, each column of which comprises the T observations on a regressor in the i'th equation of the model; β_i is a $(K_i \times 1)$ vector with typical element β_{ij}; and u_i is the corresponding $(T \times 1)$ disturbance vector.
By writing (1.2) as

$$\begin{bmatrix} y_1 \\ y_2 \\ \vdots \\ y_M \end{bmatrix} = \begin{bmatrix} X_1 & 0 & \cdots & 0 \\ 0 & X_2 & \cdots & 0 \\ \vdots & \vdots & & \vdots \\ 0 & 0 & \cdots & X_M \end{bmatrix} \begin{bmatrix} \beta_1 \\ \beta_2 \\ \vdots \\ \beta_M \end{bmatrix} + \begin{bmatrix} u_1 \\ u_2 \\ \vdots \\ u_M \end{bmatrix},$$

the model may be expressed in the compact form

$$y = X\beta + u, \qquad (1.3)$$

where y is $(TM \times 1)$, X is $(TM \times K^*)$, β is $(K^* \times 1)$, u is $(TM \times 1)$, and $K^* = \Sigma_i K_i$.

Treating each of the M equations as classical linear regression relationships, we make the conventional assumptions about the regressors:

$$X_i \text{ is fixed, with rank } (X_i) = K_i \qquad (1.4)$$

$$\lim_{T\to\infty} \left(\frac{1}{T} X_i' X_i\right) = Q_{ii} \qquad (1.5)$$

where Q_{ii} is non-singular with fixed and finite elements (i = 1, ..., M).

Further, we assume that the elements of the disturbance vector, u_i, follow a multivariate probability distribution with

$$E(u_i) = 0 \tag{1.6}$$

$$E(u_i u_i') = \sigma_{ii} I_T \quad (i = 1, \ldots, M). \tag{1.7}$$

Here, σ_{ii} represents the variance of the random disturbance in the i'th equation for each observation in the sample, I_T is an identity matrix of order T, and $E(\cdot)$ denotes the usual expectation operation.

Considering the interactions between the M equations of the model, it is assumed that

$$\lim_{T \to \infty} \left(\frac{1}{T} X_i' X_j\right) = Q_{ij} \tag{1.8}$$

$$E(u_i u_j') = \sigma_{ij} I_T \quad (i, j = 1, \ldots, M) \tag{1.9}$$

where Q_{ij} is non-singular with fixed and finite elements, and σ_{ij} represents the covariance between the disturbances of the i'th and j'th equations for each observation in the sample.

Writing (1.6), (1.7) and (1.9) more compactly, we have

$$E(u) = 0 \tag{1.10}$$

$$E(uu') = \begin{bmatrix} \sigma_{11} I_T & \cdots & \sigma_{1M} I_T \\ \vdots & & \vdots \\ \sigma_{M1} I_T & \cdots & \sigma_{MM} I_T \end{bmatrix} \tag{1.11}$$

$$= (\Sigma \otimes I_T) = \Psi,$$

where $\otimes$ denotes the usual Kronecker product (so that Ψ is $(MT \times MT)$), and $\Sigma = ((\sigma_{ij}))$ is an $(M \times M)$ positive definite symmetric matrix. The definiteness of Σ precludes the possibility of any linear dependencies among the contemporaneous disturbances in the M equations of the model.

From the form of (1.11) we see that the variance of u_{ti} is assumed to be constant for all t; the contemporaneous covariance between u_{ti} and u_{tj} is constant for all t; and the intertemporal covariances between u_{ti} and $u_{\tau j}$ $(t \neq \tau)$ are zero for all i and j. In adopting the conventional terminology of "contemporaneous" and "intertemporal" covariances we are implicitly assuming that the data are available in time-series form. However, this is not restrictive. The assumptions of the model and the results which follow apply equally if cross-section data are employed. The constancy of the contemporaneous covariances across sample points is a natural generalization of the usual assumption of homoscedastic disturbances in the single-equation linear model, and the zero intertemporal covariances generalize the usual serial independence assumption.

The reason why model (1.1), under these assumptions, is usually referred to as the seemingly unrelated regression equations model is now clear. Although the M equations of the model may appear to be unrelated, in the sense that there is no simultaneity between the variables in the system and each equation purports to explain its own dependent variable by means of some (generally different) set of regressors, none the less the equations are related stochastically through their disturbance terms. These disturbances are temporally (serially) uncorrelated both within and across equations, but they are contemporaneously correlated across the equations of the model.

The relationship between the SURE model and other types of statistical or econometric models is readily seen. First, if in fact the disturbances in different equations are uncorrelated, then the model amounts to a collection of individual multiple regression equations, each of which may be estimated separately. This situation is discussed in the next chapter.

Secondly, the SURE model is a special case of the simultaneous equations model, one involving M structural equations with M jointly dependent and K ($\geq K_i$, for all i) distinct exogenous variables, and in which neither current nor lagged endogenous variables appear as regressors in any of the structural equations. Alternatively, the SURE model may be seen to be a special example of the unrestricted reduced form of a simultaneous equations model—one in which all of the predetermined variables are strictly exogenous.

Finally, the SURE model has a close link with the conventional multivariate regression model found in the standard statistical literature. To explore this last relationship, let Y and U be $(T \times M)$ matrices with typical column vectors y_i and u_i $(i = 1, \ldots, M)$ respectively. Further, let Z denote the $(T \times K)$ matrix of T observations on each of the K distinct explanatory variables in the model. If B is a $(K \times M)$ matrix of unknown regression coefficients, then the multivariate regression model is defined by

$$Y = ZB + U, \tag{1.12}$$

$$E(U) = 0 \tag{1.13}$$

$$E\left(\frac{1}{T} U'U\right) = \Sigma. \tag{1.14}$$

If u_i is the i'th column vector of U and b_i is the i'th column vector of the coefficient matrix B, the M equations in (1.12) may be written as

$$y_i = Zb_i + u_i \quad (i = 1, \ldots, M). \tag{1.15}$$

Noticing that all X_i's are submatrices of Z, we have

$$X_i = ZJ_i \quad (i = 1, \ldots, M) \tag{1.16}$$

where J_i is a selection matrix of order $(K \times K_i)$, with elements taking the value zero or unity, as appropriate. Finally, let $\bar{X}_i$ denote the $[T \times (K - K_i)]$ matrix of T observations on the $(K - K_i)$ explanatory variables deleted from the i'th equation, so that

$$\bar{X}_i = Z\bar{J}_i \quad (i = 1, \ldots, M) \tag{1.17}$$

where $\bar{J}_i$ is a selection matrix of order $[K \times (K - K_i)]$.

The M equations of the SURE model may now be written as

$$y_i = ZJ_i\beta_i + Z\bar{J}_i 0 + u_i . \tag{1.18}$$

Comparing (1.15) and (1.18) we see that $b_i = J_i\beta_i$, which indicates that $(K - K_i)$ elements of b_i are zero. Thus, the SURE model differs from the multivariate regression model only in that it takes account of prior information concerning the absence of certain explanatory variables from certain equations of the model. Such exclusions are highly realistic, especially in the modelling of economic data, where each equation represents an "explanation" of some aspect of economic behaviour and different economic variables may be expected to be relevant in the determination of each equation. If $K = K_i$ for all i, the SURE and multivariate regression models coincide.

1.3 OUTLINE

Although it is related to other statistical models, the SURE model is of direct interest in its own right. As noted already, over two decades have passed since Zellner's seminal treatment of the SURE model and since then a substantial literature concerning this model has emerged. Srivastava and Dwivedi (1979) offered a relatively brief survey article of this literature, and Judge, Griffiths, Hill and Lee (1980) devoted a full chapter of their book to a discussion of the SURE model. However, despite the growth of the associated literature over the past decade, a comprehensive and systematic survey is not available.

This book is concerned essentially with the related problems of estimating the SURE model's parameters and of testing hypotheses about them. Various tests and estimators are examined and compared under the assumptions of the last section, and under various generalizations of these assumptions. Greater attention has been given to estimation than to hypothesis testing in the SURE model literature and this is reflected in our own presentation. Chapters 2 to 6 deal with the various estimators which have been proposed for the conventional SURE model under our basic assumptions, and discuss their finite-sample and asymptotic properties. In Chapter 7 we relax the assumption of serial independence of the disturbances in the model and allow for the possibility of temporal autocorrelation. Similarly, in Chapter 8 the assumption of error homoscedasticity is relaxed in favour of possibly heteroscedastic disturbances. The utilization of prior information relating to the covariance structure of the model's disturbances is the subject of Chapter 9, and in Chapter 10 we consider both classical and Bayesian analyses of the incorporation of prior information about the coefficient parameters when estimating the SURE model. Finally, Chapter 11 is devoted to a variety of additional topics and issues which arise in the context of the SURE model. These include varying coefficients, missing observations, goodness-of-fit measures, dynamic models and nonlinear specifications. This final chapter rounds off the overall discussion of the SURE model, and points to some directions for future research.

EXERCISES

1.1 Identify the multivariate regression model, the simultaneous equation model and the SURE model in the following diagrams:

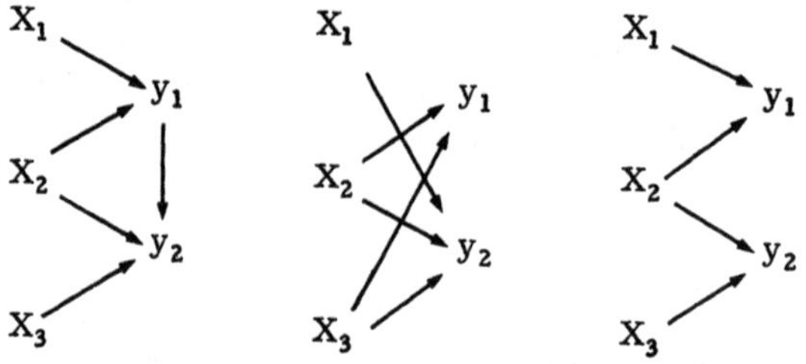

where an arrow depicts the nature of dependence. Also in each case write down the equations of the models.

1.2 Highlight the points of difference between the SURE model and the multivariate regression model. How can one model be viewed as a particular case of the other?

1.3 For the simultaneous equation model

$$y_1 + \beta_1 x_1 = u_1$$

$$\gamma_{21} y_1 + y_2 + \beta_2 x_1 = u_2$$

$$\gamma_{31} y_1 + \gamma_{32} y_2 + y_3 + \beta_{31} x_1 + \beta_{32} x_2 = u_3,$$

obtain the reduced form and observe that it is in the form of a SURE model. What kind of general simultaneous equation system will possess this property?

1.4 Consider the following SURE model in which each equation contains merely one and the same explanatory variable:

$$y_1 = \beta_1 x + u_1$$
$$y_2 = \beta_2 x + u_2$$
$$\vdots$$
$$y_M = \beta_M x + u_M$$

such that $(y_1 + y_2 + \cdots + y_M) = x$. Show that the variance covariance matrix of the disturbances is singular and that the sum of the coefficients $\beta_1, \beta_2, \ldots, \beta_M$ is unity.

1.5 In an econometric model, the variables may contain seasonal patterns which are determined in an exogenous manner and which are stable. Indicate how the SURE model provides a framework for studying the seasonal variations in the intercept term. [Hint: See Judge, Griffiths, Hill and Lee (1980; pp. 385-387).]

REFERENCES

Donnelly, W. A. (1982). "The regional demand for petrol in Australia," Economic Record 58, 317-327.

Giles, D. E. A. and P. Hampton (1984). "Regional production relationships during the industrialization of New Zealand, 1935-1948," Journal of Regional Science 24, 519-533.

Giles, D. E. A. and P. Hampton (1986). "A regional consumer demand model for New Zealand," forthcoming in Journal of Regional Science.

Judge, G. G., W. E. Griffiths, R. C. Hill and T. S. Lee (1980). The Theory and Practice of Econometrics (Wiley, New York).

Kmenta, J. (1971). Elements of Econometrics (Macmillan, New York).

Powell, A. A. (1969). "Aitken estimators as a tool in allocating predetermined aggregates," Journal of the American Statistical Association 64, 913-922.

Srivastava, V. K. and T. D. Dwivedi (1979). "Estimation of seemingly unrelated regression equations: a brief survey," Journal of Econometrics 10, 15-32.

Theil, H. (1971). Principles of Econometrics (Wiley, New York).

White, E. N. and G. J. D. Hewings (1982). "Space-time employment modelling: some results using seemingly unrelated regression estimators," Journal of Regional Science 22, 283-302.

Zellner, A. (1962). "An efficient method of estimating seemingly unrelated regression equations and tests for aggregation bias," Journal of the American Statistical Association 57, 348-368.

Zellner, A. (1971). An Introduction to Bayesian Inference in Econometrics (Wiley, New York).

2
The Least Squares Estimator and Its Variants

2.1 INTRODUCTION

In this chapter we begin our discussion of the various estimators that have been proposed for the estimation of the parameters of the SURE model, by considering those based on the principle of least squares. In discussing these estimators we recognise two basic procedures. First, we may estimate the parameters of the model equation by equation, perhaps using the OLS estimator; or secondly we may estimate the parameters in all of the equations jointly, perhaps using the generalized least squares (GLS) estimator, or one of its feasible variants in the (usual) case where Σ in (1.11) is unknown. The first approach ignores the essential jointness of the relationships which make up the SURE model, while the latter approach takes this jointness into account. The former approach implicitly assumes that the model (1.3) comprises a set of regression equations that are independent of one another, while the latter approach takes account of the non-zero correlations between the disturbances of the different equations of the model and so utilizes the additional prior information that is fundamental to the SURE specification. Intuitively, the use of this extra information should improve the efficiency of the latter class of estimators relative to the former class, at least asymptotically. However, when finite-sample properties are considered, and depending upon how this information is actually taken into account, we cannot be sure, a priori, that the latter approach is universally superior to the former. In this chapter we present some estimators derived from the basic principle of least squares, and discuss their sampling properties.

2.2 THE ORDINARY LEAST SQUARES AND GENERALIZED LEAST SQUARES ESTIMATORS

Consider the SURE model introduced in Section 1.2:

$$y = X\beta + u,$$
$$E(u) = 0, \quad E(uu') = (\Sigma \otimes I_T) = \Psi . \tag{2.1}$$

The OLS estimator of β in (2.1) is given by

$$b_0 = (X'X)^{-1}X'y . \tag{2.2}$$

This estimator ignores the fact that the joint prior variance covariance matrix for the model is Ψ, and treats this matrix as if it were scalar. The non-zero correlations between the u_i's are ignored.

From (2.1) and (2.2) we see that

$$E(b_0 - \beta) = 0, \tag{2.3}$$

and it also follows that the variance covariance matrix of b_0 is

$$\begin{aligned} V(b_0) &= E(b_0 - \beta)(b_0 - \beta)' \\ &= (X'X)^{-1}X'\Psi X(X'X)^{-1} . \end{aligned} \tag{2.4}$$

To take account of the form of the variance covariance matrix of the disturbances in (2.1) we may use the GLS, or Aitken, estimator of β:

$$\begin{aligned} b_G &= (X'\Psi^{-1}X)^{-1}X'\Psi^{-1}y \\ &= [X'(\Sigma^{-1} \otimes I_T)X]^{-1}X'(\Sigma^{-1} \otimes I_T)y . \end{aligned} \tag{2.5}$$

It is easily verified that

$$E(b_G - \beta) = 0 \tag{2.6}$$

and

$$\begin{aligned} V(b_G) &= E(b_G - \beta)(b_G - \beta)' \\ &= (X'\Psi^{-1}X)^{-1} \\ &= [X'(\Sigma^{-1} \otimes I_T)X]^{-1} . \end{aligned} \tag{2.7}$$

So, from (2.3) and (2.6) we see that both the OLS and GLS estimators of (2.1) are unbiased. Further, if we define

$$G = (X'X)^{-1}X' - (X'\Psi^{-1}X)^{-1}X'\Psi^{-1} ,$$

and note that $GX = 0$, we see that from (2.4) and (2.7),

$$V(b_0) - V(b_G) = G\Psi G' . \tag{2.8}$$

Because Ψ is positive definite by construction, $G\Psi G'$ is at least positive semi-definite, and so GLS is at least as efficient as is OLS when estimating β in this model. Indeed, provided that Σ is non-stochastic and observable, it follows from Aitken's theorem (Theil, 1971; pp. 238–239) that b_G is the best linear unbiased (BLU) estimator of β in the SURE model.

2.3 THE FEASIBLE GENERALIZED LEAST SQUARES ESTIMATOR

Looking at the expression for b_G in (2.5), it is clear that it is not an operational or feasible estimator of β because in general Σ (and hence Ψ) will be unobservable. Recognising this, Zellner (1962) proposed an estimator of β in the SURE model, basing this on (2.5), but with Σ replaced by an observable $(M \times M)$ matrix, S. In particular, the elements of S are chosen to be estimators of the corresponding elements of Σ.

With this replacement for Σ (and hence Ψ), we now have a feasible generalized least squares (FGLS) estimator of β in (2.1):

$$b_F = [X'(S^{-1} \otimes I_T)X]^{-1}X'(S^{-1} \otimes I_T)y . \tag{2.9}$$

We are assuming that the matrix $S = ((s_{ij}))$ is non-singular, where s_{ij} is some estimator of σ_{ij} (see Bacon, 1974, for an OLS interpretation).

Although there are many possible choices of S, two ways of obtaining the s_{ij}'s are popular. Each of these is based on residuals obtained by the preliminary application of OLS in one way or another. The first approach involves so-called "unrestricted residuals," because the restrictions on the coefficients of the SURE model which distinguish it from the multivariate regression model are ignored when obtaining the residuals to be used for constructing the s_{ij}'s. Recalling the notation of Section 1.2, let K be the total number of distinct regressors in the full model (2.1) and let Z be the corresponding (T × K) observation matrix for these variables. Regressing each of the M dependent variables in turn on the columns of Z (that is, estimating each equation of (1.15) by OLS), we obtain the (T × 1) "unrestricted" residual vectors,

$$\begin{aligned}\tilde{u}_i &= y_i - Z(Z'Z)^{-1}Z'y_i \\ &= \bar{P}_z y_i \qquad (i = 1, \ldots, M)\end{aligned}$$

where

$$\bar{P}_z = I_T - Z(Z'Z)^{-1}Z' .$$

From these residuals we may obtain consistent estimators of the σ_{ij}'s as

$$\begin{aligned}\tilde{s}_{ij} &= \frac{1}{T}\tilde{u}_i'\tilde{u}_j \\ &= \frac{1}{T}y_i'\bar{P}_z y_j \qquad (i, j = 1, \ldots, M) .\end{aligned} \tag{2.10}$$

Because X_i is a sub-matrix of Z, we have from (1.16),

$$X_i = ZJ_i ,$$

where as before J_i is a $(K \times K_i)$ selection matrix. It is easy to see that

$$\begin{aligned}\bar{P}_z X_i &= X_i - Z(Z'Z)^{-1}Z'X_i \\ &= X_i - ZJ_i \\ &= 0 .\end{aligned}$$

Using this result, we have

$$\begin{aligned}y_i'\bar{P}_z y_j &= (\beta_i'X_i' + u_i')\bar{P}_z(X_j\beta_j + u_j) \\ &= u_i'\bar{P}_z u_j ,\end{aligned}$$

so that from (2.10),

$$\begin{aligned} E(\tilde{s}_{ij}) &= \frac{1}{T} E(u_i' \bar{P}_z u_j) \\ &= \frac{1}{T} \sigma_{ij} \operatorname{tr}(\bar{P}_z) \\ &= \left(1 - \frac{K}{T}\right) \sigma_{ij}, \end{aligned} \tag{2.11}$$

from which it follows that an unbiased estimator of σ_{ij} is obtained by replacing $1/T$ by $1/(T - K)$ in (2.10). Similarly, it can be shown that the minimum mean squared error (best invariant) estimator of σ_{ij} may be obtained if $1/(T - K + 2)$ is substituted for $1/T$ in the definition of $\tilde{s}_{ij}$.

An alternative estimator of σ_{ij} may be developed by using residuals which have been obtained by taking into account the restrictions on the coefficients which effectively distinguish the SURE model from the multivariate regression model. The "restricted" residual vectors are obtained by estimating each equation of (2.1) separately by OLS, yielding

$$\begin{aligned} \hat{u}_i &= y_i - X_i(X_i'X_i)^{-1}X_i'y_i \\ &= \bar{P}_{x_i} y_i \qquad (i = 1, \ldots, M) \end{aligned}$$

so that an alternative consistent estimator for σ_{ij} is

$$\begin{aligned} \hat{s}_{ij} &= \frac{1}{T} \hat{u}_i' \hat{u}_j \\ &= \frac{1}{T} y_i' \bar{P}_{x_i} \bar{P}_{x_j} y_j, \end{aligned} \tag{2.12}$$

where the projection matrices $\bar{P}_{x_i}$ and $\bar{P}_{x_j}$ are defined analogously to $\bar{P}_z$.

If T in the denominator of (2.12) is replaced by

$$\operatorname{tr}(\bar{P}_{x_i} \bar{P}_{x_j}) = [T - K_i - K_j + \operatorname{tr}(X_i'X_i)^{-1}X_i'X_j(X_j'X_j)^{-1}X_j'X_i] \tag{2.13}$$

then this yields an unbiased estimator of σ_{ij}. Another interesting choice of the divisor in (2.12) is $[(T - K_i)(T - K_j)]^{\frac{1}{2}}$ (see Johnston, 1984; p. 338). The latter choice of divisor when constructing $\hat{s}_{ij}$ may, however, lead to an estimator of Σ which is not positive definite. For example, consider a two equation SURE model with

$$y_1 = \begin{bmatrix} y_{11} \\ y_{21} \\ y_{31} \end{bmatrix}; \quad y_2 = \begin{bmatrix} y_{12} \\ y_{22} \\ y_{32} \end{bmatrix};$$

$$X_1 = \begin{bmatrix} 1 \\ 0 \\ 0 \end{bmatrix}; \quad X_2 = \begin{bmatrix} 1 & 0 \\ 0 & 1 \\ 0 & 0 \end{bmatrix},$$

so that we have $M = 2$, $T = 3$, $K_1 = 1$ and $K_2 = 2$.

It is easily seen that

$$(X_1'X_1) = 1, \quad (X_2'X_2) = \begin{bmatrix} 1 & 0 \\ 0 & 1 \end{bmatrix},$$

$$(X_1'X_2) = [1, 0], \quad \mathrm{tr}(\bar{P}_{x_1}\bar{P}_{x_2}) = 1,$$

so that, from (2.12) and (2.13),

$$\begin{bmatrix} \hat{s}_{11} & \hat{s}_{12} \\ \hat{s}_{21} & \hat{s}_{22} \end{bmatrix} = \begin{bmatrix} \frac{1}{2}(y_{21}^2 + y_{31}^2) & y_{31}y_{32} \\ y_{31}y_{32} & y_{32}^2 \end{bmatrix},$$

which is singular if y_{21} and y_{31} are equal in value. Further, this matrix is indefinite if $|y_{21}|$ is smaller in value than is $|y_{31}|$ (see Theil, 1971; p. 326).

Clearly, the assumed non-singularity of Σ in no way ensures that its estimator, S, also will be non-singular. In particular, if the number of equations in the SURE model exceeds the number of data observations ($M > T$) then S will be singular if it is estimated by either of the above approaches. To see this, consider the $(T \times M)$ matrix U which has u_i as its i'th column vector:

$$U = [u_1, \ldots, u_M],$$

and let U^* be the $(T \times M)$ matrix obtained by replacing the u_i's in U by some residual vectors such as the $\tilde{u}_i$'s or the $\hat{u}_i$'s. If S is obtained as any scalar multiple of $(U^{*\prime}U^*)$, then this estimator of Σ is singular if U^* has less than full column rank, which will be the case if $M > T$. Note, however, that $M \leq T$ is a necessary but not sufficient condition for the non-singularity of S, as the earlier two equation example illustrates.

We now have $\tilde{S} = ((\tilde{s}_{ij}))$ and $\hat{S} = ((\hat{s}_{ij}))$ as two explicit choices of S as an estimator for Σ in the construction of an FGLS estimator of β in the SURE model. These two choices lead to the seemingly unrelated unrestricted residuals (SUUR) and seemingly unrelated restricted residuals (SURR) estimators proposed by Zellner (1962, 1963):

$$\tilde{\beta}_{SU} = [X'(\tilde{S}^{-1} \otimes I_T)X]^{-1}X'(\tilde{S}^{-1} \otimes I_T)y$$
$$\hat{\beta}_{SR} = [X'(\hat{S}^{-1} \otimes I_T)X]^{-1}X'(\hat{S}^{-1} \otimes I_T)y\,. \tag{2.14}$$

Of course, choosing $S = I_M$ collapses the FGLS estimator of β to the OLS estimator, (2.2). Thus, the OLS estimator may be regarded as a particular FGLS estimator of β in the SURE model, although intuitively not a very appealing one in view of the fact that it ignores the prior information about the model's variance covariance structure.

2.4 OPTIMALITY OF ORDINARY LEAST SQUARES

In some situations the OLS and FGLS estimators are identical. Suppose that the same explanatory variables appear in each of the M equations of (2.1), so that $X_1 = X_2 = \cdots = X_M = Z$ and $K_1 = K_2 = \cdots = K_M = K$, and the SURE model collapses to the multivariate regression model. Then

$$X = \begin{bmatrix} Z & 0 & \cdots & 0 \\ 0 & Z & \cdots & 0 \\ \vdots & \vdots & & \vdots \\ 0 & 0 & \cdots & Z \end{bmatrix} = (I_M \otimes Z)\,,$$

so from (2.2),

$$\begin{aligned} b_0 &= (X'X)^{-1}X'y \\ &= (I_M \otimes Z'Z)^{-1}(I_M \otimes Z')y \\ &= [I_M \otimes (Z'Z)^{-1}Z']y\,, \end{aligned} \tag{2.15}$$

and from (2.9),

$$\begin{aligned} b_F &= [(I_M \otimes Z')(S^{-1} \otimes I_T)(I_M \otimes Z)]^{-1}(I_M \otimes Z')(S^{-1} \otimes I_T)y \\ &= (S^{-1} \otimes Z'Z)^{-1}(S^{-1} \otimes Z')y \\ &= [I_M \otimes (Z'Z)^{-1}Z']\,y\,. \end{aligned} \tag{2.16}$$

So, in this case, b_F, and in particular the SUUR and SURR estimators of β, collapse to the OLS estimator. As (2.16) does not involve S, it follows that b_G is also identical to b_0 when each equation of the SURE model has identical regressors.

As a second special case, consider the situation when Σ is known to be diagonal, so that only its diagonal elements are estimated when constructing an FGLS estimator of β. In this case S is also diagonal and it follows that

$$\begin{aligned} b_F &= [X'(S^{-1} \otimes I_T)X]^{-1}X'(S^{-1} \otimes I_T)y \\ &= [(S^{-1} \otimes I_T)X'X]^{-1}(S^{-1} \otimes I_T)X'y \\ &= (X'X)^{-1}(S \otimes I_T)(S^{-1} \otimes I_T)X'y \\ &= (X'X)^{-1}X'y , \end{aligned}$$

and so b_F collapses to b_0. This result is less surprising than the previous one—if Σ is diagonal then the individual equations of the SURE model have disturbance terms which are pairwise uncorrelated. Thus, the model comprises M separate equations and OLS should be optimal, in the BLU sense.

These two special situations in which the GLS and FGLS estimators for the parameters of the SURE model collapse to OLS were indicated by Zellner (1962). It is interesting to ask whether or not there are more general situations for which this equivalence may be established. In fact, various conditions for the equivalence of the OLS and GLS estimators in the general linear model have been established by several authors, such as Gourieroux and Monfort (1980), Krämer (1980), Kruskal (1968), McElroy (1967), Mitra and Moore (1973), Norlén (1975), Rao (1968), Watson (1967) and Zyskind (1967). Using these results in the context of the SURE model a number of situations may be isolated in which the estimators b_G and b_0 coincide. For example, consider the necessary and sufficient condition given by Rao (1968; p. 260, Lemma 1). Defining $K^* = \Sigma_i K_i$, let W be an $[MT \times (MT - K^*)]$ matrix of full column rank and orthogonal to X; i.e., $X'W = 0$. Then it may be shown that a necessary and sufficient condition for the optimality (in the BLU sense) of the OLS estimator of β in the SURE model is

$$X'(\Sigma \otimes I_T)W = 0 . \qquad (2.17)$$

Dwivedi and Srivastava (1978) took W to be block-diagonal:

$$W = \begin{bmatrix} W_1 & 0 & \cdots & 0 & 0 \\ 0 & W_2 & \cdots & 0 & 0 \\ \vdots & \vdots & & \vdots & \vdots \\ 0 & 0 & \cdots & 0 & W_M \end{bmatrix}$$

with W_i of order $[T \times (T - K_i)]$ and of full column rank. In this case the orthogonality of X and W, and (2.17), imply

$$X_i' W_i = 0 \qquad (i = 1, \ldots, M)$$

and

$$\sigma_{ij} X_i' W_j = 0 \qquad (i, j = 1, \ldots, M)$$

so that (2.17) is equivalent to the conditions

$$\sigma_{ij} X_i' W_j = 0 \qquad (i, j = 1, \ldots, M;\ i \neq j)\,. \tag{2.18}$$

When Σ is diagonal, $\sigma_{ij} = 0$ for all $i \neq j$, and in this case (2.18) obviously is satisfied. On the other hand, if Σ is known to be non-diagonal, then the conditions (2.18) imply that b_0 and b_G will be identical if and only if

$$X_i' W_j = 0 \qquad (i, j = 1, \ldots, M;\ i \neq j)\,. \tag{2.19}$$

Because (2.19) does not involve the σ_{ij}'s or any estimators of these parameters, these conditions also characterise the situations under which the FGLS estimators of β will coincide with b_0 (given that Σ is non-diagonal, so that a genuine SURE framework holds). To specify these conditions more fully, let us suppose that X* and W* are two non-stochastic matrices of order $(T \times L)$ and $[T \times (T - L)]$ respectively, each of full column rank. Further, let us assume that these two matrices are orthogonal. If $\mathcal{M}(X^*)$ and $\mathcal{M}(W^*)$ denote the column spaces of X* and W* respectively, and if $\mathcal{V}$ is the vector space of all T-tuple vectors with real or complex elements, then $\mathcal{V}$ is the direct sum of the sub-spaces $\mathcal{M}(X^*)$ and $\mathcal{M}(W^*)$. As the columns of X* and W* generate the entire space, there exist matrices C_{1i}, C_{2i}, C_{3i} and C_{4i} such that

$$\begin{aligned} X_i &= X^*C_{1i} + W^*C_{2i} \qquad (i = 1, \ldots, M) \\ W_i &= X^*C_{3i} + W^*C_{4i} \qquad (i = 1, \ldots, M)\,. \end{aligned} \tag{2.20}$$

If X_i ($i = 1, \ldots, M$) all lie in $\mathcal{M}(X^*)$ and W_i ($i = 1, \ldots, M$) all lie in $\mathcal{M}(W^*)$, then the matrices C_{2i} and C_{3i} ($i = 1, \ldots, M$) will be null, so that from (2.20) we have the following:

$$X_i = X^*C_{1i} \qquad (i = 1, \ldots, M)$$

$$W_i = W^*C_{4i} \qquad (i = 1, \ldots, M)$$

which, by the orthogonality of X^* and W^*, satisfy $X_i'W_j = 0$ ($i, j = 1, \ldots, M$). This characterises the situations in which b_F and b_0 coincide, of which one interesting special case is when $X_1 = X_2 = \cdots = X_M$.

Now, let us relax the assumption that W is block-diagonal and, following Dwivedi and Srivastava (1978; pp. 394–395), partition W as

$$W = \begin{bmatrix} W_{11} & \cdots & W_{1M} \\ \vdots & & \vdots \\ W_{M1} & \cdots & W_{MM} \end{bmatrix},$$

with W_{ij} of order $[T \times (T - K_j)]$. Then the orthogonality of X and W implies that

$$X_i'W_{ij} = 0 \qquad (i, j = 1, \ldots, M)$$

and these conditions are satisfied if W_{ij} is of the form

$$\begin{aligned} W_{ij} &= W_i C_{ij}^* \\ X_i'W_i &= 0 \qquad (i, j = 1, \ldots, M) \end{aligned} \tag{2.21}$$

where C_{ij}^* is a non-stochastic matrix of order $[(T - K_i) \times (T - K_j)]$. Further, (2.17) yields the following set of conditions:

$$\sum_{m=1}^{M} \sigma_{im} X_i'W_{mj} = 0 \qquad (i, j = 1, \ldots, M)$$

which are satisfied when the constraints in (2.21) hold, together with either a diagonal Σ or the additional conditions $X_i'W_j = 0$ ($i \neq j$; $i, j = 1, \ldots, M$).

When W is not block-diagonal, a more useful set of conditions than these for the equivalence of the FGLS and OLS estimators of β may be derived from (2.17). As W is orthogonal to X, condition (2.17) holds if and only if

$(\Sigma \otimes I_T)X$ belongs to the space generated by the columns of X. That is, the OLS estimator of β in the SURE model is optimal if and only if

$$(\Sigma \otimes I_T)X = XD, \tag{2.22}$$

for some matrix D of order $(K^* \times K^*)$. This is just the condition derived by Kruskal (1968).

If we partition D into

$$D = \begin{bmatrix} D_{11} & \cdots & D_{1M} \\ \vdots & & \vdots \\ D_{M1} & \cdots & D_{MM} \end{bmatrix}$$

with D_{ij} of order $(K_i \times K_j)$, then condition (2.22) implies the following set of necessary and sufficient conditions for the optimality of b_0:

$$\sigma_{ij}X_j = X_iD_{ij} \qquad (i, j = 1, \ldots, M). \tag{2.23}$$

For $i = j$, these conditions obviously are satisfied if we choose $D_{ii} = \sigma_{ii}I_{K_i}$. For $i \neq j$, a simple choice of D_{ij} is a null matrix if $\sigma_{ij} = 0$. If $\sigma_{ij} \neq 0$, we have

$$\begin{aligned} \sigma_{ij}X_j &= X_iD_{ij} \\ \sigma_{ji}X_i &= X_jD_{ji} \qquad (i \neq j) \end{aligned} \tag{2.24}$$

which means that X_i and X_j span the same column space. That is, the OLS and GLS estimators (and the FGLS estimators) will be identical if all of the X_i's span the same column space (see Kapteyn and Fiebig, 1981, for details).

We have derived two sets of conditions under which b_0 and b_G are identical and the optimality of the OLS estimator of β in the SURE model is assured. The first set includes the special constraint that Σ is diagonal, while the second set does not place any constraints on Σ, but places restrictions on the observation matrix, X. Obviously, the second set of conditions equally characterizes the equivalence of b_0 and the b_F estimators, including the SURR and SUUR estimators. These results are of some practical interest particularly when a simple inspection of the structure of the regressor matrix may indicate whether or not there is any need to adopt an FGLS estimator for the SURE model, or if the OLS estimator will in fact be optimal in the sense of the Gauss-Markov theorem.

2.5 SIMPLIFICATIONS OF FEASIBLE GENERALIZED LEAST SQUARES

The results of the preceding section provide important clues to enable us to simplify the computation of FGLS estimators for models in which the regressor matrices have specific structures. For example, consider a SURE model in which the regressors are pairwise orthogonal across equations; i.e., $X_i'X_j = 0$ $(i, j = 1, \ldots, M;\ i \neq j)$. In this case, the FGLS estimator of β given in (2.9) simplifies to

$$b_F = \begin{bmatrix} (s^{11})^{-1} \sum_{m=1}^{M} s^{1m}(X_1'X_1)^{-1}X_1'y_m \\ \vdots \\ (s^{MM})^{-1} \sum_{m=1}^{M} s^{Mm}(X_M'X_M)^{-1}X_M'y_m \end{bmatrix}$$

where s^{ij} is the (i, j)'th element of S^{-1}. Looking at each element of this expression for b_F we see that under such orthogonality of the regressor matrices the FGLS estimator of β_i is just the OLS estimator of β_i, plus a linear combination of the OLS estimators of the coefficient vectors in $(M-1)$ linear models, these being formed by regressing each of $y_1, y_2, \ldots, y_M$ (except for y_i) on the columns of X_i.

A second special SURE model arises when some of the equations include the full set of regressors, as discussed by Revankar (1974) and Schmidt (1978), and others. Without loss of generality, assume that Z is the regressor matrix for each of the first m equations, while in each of the remaining equations one or more of the full set of regressors is excluded. Then the SURE model may be partitioned into two sub-systems and written as:

$$\begin{bmatrix} y_1 \\ \vdots \\ y_m \\ \\ y_{m+1} \\ \vdots \\ y_M \end{bmatrix} = \begin{bmatrix} Z & \cdots & 0 & 0 & \cdots & 0 \\ \vdots & & \vdots & \vdots & & \vdots \\ 0 & \cdots & Z & 0 & \cdots & 0 \\ \\ 0 & \cdots & 0 & X_{m+1} & \cdots & 0 \\ \vdots & & \vdots & \vdots & & \vdots \\ 0 & \cdots & 0 & 0 & \cdots & X_M \end{bmatrix} \begin{bmatrix} \beta_1 \\ \vdots \\ \beta_m \\ \\ \beta_{m+1} \\ \vdots \\ \beta_M \end{bmatrix} + \begin{bmatrix} u_1 \\ \vdots \\ u_m \\ \\ u_{m+1} \\ \vdots \\ u_M \end{bmatrix}$$

or

$$\begin{bmatrix} y_* \\ y_{**} \end{bmatrix} = \begin{bmatrix} (I_m \otimes Z) & 0 \\ 0 & X_{**} \end{bmatrix} \begin{bmatrix} \beta_* \\ \beta_{**} \end{bmatrix} + \begin{bmatrix} u_* \\ u_{**} \end{bmatrix} . \tag{2.25}$$

If we partition S and S^{-1} conformably as

$$S = \begin{bmatrix} S_{*,*} & S_{*,**} \\ S_{**,*} & S_{**,**} \end{bmatrix} \begin{matrix} (m) \\ (M-m) \end{matrix}$$

$$\qquad (m) \qquad (M-m)$$

and

$$S^{-1} = \begin{bmatrix} S^{*,*} & S^{*,**} \\ S^{**,*} & S^{**,**} \end{bmatrix} \begin{matrix} (m) \\ (M-m) \end{matrix}$$

$$\qquad (m) \qquad (M-m)$$

and apply FGLS to estimate the elements of β_* and β_{**}, we obtain the following estimating equations:

$$\begin{bmatrix} (I_m \otimes Z')(S^{*,*} \otimes I_T)(I_m \otimes Z) & (I_m \otimes Z')(S^{*,**} \otimes I_T)X_{**} \\ X'_{**}(S^{**,*} \otimes I_T)(I_m \otimes Z) & X'_{**}(S^{**,**} \otimes I_T)X_{**} \end{bmatrix} \begin{bmatrix} b_* \\ b_{**} \end{bmatrix}$$

$$= \begin{bmatrix} (I_m \otimes Z')(S^{*,*} \otimes I_T)y_* + (I_m \otimes Z')(S^{*,**} \otimes I_T)y_{**} \\ X'_{**}(S^{**,*} \otimes I_T)y_* + X'_{**}(S^{**,**} \otimes I_T)y_{**} \end{bmatrix}$$

or

$$(S^{*,*} \otimes Z'Z)b_* + (S^{*,**} \otimes Z')X_{**}b_{**} = (S^{*,*} \otimes Z')y_* + (S^{*,**} \otimes Z')y_{**}$$

$$X'_{**}(S^{**,*} \otimes Z)b_* + X'_{**}(S^{**,**} \otimes I_T)X_{**}b_{**} \tag{2.26}$$

$$= X'_{**}(S^{**,*} \otimes I_T)y_* + X'_{**}(S^{**,**} \otimes I_T)y_{**} .$$

Pre-multiplying the first equation in (2.26) by $X'_{**}(S^{**,*}(S^{*,*})^{-1} \otimes Z(Z'Z)^{-1})$, we obtain:

$$X'_{**}(S^{**,*} \otimes Z)b_* + X'_{**}(S^{**,*}(S^{*,*})^{-1}S^{*,**} \otimes P_z)X_{**}b_{**}$$

$$= X'_{**}(S^{**,*} \otimes P_z)y_* + X'_{**}(S^{**,*}(S^{*,*})^{-1}S^{*,**} \otimes P_z)y_{**}, \quad (2.27)$$

where $P_z = Z(Z'Z)^{-1}Z'$.

Now, subtracting (2.27) from the second equation in (2.26), and using the results

$$S^{-1}_{**,**} = S^{**,**} - S^{**,*}(S^{*,*})^{-1}S^{*,**}$$

$$X'_i P_z = X_i \qquad (i = m+1,\ m+2,\ \ldots,\ M)$$

we have

$$X'_{**}(S^{-1}_{**,**} \otimes I_T)X_{**}b_{**} = X'_{**}(S^{-1}_{**,**} \otimes I_T)y_{**},$$

which may be solved to yield:

$$b_{**} = [X'_{**}(S^{-1}_{**,**} \otimes I_T)X_{**}]^{-1}X'_{**}(S^{-1}_{**,**} \otimes I_T)y_{**}. \quad (2.28)$$

This is the FGLS estimator of β_{**} which would have been obtained had the second sub-system of the full SURE model been treated as a separate SURE model in its own right. Also, from the second equation in (2.26) we get:

$$b_* = (I_m \otimes (Z'Z)^{-1}Z')y_* + [(S^{*,*})^{-1}S^{*,**} \otimes (Z'Z)^{-1}Z'](y_{**} - X_{**}b_{**})$$

$$= (I_m \otimes (Z'Z)^{-1}Z')y_* - [S_{*,**}S^{-1}_{*,*} \otimes (Z'Z)^{-1}Z'](y_{**} - X_{**}b_{**}), \quad (2.29)$$

which shows that the FGLS estimator of the coefficient vector β_* in the first sub-system of the SURE model differs from the OLS estimator of β_* by a vector which is a linear combination of the FGLS residuals from the second sub-system. So, in this second special case involving model (2.25) we see that taking account of the special structure of the system may be computationally advantageous. If we apply the FGLS estimator directly to the full SURE model, then this involves the inversion of the $(K^* \times K^*)$ matrix $[X'(S^{-1} \otimes I_T)X]$. Not only may this be computationally burdensome, but high sample correlations among the regressors may adversely affect the numerical precision of this inversion. These difficulties may be reduced considerably if the full model may be partitioned into sub-systems so that b_{**} and b_* in (2.28) and (2.29) may be used to obtain the FGLS estimates of the model's parameters. In this case the dimensions of the matrices to be inverted in applying the estimators b_* and b_{**} are less than K^*.

These observations relate closely to a well-known result in the literature on the estimation of simultaneous structural equations models. As the simultaneous equations model collapses to the SURE model when no jointly dependent or lagged endogenous variables appear as regressors in any of the equations, the two stage least squares (2SLS) and three stage least squares (3SLS) estimators are easily shown to collapse to the OLS and SURR estimators respectively in this case. If we consider partitioning a simultaneous equations model into two sub-structures, one containing overidentified structural equations and the other containing only the just-identified equations, then as Narayanan (1969) showed, the estimation of the entire simultaneous equations model by 3SLS amounts to the separate estimation of the overidentified subsystem by 3SLS while ignoring the just-identified sub-system. The estimates of the parameters in the just-identified sub-system obtained by applying 3SLS to the full model are just those obtained by an estimator amounting to 2SLS plus a linear function of 3SLS residuals associated with the over-identified set of equations. Relating the specification of a simultaneous equations model to that of the SURE model, Schmidt (1978) considered the concept of identification in the context of seemingly unrelated regressions. The i'th equation in the SURE model is said to be just-identified if $K_i = K$, and overidentified if $K_i < K$, where it will be recalled that K is the total number of distinct regressors in the SURE model. Using this idea, Schmidt obtained the results attributed above to Narayanan.

As a third special form of the regressor matrices in the SURE model, let us consider what might be termed a "subset system" of seemingly unrelated regressions. By this, we mean a SURE model in which the regressors are constrained so that the regressors in each equation form a (not necessarily proper) subset of the regressors in the previous equation. In other words, the columns of X_{i+1} are contained in the columns of X_i $(i = 1, \ldots, M-1)$, and $X_1 = Z$. For such a model, the FGLS estimator again simplifies, with corresponding computational advantages. To see this, consider a two equation model. Setting $M = 2$, $m = 1$ and $Z = X_1$ and using (2.28) and (2.29), it is easily seen that the FGLS estimator of β_2 is just the OLS estimator, while the FGLS estimator of β_1 is the OLS estimator of β_1 adjusted by a linear function of the OLS residuals from the second equation of the model. When $M > 2$ we may still set $m = 1$ and $X_1 = Z$ in (2.28) and (2.29), and observe that the FGLS estimators of the coefficients (except for those in the first equation) derived from the estimation of all M equations of the model, are identical to the FGLS estimators obtained when the last $(M - 1)$ equations of the system are treated as a self-contained SURE model. Similar results may be obtained by considering $m = 2, 3, \ldots$, etc. To illustrate this point further, consider a three equation SURE model:

$$y_1 = X_1\beta_1 + u_1$$
$$y_2 = X_2\beta_2 + u_2 \quad (2.30)$$
$$y_3 = X_3\beta_3 + u_3,$$

where X_3 is a sub-matrix of X_2, which in turn is a sub-matrix of X_1.

Setting $m = 1$, $M = 3$ and $Z = X_1$ in (2.28) and (2.29) yields the FGLS estimators $b_{(1)F}$, $b_{(2)F}$ and $b_{(3)F}$ for β_1, β_2 and β_3 respectively, as follows:

$$b_{(1)F} = (X_1'X_1)^{-1}X_1'y_1 - (s_{12}/s_{11})(X_1'X_1)^{-1}X_1'(y_2 - X_2b_{(2)F}) - (s_{13}/s_{11})(X_1'X_1)^{-1}X_1'(y_3 - X_3b_{(3)F}) \quad (2.31)$$

$$\begin{bmatrix} b_{(2)F} \\ b_{(3)F} \end{bmatrix} = \left\{ \begin{bmatrix} X_2' & 0 \\ 0 & X_3' \end{bmatrix} \begin{bmatrix} s_{22}I_T & s_{23}I_T \\ s_{32}I_T & s_{33}I_T \end{bmatrix}^{-1} \begin{bmatrix} X_2 & 0 \\ 0 & X_3 \end{bmatrix} \right\}^{-1} \cdot \begin{bmatrix} X_2' & 0 \\ 0 & X_3' \end{bmatrix} \begin{bmatrix} s_{22}I_T & s_{23}I_T \\ s_{32}I_T & s_{33}I_T \end{bmatrix}^{-1} \begin{bmatrix} y_2 \\ y_3 \end{bmatrix}$$

$$= \begin{bmatrix} s_{33}X_2'X_2 & -s_{23}X_2'X_3 \\ -s_{32}X_3'X_2 & s_{22}X_3'X_3 \end{bmatrix}^{-1} \begin{bmatrix} s_{33}X_2'y_2 - s_{23}X_2'y_3 \\ -s_{32}X_3'y_2 + s_{22}X_3'y_3 \end{bmatrix}.$$

To obtain explicit expressions for $b_{(2)F}$ and $b_{(3)F}$, we can again use (2.28) and (2.29) with $m = 1$ and $M = 2$. This yields:

$$b_{(2)F} = (X_2'X_2)^{-1}X_2'y_2 - (s_{23}/s_{33})(X_2'X_2)^{-1}X_2'(y_3 - X_3b_{(3)F}) \quad (2.33)$$

$$b_{(3)F} = (X_3'X_3)^{-1}X_3'y_3. \quad (2.34)$$

Finally, denoting the OLS estimator of β_i by

$$b_{(i)0} = (X_i'X_i)^{-1}X_i'y_i \quad (i = 1, 2, 3) \quad (2.35)$$

the following expressions may be obtained from (2.31), (2.33) and (2.34):

$$
\begin{aligned}
b_{(1)F} &= b_{(1)0} - (s_{12}/s_{11})(X_1'X_1)^{-1}X_1'(y_2 - X_2 b_{(2)0}) \\
&\quad - (1/s_{11})(X_1'X_1)^{-1}X_1'[s_{13}I_T - (s_{12}s_{23}/s_{33})P_{X_2}](y_3 - X_3 b_{(3)0}) \\
b_{(2)F} &= b_{(2)0} - (s_{23}/s_{33})(X_2'X_2)^{-1}X_2'(y_3 - X_3 b_{(3)0}) \\
b_{(3)F} &= b_{(3)0} \, .
\end{aligned}
\tag{2.36}
$$

Again, we see that the FGLS estimator of the coefficient vector in the last equation of a subset system of seemingly unrelated regressions is just the estimator which would be obtained by applying OLS to the last equation itself.

We have seen three examples which illustrate that in certain cases the regressor matrices in the equations of a SURE model may be so structured that the computation of the FGLS estimator may be substantially simplified. In each of these cases the FGLS estimator may be expressed in terms of the OLS estimator, plus some simple adjustment terms, and the computational burden may be reduced and the computational precision generally improved, if these features are taken into account explicitly. Of course, often no such special constraints will be present on the regressor matrices of the equations of a SURE model. In this case, care must be taken that the computational procedures used to obtain the FGLS estimates of the model's parameters ensure a high level of numerical accuracy (see Ruble, 1968; Section VII.B, for a layout of the computation of SURR estimates).

2.6 SOME ASYMPTOTIC PROPERTIES

In order to consider the asymptotic properties of the various estimators for the SURE model, recall assumptions (1.5) and (1.8):

$$
\lim_{T\to\infty} \left(\frac{1}{T} X_i'X_j\right) = Q_{ij} \qquad (i, j = 1, \ldots, M)
$$

where Q_{ij} is $(K_i \times K_j)$ with finite and fixed elements, and Q_{ii} is non-singular $(i, j = 1, \ldots, M)$. It should be noted that these assumptions about the regressor matrices rule out certain data features, such as the presence of a trend variable as a regressor. To illustrate this, suppose that the first equation of the SURE model has two regressors—an intercept and a linear time trend. So, X_1 is $(T \times 2)$ with its T row vectors being $(1, 1)$, $(1, 2)$, ..., $(1, T)$. In this case,

$$\left(\frac{1}{T} X_1' X_1\right) = \begin{bmatrix} 1 & \frac{1}{2}(T+1) \\ \frac{1}{2}(T+1) & \frac{1}{6}(T+1)(2T+1) \end{bmatrix},$$

and clearly as $T \to \infty$ the limiting value of this matrix does not have finite elements.

Turning to the asymptotic properties of the estimators, first consider the asymptotic distributions of the OLS and GLS estimators for the SURE model. From Theil (1971; pp. 398–402) we observe that the asymptotic distribution of $\sqrt{T}(b_0 - \beta)$ is multivariate Normal with mean vector 0 and variance covariance matrix $(Q^{-1}Q_\Sigma Q^{-1})$, where

$$Q = \begin{bmatrix} Q_{11} & \cdots & 0 \\ \vdots & & \vdots \\ 0 & \cdots & Q_{MM} \end{bmatrix},$$

$$Q_\Sigma = \begin{bmatrix} \sigma_{11}Q_{11} & \cdots & \sigma_{1M}Q_{1M} \\ \vdots & & \vdots \\ \sigma_{M1}Q_{M1} & \cdots & \sigma_{MM}Q_{MM} \end{bmatrix}, \qquad (2.37)$$

Q being block-diagonal. Similarly, if we define

$$Q_{\underline{\Sigma}} = \begin{bmatrix} \sigma^{11}Q_{11} & \cdots & \sigma^{1M}Q_{1M} \\ \vdots & & \vdots \\ \sigma^{M1}Q_{M1} & \cdots & \sigma^{MM}Q_{MM} \end{bmatrix},$$

where it will be recalled that $\Sigma^{-1} = ((\sigma^{ij}))$, then the asymptotic distribution of $\sqrt{T}(b_G - \beta)$ is multivariate Normal with mean vector 0 and variance covariance matrix $Q_{\underline{\Sigma}}^{-1}$.

Next, consider the family of FGLS estimators of β with S chosen as a consistent estimator of Σ. From Zellner (1962), the asymptotic distribution of $\sqrt{T}(b_F - \beta)$ is also multivariate Normal with mean vector 0 and variance covariance matrix $Q_{\underline{\Sigma}}^{-1}$ (see also Theil, 1971; pp. 401–402). It should be noted that the consistency of S is not necessary for the consistency of the

FGLS estimator. For example, $S = I_M$ is not a consistent estimator of Σ, and yet this choice of S implies the OLS estimator of β, which is consistent. Further, we see that the choice of a consistent estimator for Σ ensures that the FGLS estimators (including SURR and SUUR) all have the same asymptotic distributional properties, and these properties also are shared by the GLS estimator of β in the SURE model. The matrix $Q_{\underline{\Sigma}}^{-1}$ is, of course, unobservable. It may be estimated consistently by $[X'(S^{-1} \otimes I_T)X]^{-1}$ if S is consistent for Σ. Square roots of the diagonal elements of this estimator provide asymptotic standard errors for the FGLS estimators of the elements of β. The finite sample properties of $[X'(S^{-1} \otimes I_T)X]^{-1}$ as an estimator of $Q_{\underline{\Sigma}}^{-1}$ are, however, unexplored.

To evaluate the potential gain in asymptotic efficiency to be obtained by using the FGLS (rather than OLS) estimator, consider the ratio of their generalized variances (i.e., the determinants of their variance covariance matrices):

$$\begin{aligned} \eta &= |Q_{\underline{\Sigma}}^{-1}| / |Q^{-1} Q_{\Sigma} Q^{-1}| \\ &= |Q^2| / (|Q_{\underline{\Sigma}}| \cdot |Q_{\Sigma}|) . \end{aligned} \tag{2.38}$$

The positive semi-definiteness of (2.8) implies that

$$|Q^{-1} Q_{\Sigma} Q^{-1}| \geq |Q_{\underline{\Sigma}}^{-1}| , \tag{2.39}$$

so that η cannot exceed unity. Noting that

$$|Q| = \prod_{i=1}^{M} |Q_{ii}| ,$$

we observe from Rao (1973; p. 74) that

$$|Q_{\underline{\Sigma}}| \leq \prod_{i=1}^{M} (\sigma^{ii})^{K_i} |Q_{ii}| \tag{2.40}$$

$$|Q_{\Sigma}| \leq \prod_{i=1}^{M} (\sigma_{ii})^{K_i} |Q_{ii}| , \tag{2.41}$$

so that a lower bound for η is $\Pi_{i=1}^{M} (\sigma^{ii} \sigma_{ii})^{-K_i}$.

To summarize, we have

$$\prod_{i=1}^{M} (\sigma^{ii}\sigma_{ii})^{-K_i} \leq \eta \leq 1. \tag{2.42}$$

This result was obtained by Zellner and Huang (1962) when comparing the OLS and GLS estimators of β in the SURE model, and applies equally when making an asymptotic comparison between OLS and FGLS. The lower bound for η given in (2.42) is met when $\sigma_{ij} \neq 0$ and $Q_{ij} = 0$ for $i \neq j$. This situation arises when the disturbances are contemporaneously correlated across equations and the regressor matrices are (asymptotically) orthogonal across equations. In this case the asymptotic efficiency gain in using FGLS rather than OLS is maximized. For a two equation model with $X_1'X_2 = 0$, we have (Zellner and Huang, 1962; p. 309):

$$\eta = \frac{1}{(\sigma^{11}\sigma_{11})^{K_1}(\sigma^{22}\sigma_{22})^{K_2}} = \frac{1}{(1 - \rho_{12}^2)^{K_1+K_2}}, \tag{2.43}$$

where ρ_{12} is the coefficient of correlation between the disturbances in the two equations of the model. Thus we see that the orthogonality of X_1 and X_2 ensures that the relative asymptotic efficiency of FGLS over OLS is maximized, and that this gain in efficiency increases quite rapidly as the correlation between the disturbances of the model's two equations approaches unity.

Having obtained the asymptotic distributions of these estimators of β we may also consider asymptotic tests of hypotheses about the elements of this vector. The asymptotic distribution of the vector

$$\sqrt{T}\left[\frac{1}{T} X'(S^{-1} \otimes I_T)X\right]^{\frac{1}{2}}(b_F - \beta) \tag{2.44}$$

is multivariate Normal with mean vector 0 and variance covariance matrix I_{K^*}, where S is any consistent estimator of Σ, and $K^* = \Sigma_i K_i$. Thus, the asymptotic distribution of the scalar quantity

$$T(b_F - \beta)'\left[\frac{1}{T}X'(S^{-1} \otimes I_T)X\right](b_F - \beta)$$

$$= u'(S^{-1} \otimes I_T)X[X'(S^{-1} \otimes I_T)X]^{-1}X'(S^{-1} \otimes I_T)u \tag{2.45}$$

is χ^2 with K* degrees of freedom.

Similarly, the asymptotic distribution of the scalar quantity

$$(y - Xb_F)'(S^{-1} \otimes I_T)(y - Xb_F)$$

$$= u'\{(S^{-1} \otimes I_T) - (S^{-1} \otimes I_T)X[X'(S^{-1} \otimes I_T)X]^{-1}X'(S^{-1} \otimes I_T)\}u \tag{2.46}$$

is χ^2 with (MT - K*) degrees of freedom. Moreover, it may be shown that the quantities in (2.45) and (2.46) are stochastically independent. Using a convergence theorem of Cramér (1946 p. 254), the asymptotic distribution of the ratio

$$\tilde{F} = \left[\frac{MT - K^*}{K^*}\right]\left[\frac{(b_F - \beta)'X'(S^{-1} \otimes I_T)X(b_F - \beta)}{(y - Xb_F)'(S^{-1} \otimes I_T)(y - Xb_F)}\right] \tag{2.47}$$

may be shown to be the same as the asymptotic distribution of a statistic having the same form as (2.47) but with S replaced by Σ, viz.,

$$F^* = \left[\frac{MT - K^*}{K^*}\right]\left[\frac{(b_F - \beta)'X'(\Sigma^{-1} \otimes I_T)X(b_F - \beta)}{(y - Xb_F)'(\Sigma^{-1} \otimes I_T)(y - Xb_F)}\right].$$

Assuming that u is normally distributed, F* (but not $\tilde{F}$) has a finite-sample distribution which is Snedecor's F with K* and (MT - K*) degrees of freedom. Further, the asymptotic distribution of $K^*\tilde{F}$ is χ^2 with K* degrees of freedom.

The asymptotic testing of hypotheses about the elements of β in the SURE model is now easily illustrated. For example, consider testing the null hypothesis

$$H_0: R\beta = 0$$

against the alternative hypothesis

$$H_1: R\beta \neq 0,$$

where R is any known and fixed matrix of order $(L \times K^*)$. This hypothesis involves the imposition of $L(< K^*)$ linear homogeneous restrictions on the elements of β. The homogeneity of these restrictions is not limiting as non-homogeneous restrictions may be readily expressed in this form through a simple transformation of the model. A more general discussion in terms of non-homogeneous restrictions is given in Section 10.2. The present discussion serves as an introduction to some basic aspects of inference with the SURE model.

Two common examples of such hypotheses involve, first, where we wish to test the equality of the coefficient vectors β_i (i = 1, ..., M). In this case, R is taken as the $[(M-1)K_1 \times MK_1]$ matrix

$$R = \begin{bmatrix} I_{K_1} & -I_{K_1} & 0 & \cdots & & 0 \\ 0 & I_{K_1} & -I_{K_1} & \cdots & & 0 \\ \vdots & \vdots & \vdots & & & \vdots \\ 0 & 0 & 0 & \cdots & I_{K_1} & -I_{K_1} \end{bmatrix} . \tag{2.48}$$

Notice that this hypothesis makes sense only if $K_i = K_j$ (i,j = 1, 2, ..., M) so that $K^* = MK_1$. Secondly, if we wish to test the significance of any specific individual coefficient element, say β_{ih} (the h'th element of β_i), then R is chosen to be $(1 \times K^*)$, of the form:

$$R = [0, \ldots, 0, 1, 0, \ldots, 0], \tag{2.49}$$

where the unit element is in the $(\Sigma_{j=1}^{i-1} K_j + h)$'th position.

Zellner (1962) used the above-mentioned convergence theorem of Cramér to show that the asymptotic distribution of $L\tilde{F}_R$, with

$$\tilde{F}_R = \left[\frac{MT - K^*}{L}\right]\left[\frac{(Rb_F)'R[X'(S^{-1} \otimes I_T)X]^{-1}R'(Rb_F)}{(y - Xb_F)'(S^{-1} \otimes I_T)(y - Xb_F)}\right], \tag{2.50}$$

is χ^2 with L degrees of freedom when H_0 is true. Thus, we have a suitable statistic for testing $H_0: R\beta = 0$ against $H_1: R\beta \neq 0$.

It might be noted that in the special case of testing the equality of the coefficient vectors in the different equations of the model, Zellner (1962) showed that the likelihood ratio test (LRT) approach leads to a statistic which is asymptotically equivalent to $L\tilde{F}_R$. He assumed normality of u in formulating the likelihood function for (2.1), but at this stage we are not placing any such limitations on the formulation of the basic SURE model.

Finally, note that in the special case of testing $H_0: \beta_{ih} = 0$ against $H_1: \beta_{ih} \neq 0$, where R is chosen as in (2.49), then (2.50) collapses to

$$\tilde{F}_R = \frac{(MT - K^*)(b_{ih} - \beta_{ih})^2}{q_{ih}(y - Xb_F)'(S^{-1} \otimes I_T)(y - Xb_F)}, \tag{2.51}$$

where b_{ih} is the FGLS estimator of β_{ih} and q_{ih} is the $(\Sigma_{j=1}^{i-1} K_j + h)$'th diagonal element of the matrix $[X'(S^{-1} \otimes I_T)X]^{-1}$.

Next, consider testing for structural change in the SURE model, as discussed by Richmond (1982). Suppose that the hypothesized structural change occurs after $\overset{*}{T}$ observations. Splitting the T observations into $\overset{*}{T}$ and $\overset{**}{T} = (T - \overset{*}{T})$ observations, we introduce the following notation:

$$y_i = \begin{bmatrix} \overset{*}{y}_i \\ \overset{**}{y}_i \end{bmatrix} \begin{matrix} (\overset{*}{T}) \\ (\overset{**}{T}) \end{matrix}, \quad X_i = \begin{bmatrix} \overset{*}{X}_i \\ \overset{**}{X}_i \end{bmatrix} \begin{matrix} (\overset{*}{T}) \\ (\overset{**}{T}) \end{matrix}, \quad u_i = \begin{bmatrix} \overset{*}{u}_i \\ \overset{**}{u}_i \end{bmatrix} \begin{matrix} (\overset{*}{T}) \\ (\overset{**}{T}) \end{matrix}$$

$$(1) \qquad\qquad (K_i) \qquad\qquad (1)$$

$$\underset{(M\overset{*}{T}\times 1)}{\overset{*}{y}} = \begin{bmatrix} \overset{*}{y}_1 \\ \overset{*}{y}_2 \\ \vdots \\ \overset{*}{y}_M \end{bmatrix}, \quad \underset{(M\overset{*}{T}\times \overset{*}{K})}{\overset{*}{X}} = \begin{bmatrix} \overset{*}{X}_1 & 0 & \cdots & 0 \\ 0 & \overset{*}{X}_2 & \cdots & 0 \\ \vdots & \vdots & & \vdots \\ 0 & 0 & \cdots & \overset{*}{X}_M \end{bmatrix}, \quad \underset{(M\overset{*}{T}\times 1)}{\overset{*}{u}} = \begin{bmatrix} \overset{*}{u}_1 \\ \overset{*}{u}_2 \\ \vdots \\ \overset{*}{u}_M \end{bmatrix}$$

$$\underset{(M\overset{**}{T}\times 1)}{\overset{**}{y}} = \begin{bmatrix} \overset{**}{y}_1 \\ \overset{**}{y}_2 \\ \vdots \\ \overset{*}{y}_M \end{bmatrix}, \quad \underset{(M\overset{**}{T}\times 1)}{\overset{**}{X}} = \begin{bmatrix} \overset{**}{X}_1 & 0 & \cdots & 0 \\ 0 & \overset{**}{X}_2 & \cdots & 0 \\ \vdots & \vdots & & \vdots \\ 0 & 0 & \cdots & \overset{**}{X}_M \end{bmatrix}, \quad \underset{(M\overset{**}{T}\times 1)}{\overset{**}{u}} = \begin{bmatrix} \overset{**}{u}_1 \\ \overset{**}{u}_2 \\ \vdots \\ \overset{**}{u}_M \end{bmatrix}.$$

Thus, the SURE model exhibiting some structural change can be written as

$$\overset{*}{y} = \overset{*}{X}\overset{*}{\beta} + \overset{*}{u}$$

$$\overset{**}{y} = \overset{**}{X}\,\overset{**}{\beta} + \overset{**}{u}, \tag{2.52}$$

where $\overset{*}{\beta}$ and $\overset{**}{\beta}$ refer to the coefficient vector in the two subsamples of $\overset{*}{T}$ and $\overset{**}{T}$ observations respectively, and the null hypothesis of no structural change is $H_0: \overset{*}{\beta} = \overset{**}{\beta}$. Obviously, the FGLS estimator of $\overset{*}{\beta} = \overset{**}{\beta} = \beta$ (say) under H_0 is

$$b_F = [X'(S^{-1} \otimes I_T)X]^{-1}X'(S^{-1} \otimes I_T)y\ , \tag{2.53}$$

while the FGLS estimators of $\overset{*}{\beta}$ and $\overset{**}{\beta}$ under the maintained hypothesis are given by

$$\begin{aligned} \overset{*}{b}_F &= [\overset{*}{X}'(\overset{*}{S}^{-1} \otimes I_{\overset{*}{T}})\overset{*}{X}]^{-1}\overset{*}{X}'(\overset{*}{S}^{-1} \otimes I_{\overset{*}{T}})\overset{*}{y} \\ \overset{**}{b}_F &= [\overset{**}{X}'(\overset{**}{S}^{-1} \otimes I_{\overset{**}{T}})\overset{**}{X}]^{-1}\overset{**}{X}'(\overset{**}{S}^{-1} \otimes I_{\overset{**}{T}})\overset{**}{y} \end{aligned} \tag{2.54}$$

where $\overset{*}{S}$ and $\overset{**}{S}$ denote consistent estimators of Σ based on $\overset{*}{T}$ and $\overset{**}{T}$ observations.

Estimating the elements of Σ by the appropriate residual sums of squares and cross-products, divided by the number of observations, suppose that Σ_F, $\overset{*}{\Sigma}_F$ and $\overset{**}{\Sigma}_F$ are the estimators of Σ based on b_F, $\overset{*}{b}_F$ and $\overset{**}{b}_F$ respectively. If S is defined by

$$S = \frac{1}{T}(\overset{*}{T}\overset{*}{S} + \overset{**}{T}\overset{**}{S}) \tag{2.55}$$

then a suitable test statistic for H_0 is

$$\phi = T\,\mathrm{tr}(S^{-1}\Sigma_F) - \overset{*}{T}\,\mathrm{tr}(S^{-1}\overset{*}{\Sigma}_F) - \overset{**}{T}\,\mathrm{tr}(S^{-1}\overset{**}{\Sigma}_F)\ , \tag{2.56}$$

which (see Richmond, 1982) is a general form of the test statistic due to Chow (1960). Under H_0 the asymptotic distribution of ϕ is χ^2 with K* degrees of freedom, provided that all of the regressor matrices $\overset{*}{X}_i$'s and $\overset{**}{X}_i$'s have full column rank.

These asymptotic tests, and the asymptotic properties of the various estimators considered in this chapter are of some help in as much as they provide a basis for drawing inferences about the parameters of the SURE model if the available sample is sufficiently large. In general, of course, limited data availability will severely weaken the applicability of these results, and the emphasis will have to be on tests and estimators which have known and desirable properties in finite samples. We turn to a detailed discussion of these issues in the next two chapters and in Chapter 10 we consider some further aspects of hypothesis testing in the general context of prior information about the SURE model's parameters.

EXERCISES

2.1 Show that any FGLS estimator can be interpreted as the OLS estimator applied to a suitably transformed version of the SURE model.

2.2 Consider the following two equation SURE model in which each equation contains merely one regressor:

$$\begin{aligned} y_1 &= \beta_1 x_1 + u_1 \\ y_2 &= \beta_2 x_2 + u_2 \\ E(u_1) &= E(u_2) = 0 \end{aligned}$$

$$E\begin{pmatrix} u_1 \\ u_2 \end{pmatrix}(u_1' u_2') = \begin{bmatrix} \dfrac{\sigma^2}{x_1'x_1} I_T & \dfrac{\sigma^2}{x_1'x_2} I_T \\[2ex] \dfrac{\sigma^2}{x_1'x_2} I_T & \dfrac{\sigma^2}{x_2'x_2} I_T \end{bmatrix}$$

where σ^2 is an unknown positive scalar. Show that the GLS estimators of β_1 and β_2 do not exist, while the OLS estimators do exist.

2.3 In the two equation SURE model

$$\begin{aligned} y_1 &= \beta_1 x_1 + u_1 \\ y_2 &= \beta_2 x_2 + u_2 \end{aligned}$$

the coefficients β_1 and β_2 are estimated by OLS and GLS in two ways, first by using T observations and then using (T + 1) observations. Obtain recursion relationships which may be used to update both the OLS and GLS estimators for this model as additional data become available.

2.4 Consider the following two SURE models:

Model A:

$$\begin{aligned} y_1 &= \beta_{11} x_1 + \beta_{12} t + u_1 \\ y_2 &= \beta_{21} x_2 + \beta_{22} t + u_2 \end{aligned}$$

Model B:

$$\begin{aligned} y_1^* &= \beta_{11} x_1^* + v_1 \\ y_2^* &= \beta_{21} x_2^* + v_2 \end{aligned}$$

where y_1^*, y_2^*, x_1^* and x_2^* are the residual vectors obtained by regressing the variables y_1, y_2, x_1 and x_2 respectively on the time-trend variable t.

What is the relationship between the OLS estimators of β_{11} (and of β_{21}) obtained from the two models?

2.5 Consider a SURE model in which the disturbances across the equations are equicorrelated and have identical variances, so that

$$\Sigma = \sigma^2[(1-\rho)I_M + \rho e_M e_M']$$

where σ^2 (unknown) and ρ (known) are positive scalar quantities, and e_M denotes an $(M \times 1)$ vector with all elements equal to unity. Prove that

$$\Sigma^{-1} = \frac{1}{\sigma^2(1-\rho)}\left[I_M - \frac{1}{1+(M-1)\rho} e_M e_M'\right].$$

Obtain an explicit expression for the GLS estimator of β and compare its variance covariance matrix with that of the OLS estimator. Comment on the relative efficiencies of these estimators.

2.6 Obtain the variance covariance matrices of the residual vectors $u_0^* = (y - Xb_0)$ and $u_G^* = (y - Xb_G)$. What can you say about the difference between these two matrices?

2.7 A SURE model with no intercept term in any equation is estimated by OLS. The estimator obtained is to be compared with the OLS estimator constructed for a SURE model in which the regressor observations are taken to be deviations from the corresponding sample means. Compare the efficiency of these two estimators. How does this efficiency change when the GLS estimator (assuming Σ to be known) is employed instead of the OLS estimator?

2.8 Work out the details of the derivation of the asymptotic distributions of $\sqrt{T}(b_0 - \beta)$ and $\sqrt{T}(b_F - \beta)$, as stated in Section 2.6.

2.9 Prove that a set of sufficient conditions for the FGLS estimator to be consistent is

(i) $\text{plim}\left[\frac{1}{T}X'(S^{-1} \otimes I_T)X\right]$ is finite and invertible,

(ii) $\text{plim}\left[\frac{1}{T}X'(S^{-1} \otimes I_T)u\right]$ is a null vector.

Give an example to demonstrate that these conditions are not necessary.

2.10 Prove that for the GLS and FGLS estimators of the SURE model to have identical asymptotic distributions, it is merely sufficient to have

$$\text{plim}\left[\frac{1}{T}X'[(S^{-1} - \Sigma^{-1}) \otimes I_T]X\right] = 0$$

$$\text{plim}\left[\frac{1}{\sqrt{T}}X'[(S^{-1} - \Sigma^{-1}) \otimes I_T]u\right] = 0\ .$$

[Hint: See Theil (1971; p. 399).]

2.11 Assuming that the SURE model specification continues to be applicable at the (T + 1)'th time period, suggest an expression for the forecast of the model's dependent variables on the basis of FGLS estimation, and work out the asymptotic distribution of the forecast vector.

2.12 The SURE model can be formulated as a multivariate regression model accompanied by a set of constraints, $R\beta = 0$. Specify the matrix R and suggest an asymptotic test for the hypothesis that $R\beta = 0$. [Hint: See Harvey (1981; pp. 337-338).]

2.13 Using the data associated with the two equation SURE model considered by Zellner (1962) (see also Theil, 1971; p. 295), compute the OLS, SUUR and SURR estimates of the regression coefficients. Also provide their (consistently) estimated asymptotic standard errors.

2.14 In a two equation SURE model with nonzero correlation between the disturbances across the equations, prove that a necessary and sufficient condition for the OLS and GLS estimators to be identical is $P_{x_1} = P_{x_2}$. [Hint: See Kariya (1981).]

REFERENCES

Bacon, R. W. (1974). "A simplified exposition of seemingly unrelated regression and three stage least squares," Oxford Bulletin of Economics and Statistics 36, 229-233.

Chow, G. C. (1960). "Tests of equality between sets of coefficients in two linear regressions," Econometrica 28, 591-605.

Cramér, H. (1946). Mathematical Methods of Statistics (Princeton University Press, Princeton).

Dwivedi, T. D. and V. K. Srivastava (1978). "Optimality of least squares in the seemingly unrelated regression equation model," Journal of Econometrics 7, 391-395.

Gourieroux, C. and A. Monfort (1980). "Sufficient linear structures: econometric applications," Econometrica 48, 1083-1097.

Harvey, A. C. (1981). The Econometric Analysis of Time Series (Philip Allan, Oxford).

Johnston, J. (1984). Econometric Methods (McGraw-Hill, New York).

Kapteyn, A. and D. G. Fiebig (1981). "When are two-stage and three-stage least squares estimates identical," Economics Letters 8, 53-57.

Kariya, T. (1981). "Tests for the independence between two seemingly unrelated regression equations," Annals of Statistics 9, 381-390.

Krämer, W. (1980). "A note on the equality of ordinary least squares and Gauss-Markov estimates in the general linear model," Sankhyā A 42, 130-131.

Kruskal, W. (1968). "When are Gauss-Markov and least squares estimators identical? A coordinate-free approach," Annals of Mathematical Statistics 39, 70-75.

McElroy, F. W. (1967). "A necessary and sufficient condition that ordinary least-squares be best linear unbiased," Journal of the American Statistical Association 62, 1302-1304.

Mitra, S. K. and B. J. Moore (1973). "Gauss-Markov estimation with an incorrect dispersion matrix," Sankhyā A 35, 139-152.

Narayanan, R. (1969). "Computation of Zellner-Theil's three stage least squares estimates," Econometrica 37, 298-306.

Norlén, U. (1975). "The covariance matrices for which least squares is best linear unbiased," Scandinavian Journal of Statistics 2, 85-90.

Rao, C. R. (1968). "A note on previous lemma in the theory of least squares and some further results," Sankhyā A 30, 259-266.

Rao, C. R. (1973). Linear Statistical Inference and Its Applications (Wiley, New York).

Revankar, N. S. (1974). "Some finite sample results in the context of two seemingly unrelated regression equations," Journal of the American Statistical Association 69, 187-190.

Richmond, J. (1982). "Testing for structural change in the SURE model," Discussion Paper No. 206, Department of Economics, University of Essex, Colchester.

Ruble, W. (1968). "Improving the computation of simultaneous stochastic linear equations estimates," Agricultural Economics Report No. 116, Michigan State University, East Lansing.

Schmidt, P. (1978). "A note on the estimation of seemingly unrelated regression systems," Journal of Econometrics 7, 259-261.

Theil, H. (1971). Principles of Econometrics (Wiley, New York).

Watson, G. S. (1967). "Linear least squares regression," Annals of Mathematical Statistics 38, 1679-1699.

Zellner, A. (1962). "An efficient method of estimating seemingly unrelated regression equations and tests for aggregation bias," Journal of the American Statistical Association 57, 348-368.

Zellner, A. (1963). "Estimators for seemingly unrelated regression equations: some exact finite sample results," Journal of the American Statistical Association 58, 977-992.

Zellner, A. and D. S. Huang (1962). "Further properties of efficient estimators for seemingly unrelated regression equations," International Economic Review 3, 300-313.

Zyskind, G. (1967). "On canonical forms, non-negative covariance matrices and best and simple least squares linear estimators in linear models," Annals of Mathematical Statistics 38, 1092-1109.

3

Approximate Distribution Theory for Feasible Generalized Least Squares Estimators

3.1 INTRODUCTION

In the preceding chapter we noted that the GLS and FGLS estimators have identical asymptotic distributions, provided that the matrix S is a consistent estimator of Σ. Thus, it is not possible to distinguish between the performances of SUUR and SURR (or any other FGLS estimators based on a consistent S) by considering their asymptotic distributions. Moreover, a consideration of the asymptotic distribution of any of these estimators is of limited help when in practice the available sample size, T, is finite and perhaps quite small.

Knowledge of an estimator's asymptotic properties is a starting point. However, information is also needed about the rate at which these properties are approximated as T increases, and whether or not any known asymptotic rankings of different estimators still apply in finite samples. For instance, a consistent estimator which has smallest asymptotic variance may not continue to exhibit smallest variability in samples of finite size. For these reasons, it is imperative to investigate the properties of different estimators of the parameters in the SURE model when the sample size is finite. To this end, two types of approximate distribution theory are presented and used in this chapter. The first of these is based on the large-sample asymptotic approximation to the sampling distribution, while the second deals with the numerical evaluation of the sampling distribution of an estimator by means of the Monte Carlo technique. In the next three sections the first of these approaches is applied in turn to bias vectors, variance covariance matrices, and complete density and distribution functions for estimators of the SURE model. Monte Carlo results are discussed in Section 3.5; and the last section of the chapter is devoted to a discussion of the relative merits of these two approaches to approximating the finite-sample properties of these estimators.

3.2 ASYMPTOTIC APPROXIMATIONS FOR BIAS VECTORS

Consider the SURE model:

$$\begin{aligned} y &= X\beta + u \\ E(u) &= 0, \qquad E(uu') = (\Sigma \otimes I_T) = \Psi \end{aligned} \tag{3.1}$$

and the consistent FGLS estimators of β:

$$b_F = [X'(S^{-1} \otimes I_T)X]^{-1}X'(S^{-1} \otimes I_T)y \tag{3.2}$$

where S is any consistent estimator of Σ. We now consider the bias of b_F.

In order to study the large-sample properties of b_F we maintain the assumption that the limit of the matrix $(T^{-1}X_i'X_i)$ as $T \to \infty$ is finite and invertible $(i = 1, 2, \ldots, M)$ while the limit of the matrix $(T^{-1}X_i'X_j)$ as $T \to \infty$ is finite $(i, j = 1, 2, \ldots, M)$. This implies that the elements of the matrix $(X'X)$ are of order $O(T)$ where $O(\cdot)$ denotes the order arising from mathematical convergence. Similarly, $(X_i'u_j)$ $(i, j = 1, 2, \ldots, M)$ is of order $O_p(T^{\frac{1}{2}})$ where $O_p(\cdot)$ denotes the order of convergence in probability.

Substituting (3.1) in (3.2), we get the estimation error of the FGLS estimator:

$$(b_F - \beta) = [X'(S^{-1} \otimes I_T)X]^{-1}X'(S^{-1} \otimes I_T)u\,. \tag{3.3}$$

Let us write

$$\begin{aligned} \Delta &= (S \otimes I_T) - (\Sigma \otimes I_T) \\ &= (S \otimes I_T) - \Psi\,, \end{aligned}$$

where the elements of the (MT × MT) matrix Δ are of order $O_p(T^{-\frac{1}{2}})$ which, for brevity, is stated as "Δ is $O_p(T^{-\frac{1}{2}})$."

Observing that Ψ is O(1), we can write

$$
\begin{aligned}
(S^{-1} \otimes I_T) &= (\Psi + \Delta)^{-1} \\
&= [\Psi(I + \Psi^{-1}\Delta)]^{-1} \\
&= \Psi^{-1} - \Psi^{-1}\Delta\Psi^{-1} + \Psi^{-1}\Delta\Psi^{-1}\Delta\Psi^{-1} - \Psi^{-1}\Delta\Psi^{-1}\Delta\Psi^{-1}\Delta\Psi^{-1} + \cdots
\end{aligned}
\tag{3.4}
$$

so that

$$X'(S^{-1} \otimes I_T)u = X'\Psi^{-1}u - X'\Psi^{-1}\Delta\Psi^{-1}u + X'\Psi^{-1}\Delta\Psi^{-1}\Delta\Psi^{-1}u - \cdots$$

$$
\begin{aligned}
[X'(S^{-1} \otimes I_T)X]^{-1} &= [X'\Psi^{-1}X - X'\Psi^{-1}\Delta\Psi^{-1}X + X'\Psi^{-1}\Delta\Psi^{-1}\Delta\Psi^{-1}X - \cdots]^{-1} \\
&= \Omega[I - X'\Psi^{-1}\Delta\Psi^{-1}X\Omega + X'\Psi^{-1}\Delta\Psi^{-1}\Delta\Psi^{-1}X\Omega - \cdots]^{-1} \\
&= \Omega + \Omega X'\Psi^{-1}\Delta\Psi^{-1}X\Omega - \Omega X'\Psi^{-1}\Delta Q\Delta\Psi^{-1}X\Omega \cdots
\end{aligned}
\tag{3.5}
$$

where

$$\Omega = [X'\Psi^{-1}X]^{-1} = [X'(\Sigma^{-1} \otimes I_T)X]^{-1}$$

$$Q = \Psi^{-1} - \Psi^{-1}X\Omega X'\Psi^{-1} .$$

Using (3.5) in (3.3), we find (Srivastava, 1970; p. 487):

$$(b_F - \beta) = \xi_{-1/2} + \xi_{-1} + \xi_{-3/2} + O_p(T^{-g}) \qquad (g \geq 2) \tag{3.6}$$

where

$$
\begin{aligned}
\xi_{-1/2} &= \Omega X'\Psi^{-1}u \\
\xi_{-1} &= -\Omega X'\Psi^{-1}\Delta Q u \\
\xi_{-3/2} &= \Omega X'\Psi^{-1}\Delta Q\Delta Q u .
\end{aligned}
\tag{3.7}
$$

Here the subscripts of ξ indicate the order of magnitude in probability; i.e., ξ_g is $O_p(T^g)$ with $g = -1/2,\ -1,\ -3/2$.

If we approximate the estimation error by the leading term, i.e.,

$$(b_F - \beta) = \xi_{-1/2} \tag{3.8}$$

we observe that

$$E(b_F - \beta) = E(\xi_{-1/2}) = 0 \tag{3.9}$$

from which it follows that, to a first-order asymptotic approximation, the FGLS estimator is unbiased.

In order to improve on this first-order asymptotic approximation, let us consider the bias vector to order $O(T^{-1})$ so that

$$E(b_F - \beta) = E(\xi_{-1/2}) + E(\xi_{-1}) \ . \tag{3.10}$$

Using the result

$$\begin{aligned} X'Q &= X'[\Psi^{-1} - \Psi^{-1}X\Omega X'\Psi^{-1}] \\ &= X'\Psi^{-1} - X'\Psi^{-1}X\Omega X'\Psi^{-1} \\ &= X'\Psi^{-1} - X'\Psi^{-1} = 0 \end{aligned} \tag{3.11}$$

we observe that

$$\begin{aligned} \xi_{-1} &= -\Omega X'\Psi^{-1}\Delta Qu \\ &= -\Omega X'\Psi^{-1}(S \otimes I_T)Qu + \Omega X'Qu \\ &= -\Omega X'\Psi^{-1}(S \otimes I_T)Qu \ . \end{aligned} \tag{3.12}$$

If S is an even function of the disturbances, it follows from (3.12) that $E(\xi_{-1})$ is a null vector provided that the disturbances are symmetrically distributed. Notice that no such specification about the distribution of disturbances is required for $E(\xi_{-\frac{1}{2}})$ to be a null vector. It thus follows from (3.10) that, according to a second-order asymptotic approximation, the FGLS estimator is unbiased to order $O(T^{-1})$ provided that S is an even function of disturbances which follow a symmetric probability distribution. In particular, the choices $S = \tilde{S}$ and $S = \hat{S}$ satisfy this requirement, so that the SUUR and SURR estimators are unbiased to order $O(T^{-1})$ for symmetric distributions of the disturbances. In fact, it will be demonstrated in the next chapter that these estimators are exactly unbiased under symmetry of the distribution of the disturbances in the SURE model, together with some additional conditions.

3.3 ASYMPTOTIC APPROXIMATIONS FOR VARIANCE COVARIANCE MATRICES

From the first-order asymptotic approximation (3.8) for the estimation error of the FGLS estimator, we observe that the variance covariance matrix of b_F, to order $O(T^{-1})$, is (Srivastava, 970; p. 487)

$$\begin{aligned} V(b_F) &= E(b_F - \beta)(b_F - \beta)' \\ &= E(\xi_{-1/2}\xi'_{-1/2}) \\ &= \Omega X'\Psi^{-1}E(uu')\Psi^{-1}X\Omega \\ &= \Omega X'\Psi^{-1}X\Omega \\ &= \Omega , \end{aligned} \tag{3.13}$$

which does not require symmetry of the distribution of the disturbances. So, to order $O(T^{-1})$, the asymptotic approximation for the variance covariance matrix of the FGLS estimator is equal to the exact variance covariance matrix of the GLS estimator.

In order to have a closer approximation, let us consider the variance covariance matrix of b_F to order $O(T^{-2})$:

$$\begin{aligned} V(b_F) &= E(b_F - \beta)(b_F - \beta)' \\ &= E(\xi_{-1/2}\xi'_{-1/2}) + E(\xi_{-1}\xi'_{-1/2} + \xi_{-1/2}\xi'_{-1}) \\ &\quad + E(\xi_{-3/2}\xi'_{-1/2} + \xi_{-1/2}\xi'_{-3/2} + \xi_{-1}\xi'_{-1}) \\ &= \Omega + E(\xi_{-1}\xi'_{-1/2} + \xi_{-1/2}\xi'_{-1}) \\ &\quad + E(\xi_{-3/2}\xi'_{-1/2} + \xi_{-1/2}\xi'_{-3/2} + \xi_{-1}\xi'_{-1}) . \end{aligned} \tag{3.14}$$

It is difficult to evaluate the expectations in (3.14) in general, so we assume that the disturbances follow a Normal distribution, restrict our attention to the SUUR and SURR estimators, and consider the results of Srivastava (1970), Srivastava and Upadhyaya (1978) and others.

3.3.1 The SUUR and SURR Estimators

Let us recall the definition of $\tilde{S}$. Its (i, j)'th element is

$$\tilde{s}_{ij} = \frac{1}{T} y_i'[I_T - Z(Z'Z)^{-1}Z']y_j$$

$$= \frac{1}{T} y_i' \bar{P}_z y_j$$

where Z is a $(T \times K)$ matrix of T observations on all of the K distinct regressors in the model. As X_i is a sub-matrix of Z, we have

$$X_i' \bar{P}_z = 0 \qquad (i = 1, 2, \ldots, M) \tag{3.15}$$

from which we find:

$$\tilde{s}_{ij} = \frac{1}{T}(X_i\beta_i + u_i)'\bar{P}_z(X_j\beta_j + u_j)$$

$$= \frac{1}{T} u_i' \bar{P}_z u_j \ .$$

Now, from (3.11), we see that (Srivastava and Upadhyaya, 1978; p. 93)

$$E(\xi_{-1}\xi'_{-1/2}) = -\Omega X'\Psi^{-1}E[\Delta Quu']\Psi^{-1}X\Omega$$

$$= -\Omega X'\Psi^{-1}E[(\tilde{S}\otimes I_T)Quu']\,\Psi^{-1}X\Omega + \Omega X'QE(uu')\Psi^{-1}X\Omega$$

$$= -\Omega X'\Psi^{-1}E[(\tilde{S}\otimes I_T)Quu']\,\Psi^{-1}X\Omega \ . \tag{3.16}$$

If we write $\Sigma^{-1} = ((\sigma_{ij}))^{-1} = ((\sigma^{ij}))$ and partition Q as

$$Q = \begin{bmatrix} Q_{11} & \cdots & Q_{1M} \\ \vdots & & \vdots \\ Q_{M1} & \cdots & Q_{MM} \end{bmatrix}$$

with the (i, j)'th sub-matrix Q_{ij} of order $(T \times T)$, we can express the (g, h)'th submatrix of $E[(\tilde{S} \otimes I_T)Quu']\Psi^{-1}X$ as

$$\sum_{i,j,k}^{M} \sigma^{kh} Q_{ij} E[\tilde{s}_{gi} u_j u_k'] X_h$$

$$= \frac{1}{T} \sum_{i,j,k}^{M} \sigma^{kh} Q_{ij} E[u_g' \bar{P}_z u_i u_j u_k'] X_h$$

$$= \frac{1}{T} \sum_{i,j,k}^{M} \sigma^{kh} Q_{ij} [\sigma_{gi}\sigma_{jk}(T - K) I_T + (\sigma_{gj}\sigma_{ik} + \sigma_{gk}\sigma_{ij})\bar{P}_z] X_h$$

$$= \left(1 - \frac{K}{T}\right) \sum_{i,j}^{M} \sigma_{gi} Q_{ij} \left(\sum_{k}^{M} \sigma_{jk}\sigma^{kh}\right) X_h$$

$$= \left(1 - \frac{K}{T}\right) \sum_{i}^{M} \sigma_{gi} Q_{ih} X_h$$

$$= 0 , \tag{3.17}$$

where we have used the result (B.1.1) of the Appendix along with:

$$\sum_{k}^{M} \sigma_{jk}\sigma^{kh} = \begin{cases} 1 & \text{if } j = h \\ 0 & \text{if } j \neq h \end{cases} \tag{3.18}$$

$$Q_{ih} X_h = 0 \qquad \text{(i.e., } QX = 0\text{)} .$$

From (3.17), it follows that

$$E[(\tilde{S} \otimes I_T) Q u u'] \Psi^{-1} X = 0 , \tag{3.19}$$

which, when substituted into (3.16), yields

$$E(\xi_{-1} \xi'_{-1/2}) = 0 \tag{3.20}$$

and hence its transpose

$$E(\xi_{-1/2} \xi'_{-1}) = 0 . \tag{3.21}$$

Similarly, from result (B.1.2) of the Appendix, we have

$$E[\tilde{s}_{ai}\tilde{s}_{jk}u_\ell u_b']$$

$$= \frac{1}{T^2} E[u_a'\bar{P}_z u_i u_j'\bar{P}_z u_k u_\ell u_b']$$

$$= \left(1 - \frac{K}{T}\right)\sigma_{\ell b}\left[\left(1 - \frac{K}{T}\right)\sigma_{ai}\sigma_{jk} + \frac{1}{T}(\sigma_{aj}\sigma_{ik} + \sigma_{ak}\sigma_{ij})\right] I_T$$

$$+ \frac{1}{T}\left(1 - \frac{K}{T}\right)[\sigma_{ai}(\sigma_{j\ell}\sigma_{bk} + \sigma_{jb}\sigma_{\ell k}) + \sigma_{jk}(\sigma_{a\ell}\sigma_{ib} + \sigma_{ab}\sigma_{i\ell})]\bar{P}_z$$

$$+ \frac{1}{T^2}[\sigma_{aj}(\sigma_{bk}\sigma_{i\ell} + \sigma_{ib}\sigma_{\ell k}) + \sigma_{ak}(\sigma_{ib}\sigma_{\ell j} + \sigma_{i\ell}\sigma_{jb})$$

$$+ \sigma_{a\ell}(\sigma_{ij}\sigma_{kb} + \sigma_{jb}\sigma_{ik}) + \sigma_{ab}(\sigma_{ik}\sigma_{j\ell} + \sigma_{ij}\sigma_{\ell k})]\bar{P}_z . \tag{3.22}$$

Now consider the expectation

$$E[X'\Psi^{-1}(\tilde{S} \otimes I_T)Q(\tilde{S} \otimes I_T)Quu'\Psi^{-1}X] ,$$

the (g, h)'th sub-matrix of which is given by

$$\sum_{a,i,j,k,\ell,b}^{M} \sigma^{ga}\sigma^{bh} X_g' Q_{ij} Q_{k\ell} E[\tilde{s}_{ai}\tilde{s}_{jk}u_\ell u_b'] X_h$$

$$= \left(1 - \frac{K}{T}\right) \sum_{a,i,j,k,\ell,b}^{M} \sigma^{ga}\sigma^{bh}\left[\left(1 - \frac{K}{T}\right)\sigma_{ai}\sigma_{jk}\sigma_{\ell b}\right.$$

$$\left. + \frac{1}{T}\sigma_{\ell b}(\sigma_{aj}\sigma_{ik} + \sigma_{ak}\sigma_{ij})\right] X_g' Q_{ij} Q_{k\ell} X_h$$

$$= \left(1 - \frac{K}{T}\right) \sum_{i,j,k,\ell}^{M} \left[\left(1 - \frac{K}{T}\right)\sigma_{jk}\left(\sum_a^M \sigma^{ga}\sigma_{ai}\right)\left(\sum_b^M \sigma^{bh}\sigma_{\ell b}\right) X_g' Q_{ij} Q_{k\ell} X_h\right.$$

$$+ \frac{1}{T}\sigma_{ik}\left(\sum_a^M \sigma^{ga}\sigma_{aj}\right)\left(\sum_b^M \sigma^{bh}\sigma_{\ell b}\right) X_g' Q_{ij} Q_{k\ell} X_h$$

$$\left. + \frac{1}{T}\sigma_{ij}\left(\sum_a^M \sigma^{ga}\sigma_{ak}\right)\left(\sum_b^M \sigma^{bh}\sigma_{\ell b}\right) X_g' Q_{ij} Q_{k\ell} X_h\right]$$

$$= \left(1 - \frac{K}{T}\right)\left[\left(1 - \frac{K}{T}\right)\sum_{j,k}^{M} \sigma_{jk} X_g' Q_{gj} Q_{kh} X_h + \frac{1}{T}\sum_{i,k}^{M} \sigma_{ik} X_g' Q_{ig} Q_{kh} X_h\right.$$

$$+ \frac{1}{T} \sum_{i,j}^{M} \sigma_{ij} X'_g Q_{ij} Q_{gh} X_h \Big]$$

$$= 0 ,$$

where use has been made of (3.11), (3.18) and (3.22).

Thus, we have

$$X'\Psi^{-1} E[(\tilde{S} \otimes I_T) Q (\tilde{S} \otimes I_T) Quu'] \Psi^{-1} X = 0 . \tag{3.23}$$

Using this result with (3.19), we find (Srivastava and Upadhyaya, 1978; p. 94)

$$\begin{aligned} E(\xi_{-3/2}\xi'_{-1/2}) &= \Omega X'\Psi^{-1} E(\Delta Q \Delta Q uu')\Psi^{-1} X\Omega \\ &= \Omega X'\Psi^{-1} E[(\tilde{S} \otimes I_T) Q (\tilde{S} \otimes I_T) Quu']\Psi^{-1} X\Omega \\ &\quad - \Omega X'QE[(\tilde{S} \otimes I_T) Quu']\Psi^{-1} X\Omega \\ &\quad - \Omega X'\Psi^{-1} E[(\tilde{S} \otimes I_T) Q\Psi Quu']\Psi^{-1} X\Omega \\ &\quad + \Omega X'Q\Psi QE(uu')\Psi^{-1} X\Omega \\ &= \Omega X'\Psi^{-1} E[(\tilde{S} \otimes I_T) Q (\tilde{S} \otimes I_T) Quu']\Psi^{-1} X\Omega \\ &\quad - \Omega X'\Psi^{-1} E[(\tilde{S} \otimes I_T) Quu']\Psi^{-1} X\Omega \\ &= 0 , \end{aligned} \tag{3.24}$$

where we have employed (3.11) and the following result:

$$\begin{aligned} Q\Psi Q &= Q\Psi[\Psi^{-1} - \Psi^{-1} X\Omega X'\Psi^{-1}] \\ &= Q - QX\Omega X'\Psi^{-1} \\ &= Q . \end{aligned} \tag{3.25}$$

Transposing (3.24), we get

$$E(\xi_{-1/2}\xi'_{-3/2}) = 0 . \tag{3.26}$$

Finally, we take up the last term in (3.14):

$$
\begin{aligned}
E(\xi_{-1}\xi'_{-1}) &= \Omega X'\Psi^{-1}E[\Delta Quu'Q\Delta]\Psi^{-1}X\Omega \\
&= \Omega X'\Psi^{-1}E[(\tilde{S} \otimes I_T)Quu'Q(\tilde{S} \otimes I_T)]\Psi^{-1}X\Omega \\
&\quad - \Omega X'QE[uu'Q(\tilde{S} \otimes I_T)]\Psi^{-1}X\Omega \\
&\quad - \Omega X'\Psi^{-1}E[(\tilde{S} \otimes I_T)Quu']QX\Omega + \Omega X'QE[uu']QX\Omega \\
&= \Omega X'\Psi^{-1}E[(\tilde{S} \otimes I_T)Quu'Q(\tilde{S} \otimes I_T)]\Psi^{-1}X\Omega \,. \qquad (3.27)
\end{aligned}
$$

The (g, h)'th sub-matrix of the matrix

$$X'\Psi^{-1}E[(\tilde{S} \otimes I_T)Quu'Q(\tilde{S} \otimes I_T)]\Psi^{-1}X$$

can be written as

$$\sum_{a,i,j,k,\ell,b}^{M} \sigma^{ga}\sigma^{kh}Q_{i\ell}E[\tilde{s}_{ai}\tilde{s}_{jk}u_{\ell}u'_{b}]Q_{bj}X_h \,. \qquad (3.28)$$

From the definition and partitioned form of Q, its (b, j)'th sub-matrix is

$$Q_{bj} = \sigma^{bj}I_T - \sum_{\mu,\mu^*}^{M} \sigma^{b\mu}\sigma^{\mu^* j}X_{\mu}\Omega_{\mu\mu^*}X_{\mu^*}$$

where $\Omega_{\mu\mu^*}$ is the (μ, μ^*)'th sub-matrix of Ω, which is partitioned as

$$\Omega = \begin{bmatrix} \Omega_{11} & \cdots & \Omega_{1M} \\ \vdots & & \vdots \\ \Omega_{M1} & \cdots & \Omega_{MM} \end{bmatrix} .$$

So, we see that

$$
\begin{aligned}
\bar{P}_z Q_{bj} X_h &= \sigma^{bj}\bar{P}_z X_h - \sum_{\mu,\mu^*}^{M} \sigma^{b\mu}\sigma^{\mu^* j}\bar{P}_z X_{\mu}\Omega_{\mu\mu^*}X'_{\mu^*}X_h \\
&= 0 \,,
\end{aligned}
$$

since X_h and X_μ are sub-matrices of Z.

Using this result and (3.22), we find the expression (3.28) to be equal to:

$$\left(1-\frac{K}{T}\right)\sum_{i,j,\ell,b}^{M}\sigma_{\ell b}\Bigg[\left(1-\frac{K}{T}\right)\left(\sum_{a}^{M}\sigma^{ga}\sigma_{ai}\right)\left(\sum_{k}^{M}\sigma^{kh}\sigma_{jk}\right)X_g'Q_{i\ell}Q_{bj}X_h$$

$$+\frac{1}{T}\left(\sum_{a}^{M}\sigma^{ga}\sigma_{aj}\right)\left(\sum_{k}^{M}\sigma^{kh}\sigma_{ik}\right)X_g'Q_{i\ell}Q_{bj}X_h$$

$$+\frac{1}{T}\sum_{a}^{M}\sigma_{ij}\sigma^{ga}\left(\sum_{k}^{M}\sigma^{kh}\sigma_{ak}\right)X_g'Q_{i\ell}Q_{bj}X_h\Bigg]$$

$$=\left(1-\frac{K}{T}\right)\Bigg[\left(1-\frac{K}{T}\right)\sum_{\ell,b}^{M}\sigma_{\ell b}X_g'Q_{g\ell}Q_{bh}X_h+\frac{1}{T}\sum_{\ell,b}^{M}X_g'Q_{h\ell}(\sigma_{\ell b}I_T)Q_{bg}X_h$$

$$+\frac{1}{T}\sum_{i,j,\ell,b}^{M}\sigma_{ij}\sigma^{gh}X_g'Q_{i\ell}(\sigma_{\ell b}I_T)Q_{bj}X_h\Bigg].$$

$$=\frac{1}{T}X_g'Q_{hg}X_h+\frac{1}{T}\sum_{i,j}^{M}\sigma_{ij}\sigma^{gh}X_g'Q_{ij}X_h\,. \tag{3.29}$$

Now, if we define

$$R = X\Omega X'(\Sigma^{-1}\otimes I_T) = \begin{bmatrix} R_{11} & \cdots & R_{1M} \\ \vdots & & \vdots \\ R_{M1} & \cdots & R_{MM} \end{bmatrix}$$

$$W = \sum_{i}^{M} R_{ii}$$

$$P = \begin{bmatrix} P_{11} & \cdots & P_{1M} \\ \vdots & & \vdots \\ P_{M1} & \cdots & P_{MM} \end{bmatrix}$$

with $P_{ij} = Q_{ji}$, we observe that

$$\sum_{i,j}^{M} \sigma_{ij} Q_{ij} = M I_T - \sum_{i}^{M} R_{ii}$$
$$= M I_T - W ,$$

from which we find that (3.29) equals

$$\frac{1}{T} X'_g P_{gh} X_h + \frac{1}{T} M\sigma^{gh} X'_g X_h - \frac{1}{T}\sigma^{gh} X'_g W X_h ,$$

which provides the expression for (3.28). This yields

$$X'\Psi^{-1}[(\tilde{S} \otimes I_T)Quu'Q(\tilde{S} \otimes I_T)]\Psi^{-1}X$$
$$= \frac{1}{T} X'PX + \frac{M}{T} X'(\Sigma^{-1} \otimes I_T)X - \frac{1}{T} X'(\Sigma^{-1} \otimes W)X . \tag{3.30}$$

Substituting (3.30) in (3.27), we obtain the result of Srivastava and Upadhyaya (1978; p. 95):

$$E(\xi_{-1}\xi'_{-1}) = \frac{M}{T}\Omega - \frac{1}{T}\Omega X'[(\Sigma^{-1} \otimes W) - P]X\Omega . \tag{3.31}$$

Using (3.13), (3.20), (3.21), (3.24), (3.26) and (3.31), we get from (3.14) the asymptotic approximation for the variance covariance matrix of the SUUR estimator:

$$V(\tilde{\beta}_{SU}) = \left(1 + \frac{M}{T}\right)\Omega - \frac{1}{T}\Omega X'[(\Sigma^{-1} \otimes W) - P]X\Omega \tag{3.32}$$

the order of approximation being $O(T^{-2})$ (Srivastava and Upadhyaya, 1978; p. 91).

An alternative derivation of (3.32) may be illuminating. To this end, we define a block diagonal matrix:

$$J = \begin{bmatrix} J_1 & \cdots & 0 \\ \vdots & & \vdots \\ 0 & \cdots & J_M \end{bmatrix} \tag{3.33}$$

where J_i is a selection matrix of order $(K \times K_i)$ such that $X_i = ZJ_i$. Thus we can write

$$X = (I_M \otimes Z)J \tag{3.34}$$

so that

$$\begin{aligned} X'(\tilde{S}^{-1} \otimes I_T)X &= J'(I_M \otimes Z')(\tilde{S}^{-1} \otimes I_T)(I_M \otimes Z)J \\ &= J'(\tilde{S}^{-1} \otimes Z'Z)J \end{aligned} \tag{3.35}$$

$$\begin{aligned} X'(\tilde{S}^{-1} \otimes I_T)y &= J'(I_M \otimes Z')(\tilde{S}^{-1} \otimes I_T)y \\ &= J'(\tilde{S}^{-1} \otimes Z')y \\ &= J'(\tilde{S}^{-1} \otimes I_T)(I_M \otimes Z')y \end{aligned} \tag{3.36}$$

from which the SUUR estimator can be written as

$$\begin{aligned} \tilde{\beta}_{SU} &= [X'(\tilde{S}^{-1} \otimes I_T)X]^{-1}X'(\tilde{S}^{-1} \otimes I_T)y \\ &= [J'(\tilde{S}^{-1} \otimes Z'Z)J]^{-1}J'(\tilde{S}^{-1} \otimes I_T)(I_M \otimes Z')y \,. \end{aligned} \tag{3.37}$$

It is now easy to see that the variance covariance matrix of $\tilde{\beta}_{SU}$ is

$$\begin{aligned} &E(\tilde{\beta}_{SU} - \beta)(\tilde{\beta}_{SU} - \beta)' \\ &= E\{[J'(\tilde{S}^{-1} \otimes Z'Z)J]^{-1}J'(\tilde{S}^{-1} \otimes I_T)(I_M \otimes Z')uu'(I_M \otimes Z)(\tilde{S}^{-1} \otimes I_T) \times \\ &\qquad J[J'(\tilde{S}^{-1} \otimes Z'Z)J]^{-1}\} \\ &= E\{[J'(\tilde{S}^{-1} \otimes Z'Z)J]^{-1}J'(\tilde{S}^{-1} \otimes I_T)(I_M \otimes Z')(\Sigma \otimes I_T)(I_M \otimes Z) \times \\ &\qquad (\tilde{S}^{-1} \otimes I_T)J[J'(\tilde{S}^{-1} \otimes Z'Z)J]^{-1}\} \\ &= E\{[J'(\tilde{S}^{-1} \otimes Z'Z)J]^{-1}J'(\tilde{S}^{-1}\Sigma\tilde{S}^{-1} \otimes Z'Z)J[J'(\tilde{S}^{-1} \otimes Z'Z)J]^{-1}\} \\ &= E\{[X'(\tilde{S}^{-1} \otimes I_T)X]^{-1}X'(\tilde{S}^{-1}\Sigma\tilde{S}^{-1} \otimes I_T)X[X'(\tilde{S}^{-1} \otimes I_T)X]^{-1}\} \\ &= E\{[X'(\tilde{S}^{-1} \otimes I_T)X]^{-1}X'(\tilde{S}^{-1} \otimes I_T)(\Sigma \otimes I_T)(\tilde{S}^{-1} \otimes I_T)X[X'(\tilde{S}^{-1} \otimes I_T)X]^{-1}\}, \end{aligned} \tag{3.38}$$

where the third line of (3.38) is obtained from the second line by using the fact that the elements of $\tilde{S}$ and $(I_M \otimes Z')u$ are independently distributed, due to the normality of the disturbances.

Using (3.4) and (3.5) in (3.38) and retaining terms of order $O(T^{-2})$, we get (after algebraic simplifications):

$$\begin{aligned} V(\tilde{\beta}_{SU}) &= \Omega + \Omega X'\Psi^{-1}E(\Delta Q\Delta)\Psi^{-1}X\Omega \\ &= \Omega + \Omega X'\Psi^{-1}E[(\tilde{S} \otimes I_T)Q(\tilde{S} \otimes I_T)]\Psi^{-1}X\Omega . \end{aligned} \tag{3.39}$$

Now consider the matrix

$$X'(\Sigma^{-1} \otimes I_T)E[(\tilde{S} \otimes I_T)Q(\tilde{S} \otimes I_T)](\Sigma^{-1} \otimes I_T)X , \tag{3.40}$$

the (g, h)'th sub-matrix of which is given by

$$\begin{aligned} &\sum_{i,j,k,\ell}^{M} \sigma^{gi}\sigma^{\ell h}E(\tilde{s}_{ij}\tilde{s}_{k\ell})X_g'Q_{jk}X_h \\ &= \frac{1}{T^2}\sum_{i,j,k,\ell}^{M} \sigma^{gi}\sigma^{\ell h}E(u_i'\bar{P}_Z u_j u_k'\bar{P}_Z u_\ell)X_g'Q_{jk}X_h \\ &= \sum_{i,j,k,\ell}^{M} \sigma^{gi}\sigma^{\ell h}\left(1 - \frac{K}{T}\right)\left[\left(1 - \frac{K}{T}\right)\sigma_{ij}\sigma_{k\ell} + \frac{1}{T}(\sigma_{ik}\sigma_{j\ell} + \sigma_{i\ell}\sigma_{jk})\right] X_g'Q_{jk}X_h \\ &= X_g'\left[\left(1 - \frac{K}{T}\right)^2 \sum_{j,k}^{M}\left(\sum_{i}^{M}\sigma^{gi}\sigma_{ij}\right)\left(\sum_{\ell}^{M}\sigma^{\ell h}\sigma_{k\ell}\right)Q_{jk}\right. \\ &\quad + \frac{1}{T}\left(1 - \frac{K}{T}\right)\sum_{j,k}^{M}\left(\sum_{i}^{M}\sigma^{gi}\sigma_{ik}\right)\left(\sum_{\ell}^{M}\sigma^{\ell h}\sigma_{j\ell}\right)Q_{jk} \\ &\quad \left. + \frac{1}{T}\left(1 - \frac{K}{T}\right)\sum_{j,k}^{M}\left(\sum_{i}^{M}\sigma^{gi}\sigma_{i\ell}\right)\left(\sum_{\ell}^{M}\sigma^{\ell h}\sigma_{jk}\right)Q_{jk}\right]X_h . \end{aligned} \tag{3.41}$$

Using (3.18), it reduces to the following:

$$\frac{1}{T}\left(1 - \frac{K}{T}\right)\left[X_g'Q_{hg}X_h + \sigma^{gh}X_g'\left(\sum_{j,k}^{M}\sigma_{jk}Q_{jk}\right)X_h\right].$$

Dropping terms of order $O(T^{-2})$ in the above expression yields

$$\frac{1}{T}[X_g'P_{gh}X_h + M\sigma^{gh}X_g'X_h - \sigma^{gh}X_g'WX_h], \tag{3.42}$$

which leads to the following result:

$$X'\Psi^{-1}E[(\tilde{S} \otimes I_T)Q(\tilde{S} \otimes I_T)]\Psi^{-1}X = \frac{1}{T}X'PX + \frac{M}{T}X'\Psi^{-1}X - \frac{1}{T}X'(\Sigma^{-1} \otimes W)X . \tag{3.43}$$

Substituting (3.43) in (3.39) gives the expression (3.32) for the variance covariance matrix of the SUUR estimator, to order $O(T^{-2})$.

The expression for the variance covariance matrix, to order $O(T^{-2})$, of the SURR estimator can be derived in a similar manner (see Srivastava, 1970, for details), and it is interesting to note that it turns out to be the same as (3.32). Thus, both the SUUR and SURR estimators have identical variance covariance matrices to order $O(T^{-2})$, but they may differ if we consider higher order asymptotic approximations. This result led Kariya and Maekawa (1982) to consider third-order asymptotic approximations and to derive the variance covariance matrix, to order $O(T^{-3})$, of the SUUR estimator in the context of a two equation model. As they did not obtain a corresponding expression for the SURR estimator, it is not known whether the inclusion of a term of order $O(T^{-3})$ permits us to discriminate between the SUUR and SURR estimators in terms of their relative precisions. This matter deserves further research.

3.3.2 Comparison with OLS Estimator

The exact expression for the variance covariance matrix of the OLS estimator for the SURE model is

$$\begin{aligned} V(b_0) &= E(b_0 - \beta)(b_0 - \beta)' \\ &= (X'X)^{-1}X'\Psi X(X'X)^{-1} . \end{aligned} \tag{3.44}$$

If we compare this expression with the first-order asymptotic approximation (3.13), it is easy to verify that the FGLS estimator is more efficient than the OLS estimator. In order to elicit the gain in efficiency, let us consider the case of a two equation model and compare the FGLS and the OLS estimators of β_1. For a two equation model, we have

$$\Omega = (X'\Psi^{-1}X)^{-1} = (1 - \rho_{12}^2)\begin{bmatrix} \frac{1}{\sigma_{11}} X_1'X_1 & -\frac{\rho_{12}}{\sqrt{\sigma_{11}\sigma_{22}}} X_1'X_2 \\ -\frac{\rho_{12}}{\sqrt{\sigma_{11}\sigma_{22}}} X_2'X_1 & \frac{1}{\sigma_{22}} X_2'X_2 \end{bmatrix}^{-1},$$

so that according to the first-order asymptotic approximation (3.13), the variance covariance matrix of the FGLS estimator of β_1 is

$$V_{F,1} = \sigma_{11}(1 - \rho_{12}^2)[X_1'X_1 - \rho_{12}^2 X_1'P_{x_2}X_1]^{-1} . \tag{3.45}$$

Observing that the variance covariance matrix of the OLS estimator of β_1 is

$$V_{0,1} = \sigma_{11}(X_1'X_1)^{-1} , \tag{3.46}$$

we can derive the relative efficiencies of these estimators by considering generalized variances, obtained by taking the ratio of the determinants of the variance covariance matrices:

$$\frac{|V_{F,1}|}{|V_{0,1}|} = \frac{(1 - \rho_{12}^2)^{K_1}}{\Pi_{i=1}^{K_1} (1 - \rho_{12}^2\theta_i)} , \tag{3.47}$$

where θ_1, θ_2, ..., θ_{K_1} are the squared canonical correlation coefficients between the sets of variables in X_1 and X_2. These are the nonzero roots of the determinantal equation

$$|I_{K_1} - \theta(X_1'X_1)^{-1}X_1'X_2(X_2'X_2)^{-1}X_2'X_1| = 0 .$$

The ratio (3.47) permits us to examine the role of the correlations between the disturbances as well as between the regressors across the equations. Thus, given ρ_{12}, the maximum efficiency gain arises when all θ_i's are zero while there is no gain if all θ_i's are unity. That is, the efficiency gain is a decreasing function of the canonical correlations (see Zellner and Huang, 1962; p. 308). However, this result may not be very useful, so Binkley (1982) considered the estimation of an element of β_1, as follows. Without any loss of generality, take this to be the first element β_{11} (say) of β_1. If we partition X_1 as

$$\underset{}{X_1} = \underset{(T\times 1)}{(x_{11}} \quad \underset{[T\times(K_1-1)]}{X_{11})}$$

it is easily seen from (3.45) that the variance of the FGLS estimator of β_{11} is

$$V_{F,11} = \frac{\sigma_{11}(1-\rho_{12}^2)}{x_{11}'(I_T-\rho_{12}^2P_{x_2})x_{11} - x_{11}'(I_T-\rho_{12}^2P_{x_2})X_{11}[X_{11}'(I_T-\rho_{12}^2P_{x_2})X_{11}]^{-1}X_{11}'(I_T-\rho_{12}^2P_{x_2})x_1} \tag{3.48}$$

to a first-order approximation.

If $\hat{e}$ is the OLS residual vector obtained from the regression of the variable in x_{11} on the variables in X_{11}, we can write $V_{F,11}$ as

$$V_{F,11} = \frac{\sigma_{11}(1-\rho_{12}^2)}{\hat{e}'(I_T-\rho_{12}^2P_{x_2})\hat{e} - \hat{e}'(I_T-\rho_{12}^2P_{x_2})X_{11}[X_{11}'(I_T-\rho_{12}^2P_{x_2})X_{11}]^{-1}X_{11}'(I_T-\rho_{12}^2P_{x_2})\hat{e}}$$

$$= \frac{\sigma_{11}(1-\rho_{12}^2)}{\hat{e}'\hat{e}(1-\rho_{12}^2R_{\hat{e}\cdot X_2}^2) - \alpha} \tag{3.49}$$

where

$$R_{\hat{e}\cdot X_2}^2 = \frac{\hat{e}'P_{X_2}\hat{e}}{\hat{e}'\hat{e}}$$

$$\alpha = \rho_{12}^2\hat{e}'P_{x_2}X_{11}[X_{11}'(I_T - \rho_{12}^2P_{x_2})X_{11}]^{-1}X_{11}'P_{x_2}\hat{e} .$$

Despite the fact that the relationship between x_{11} and X_2 generally will be different from the relationship between $\hat{e}$ and X_2, it is interesting to note from (3.48) and (3.49) that $V_{F,11}$ remains unaltered whether the regressand is x_{11} or $\hat{e}$. This indicates that the effect of the correlation between the variables across the equations depends upon the correlation between the variables within the equations; i.e., the multicollinearity within the equations.

As the matrix $[X_{11}'X_{11} - X_{11}'P_{x_2}X_{11}]$ is at least positive semi-definite, the matrix

$$[X_{11}'X_{11} - \rho_{12}^2X_{11}'P_{x_2}X_{11}] - [X_{11}'X_{11} - \rho_{12}^2X_{11}'X_{11}] ,$$

and therefore

$$\frac{1}{(1-\rho_{12}^2)}(X_{11}'X_{11})^{-1} - [X_{11}'(I_T - \rho_{12}^2 P_{x_2})X_{11}]^{-1} ,$$

will also be at least positive semi-definite. Using this result, we have

$$\alpha \leq \frac{1}{(1-\rho_{12}^2)} \hat{e}' P_{x_2} P_{x_{11}} P_{x_2} \hat{e} . \tag{3.50}$$

If $e_p = P_{x_2}\hat{e}$ is the vector of predicted values of $\hat{e}$ obtained by the OLS regression of $\hat{e}$ on X_2, and $R_{e_p \cdot X_{11}}$ denotes the multiple correlation coefficient between e_p and X_{11}, we have

$$\begin{aligned} \hat{e}' P_{x_2} P_{x_{11}} P_{x_2} \hat{e} &= e_p' P_{x_{11}} e_p \\ &= R^2_{e_p \cdot X_{11}} e_p' e_p \\ &= R^2_{e_p \cdot X_{11}} \hat{e}' P_{x_2} \hat{e} \leq R^2_{e_p \cdot X_{11}} e'e , \end{aligned}$$

from which we obtain

$$\alpha \leq \frac{\rho_{12}^4 R^2_{e_p \cdot X_{11}}}{(1-\rho_{12}^2)} \hat{e}'\hat{e} .$$

Using this, we find from (3.49):

$$V_{F,11} \leq \frac{\sigma_{11}(1-\rho_{12}^2)}{\hat{e}'\hat{e}[1 - \rho_{12}^2 R^2_{\hat{e}\cdot X_2} - (\rho_{12}^4/1-\rho_{12}^2) R^2_{e_p \cdot X_{11}}]} . \tag{3.51}$$

Similarly, from (3.46) the variance of the OLS estimator of β_{11} is

$$V_{0,11} = \frac{\sigma_{11}}{x_{11}' \bar{P}_{x_{11}} x_{11}} = \frac{\sigma_{11}}{\hat{e}'\hat{e}} . \tag{3.52}$$

Recalling that the first-order asymptotic approximation of the variance covariance matrix of the FGLS estimator is equal to the exact variance covariance matrix of the GLS estimator, it follows from the best linear unbiasedness property of the GLS estimator that $V_{F,11}$ cannot exceed $V_{0,11}$:

$$V_{F,11} \leq \frac{\sigma^2}{\hat{e}'\hat{e}} \cdot \tag{3.53}$$

Combining the results (3.50) and (3.53) we can write the efficiency ratio compactly as

$$\frac{V_{F,11}}{V_{0,11}} = \left[\frac{1 - \rho_{12}^2}{1 - \rho_{12}^2 R_{\hat{e}\cdot X_2}^2 - \alpha}\right], \tag{3.54}$$

where

$$0 \leq \alpha \leq \min\left[\frac{\rho_{12}^4}{1 - \rho_{12}^2} R_{e_p \cdot X_{11}}^2,\ \rho_{12}^2(1 - R_{\hat{e}\cdot X_2}^2)\right] \cdot \tag{3.55}$$

Thus, as shown by Binkley (1978), we see that the asymptotic efficiency of the FGLS estimator is not only influenced by the correlation between disturbances and correlation between variables across the equations, but also by the correlation between the variables within the equation (i.e., the multicollinearity within the equation). So, if we have two SURE models that have the same levels of correlation between the disturbances as well as between the variables across the equations, there may be different efficiency gains for the FGLS estimators over OLS. For instance, when x_{11} is not highly correlated with the variables in X_{11}, a high correlation between x_{11} and the variables in X_2 will tend to reduce the efficiency gain. However, this reduction in the efficiency gain may be very small or even zero, and consequently significant efficiency gains may arise, when the variable x_{11} is highly correlated with the variables in X_{11}. Thus, higher correlations among the variables within the equations may be favourable to the application of the FGLS estimator and may lead to significant efficiency gains arising from FGLS estimation in comparison to OLS. See Binkley (1982; pp. 894-895) for further discussion.

If we compare (3.32) and (3.44), it is difficult to draw any clear conclusion. The same holds true for the SUUR and SURR estimators if we use (3.13) and (3.32) to study the improvement obtained by taking a second-order approximation over a first-order approximation. We therefore examine these matters in two special cases of a two equation model.

3.3.3 The Two Equation Model with Orthogonal Regressors

Consider a two equation SURE model with $X_1'X_2 = 0$. Using this property and recalling the discussion in Section 2.6, especially equation (2.43) and the related material, it is easy to obtain the following results:

$$\Omega = (1 - \rho_{12}^2)\begin{bmatrix} \sigma_{11}(X_1'X_1)^{-1} & 0 \\ 0 & \sigma_{22}(X_2'X_2)^{-1} \end{bmatrix},$$

$$\rho_{12} = \sigma_{12}/(\sigma_{11}\sigma_{22})^{\frac{1}{2}}$$

$$X'PX = 0$$

$$W = (P_{X_1} + P_{X_2}) .$$

Substituting these into (3.32), we obtain

$$V(\tilde{\beta}_{SU}) = V(\hat{\beta}_{SR}) = \left(1 + \frac{1}{T}\right)\begin{bmatrix} \sigma_{11}(1 - \rho_{12}^2)(X_1'X_1)^{-1} & 0 \\ 0 & \sigma_{22}(1 - \rho_{12}^2)(X_2'X_2)^{-1} \end{bmatrix} . \tag{3.56}$$

Thus, we see that the second-order asymptotic approximation to the variance covariance matrix calls for the addition of T^{-1} times the variance covariance matrix to the expression for the first-order asymptotic approximation. This additional term also furnishes a measure, to the order of our approximation, of the increase in variability arising from the replacement of Σ in the GLS "estimator" by its consistent estimator S.

Using (3.44), the expression for the variance covariance matrix of the OLS estimator is

$$V(b_0) = \begin{bmatrix} \sigma_{11}(X_1'X_1)^{-1} & 0 \\ 0 & \sigma_{22}(X_2'X_2)^{-1} \end{bmatrix} . \tag{3.57}$$

Comparing (3.56) and (3.57), we observe that the OLS estimator will be more efficient than the SUUR or SURR estimators if

$$\rho_{12}^2 \le \frac{1}{T+1}, \quad \text{or} \quad T \le \frac{1}{\rho_{12}^2} - 1 . \tag{3.58}$$

As T is assumed to be large, the OLS estimator will be preferable to the SUUR or SURR estimators only in the presence of very weak correlation between the disturbances across the two equations of the SURE model. On the other hand, the SUUR and SURR estimators will be preferable to the

OLS estimator when ρ_{12}^2 exceeds $1/(T+1)$, i.e., when the correlation is at least moderate.

3.3.4 The Two Equation Model with Subset Regressors

Next, let us consider a two equation model in which X_2 is a sub-matrix of X_1. It is then easy to verify that

$$X_2' P_{x_1} = X_2'$$

$$\Omega = \begin{bmatrix} \sigma_{11}(X_1'X_1)^{-1} - \sigma_{11}\rho_{12}^2(X_1'X_1)^{-1}X_1'\bar{P}_{x_2}X_1(X_1'X_1)^{-1} & \rho_{12}(\sigma_{11}\sigma_{22})^{\frac{1}{2}}(X_1'X_1)^{-1}X_1'X_2(X_2'X_2)^{-1} \\ \rho_{12}(\sigma_{11}\sigma_{22})^{\frac{1}{2}}(X_2'X_2)^{-1}X_2'X_1(X_1'X_1)^{-1} & \sigma_{22}(X_2'X_2)^{-1} \end{bmatrix}$$

$$X'PX = 0$$

$$W = (P_{x_1} + P_{x_2}) .$$

Using the above results, the expression (3.32) simplifies to

$$V(\tilde{\beta}_{SU}) = V(\hat{\beta}_{SR}) = \begin{bmatrix} V_{11} & V_{12} \\ V_{21} & V_{22} \end{bmatrix} \tag{3.59}$$

where

$$V_{11} = \sigma_{11}(X_1'X_1)^{-1} - \sigma_{11}\left(\rho_{12}^2 - \frac{1-\rho_{12}^2}{T}\right)(X_1'X_1)^{-1}X_1'\bar{P}_{x_2}X_1(X_1'X_1)^{-1}$$

$$V_{12} = V_{21}' = \rho_{12}(\sigma_{11}\sigma_{22})^{\frac{1}{2}}(X_1'X_1)^{-1}X_1'X_2(X_2'X_2)^{-1}$$

$$V_{22} = \sigma_{22}(X_2'X_2)^{-1} .$$

Thus we see that retaining second-order terms in the variance covariance matrix results in the addition of a "positive" quantity (by which we mean that the matrix

$$\frac{\sigma_{11}(1-\rho_{12}^2)}{T}(X_1'X_1)^{-1}X_1'\bar{P}_{x_2}X_1(X_1'X_1)^{-1} \tag{3.60}$$

is positive definite) to the first-order asymptotic approximation for the variance covariance matrix of the SUUR and SURR estimators of the coefficients in the first equation of the model. No changes occur in the cross-moment matrix of the estimators of β_1 and β_2 or in the variance covariance matrix of the estimator of β_2. Notice that the SUUR and SURR estimators of β_2 are identical to the OLS estimator in this case, as was discussed in Section 2.5.

The quantity (3.60) also provides the increase in the variance covariance matrix to order $O(T^{-1})$ which can be attributed to the substitution of a consistent estimator $\tilde{S}$ or $\hat{S}$ in place of Σ in the GLS "estimator" in order to make the latter "feasible."

The variance covariance matrix (3.44) of the OLS estimator is

$$V(b_0) = \begin{bmatrix} \sigma_{11}(X_1'X_1)^{-1} & \rho_{12}(\sigma_{11}\sigma_{22})^{\frac{1}{2}}(X_1'X_1)^{-1}X_1'X_2(X_2'X_2)^{-1} \\ \rho_{12}(\sigma_{11}\sigma_{22})^{\frac{1}{2}}(X_2'X_2)^{-1}X_2'X_1(X_1'X_1)^{-1} & \sigma_{22}(X_2'X_2)^{-1} \end{bmatrix}. \tag{3.61}$$

Comparing (3.59) and (3.61) we see that the variance covariance matrices of the OLS and the SUUR (SURR) estimators differ only as far as the estimation of β_1 is concerned. Further, the OLS estimator performs better, in terms of its precision, than the SUUR (SURR) estimator when ρ_{12}^2 does not exceed $1/(T + 1)$, which is the condition (3.58) obtained in the case of orthogonal regressors. Conversely, the SUUR (SURR) estimator will have better precision than OLS for moderate to high correlations between the disturbances of the equations of the SURE model.

3.4 ASYMPTOTIC APPROXIMATIONS FOR DENSITY AND DISTRIBUTION FUNCTIONS

In this section we no longer consider merely the first two moments of the sampling distributions of the various estimators for the SURE model, but consider large-sample approximations to complete density and distribution functions. The basis for comparing alternative SURE estimators is concentration probability.

3.4.1 The SUUR Estimator

Here we describe results arising from work by Phillips (1977). First define

$$\nabla = \left(\frac{T}{T - K}\right)\tilde{S} - \Sigma$$

and let w be a column vector whose elements are the distinct elements of ∇. Further, let

$$\begin{aligned} p &= \frac{1}{T}(I_M \otimes Z')u \\ C &= \frac{1}{T}(Z'Z) , \end{aligned} \tag{3.62}$$

where, as previously, Z is the $(T \times K)$ matrix of observations on all of the K distinct regressors in the M-equation SURE model.

Assuming normality of the disturbances, it is easy to see that the elements of p are stochastically independent of the elements of ∇ and hence of w. Further, p follows a multivariate Normal distribution with mean vector 0 and variance covariance matrix

$$\begin{aligned} E(pp') &= \frac{1}{T^2}(I_M \otimes Z')E(uu')(I_M \otimes Z) \\ &= \frac{1}{T^2}(I_M \otimes Z')(\Sigma \otimes I_T)(I_M \otimes Z) \\ &= \frac{1}{T^2}(\Sigma \otimes Z'Z) \\ &= \frac{1}{T}(\Sigma \otimes C) . \end{aligned} \tag{3.63}$$

From (3.37), we have

$$\begin{aligned} (\tilde{\beta}_{SU} - \beta) &= [J'(\tilde{S}^{-1} \otimes Z'Z)J]^{-1}J'(\tilde{S}^{-1} \otimes I_T)(I_M \otimes Z')u \\ &= [J'((\Sigma + \nabla)^{-1} \otimes C)J]^{-1}J'[(\Sigma + \nabla)^{-1} \otimes I_T]p . \end{aligned} \tag{3.64}$$

Taking h to be any fixed vector of order $(K^* \times 1)$, consider the linear function

$$\begin{aligned} e &= h'(\tilde{\beta}_{SU} - \beta) \\ &= h'[J'[(\Sigma + \nabla)^{-1} \otimes C]J]^{-1}J'[(\Sigma + \nabla)^{-1} \otimes I_T]p \end{aligned} \tag{3.65}$$

which is a function of ∇ and p, or equivalently of w and p.

Now, note that (Phillips, 1977; p. 150):

(i) The function e reduces to 0 if we set $p = 0$ and $w = 0$, or equivalently $\nabla = 0$.

(ii) The function e has continuous derivatives, at least up to the fourth order in the neighbourhood of the origin $(p = 0,\ w = 0)$, and these derivatives are uniformly bounded in the neighbourhood of the origin as $T \to \infty$.

(iii) Putting $p = 0$ and $w = 0$ in the vector of partial derivatives of e with respect to the elements of p, we find

$$\left(\frac{\partial e}{\partial p}\right)\Bigg|_{\substack{p=0\\w=0}} = (\Sigma^{-1} \otimes I_T)J[J'(\Sigma^{-1} \otimes C)J]^{-1}h$$

so that the quantity

$$\left(\frac{\partial e}{\partial p}\right)'\Bigg|_{\substack{p=0\\w=0}} \left(\frac{\partial e}{\partial p}\right)\Bigg|_{\substack{p=0\\w=0}} = h'[J'(\Sigma^{-1} \otimes C)J]^{-1}J'(\Sigma^{-2} \otimes I_T)J[J'(\Sigma^{-1} \otimes C)J]^{-1}h$$

is bounded above zero as $T \to \infty$.

(iv) The vector of partial derivatives of e with respect to the elements of w at $p = 0$ and $w = 0$ is a null vector.

(v) As the matrix $T\tilde{S}$ follows a Wishart distribution, it follows that $\sqrt{T}$ times the elements of ∇, or equivalently w, will have bounded moments of all orders as $T \to \infty$.

(vi) Definitionally the (i, j)'th element of ∇ is

$$\left(\frac{T}{T - K}\right)\tilde{s}_{ij} - \sigma_{ij} = \frac{1}{(T - K)} u_i'\bar{P}_Z u_j - \sigma_{ij} ,$$

whose expectation is 0. In other words, $E(\nabla) = 0$, which implies that $E(w) = 0$.

In view of the above observations, it follows from the "approximation theorem" of Sargan (1975; p. 327) that the distribution of $\sqrt{T}e = a_h$ (say) admits a valid Edgeworth expansion.

As ∇ is of order $O_p(T^{-\frac{1}{2}})$, we have

$$\begin{aligned}\tilde{S}^{-1} &= \left[\left(\frac{T - K}{T}\right)(\Sigma + \nabla)\right]^{-1} \\ &= \left(1 - \frac{K}{T}\right)^{-1}(\Sigma + \nabla)^{-1} \\ &= \Sigma^{-1} - \Sigma^{-1}\nabla\Sigma^{-1} + O_p(T^{-g}) \qquad (g \geq 1) . \qquad (3.66)\end{aligned}$$

Writing

$$p_* = T^{\frac{1}{2}}p = T^{-\frac{1}{2}}(I_M \otimes Z')u$$
$$\Omega_* = T\Omega = T[X'(\Sigma^{-1} \otimes I_T)X]^{-1} = [J'(\Sigma^{-1} \otimes C)J]^{-1} , \tag{3.67}$$

if we substitute (3.66) in (3.65) and consider a first-order approximation, we have

$$a_h = h'\Omega_* J'\Psi^{-1} p_* . \tag{3.68}$$

As p_* follows a multivariate Normal distribution with mean vector 0 and variance covariance matrix $(\Sigma \otimes C) = \Psi(I_M \otimes C)$, the first-order asymptotic approximation to the exact distribution of a_h is univariate Normal with mean 0 and variance $\theta^2 = h'\Omega_* h = Th'\Omega h$. For the second-order asymptotic approximation, we use (3.6) to express a_h to order $O_p(T^{-1})$. This gives

$$a_h = T^{\frac{1}{2}}h'\Omega X'\Psi^{-1}u - T^{\frac{1}{2}}h'\Omega X'\Psi^{-1}(\nabla \otimes I_T)Qu$$
$$+ T^{\frac{1}{2}}\Omega X'\Psi^{-1}(\nabla \otimes I_T)Q(\nabla \otimes I_T)Qu + O_p(T^{-g}) \qquad (g \geq 3/2)$$
$$= h'Bp_* + O_p(T^{-g}) \qquad (g \geq 3/2) \tag{3.69}$$

where

$$B = \Omega_* J'\Psi^{-1}[\Psi - (\nabla \otimes I_T)G + (\nabla \otimes I_T)G\Psi^{-1}(\nabla \otimes I_T)G]\Psi^{-1}$$
$$G = I_{MT} - (\Sigma^{-1} \otimes C)J\Omega_* J' . \tag{3.70}$$

Using (3.69) and recalling that ∇ and p_* are independently distributed (because ∇ and p are independent), the characteristic function of a_h up to order $O(T^{-1})$ is

$$\begin{aligned} CF(\alpha) &= E[\exp\{\sqrt{-1}\,\alpha a_h\}] \\ &= E[\exp\{\sqrt{-1}\,\alpha h'Bp_*\}] \\ &= E_\nabla[E_{p_*}(\exp\{\sqrt{-1}\,\alpha h'Bp_*\})] \\ &= E_\nabla[\exp\{-\tfrac{1}{2}\alpha^2 h'B\Psi(I_M \otimes C)B'h\}] , \end{aligned} \tag{3.71}$$

where α is a real scalar quantity.

Now, we have

$$B\Psi(I_M \otimes C)B' = \Omega_* J'\Psi^{-1}[\Psi - (\nabla \otimes I_T)G + (\nabla \otimes I_T)G\Psi^{-1}(\nabla \otimes I_T)G] \times$$

$$(I_M \otimes C)\Psi^{-1}[\Psi - (\nabla \otimes I_T)G + G'(\nabla \otimes I_T)\Psi^{-1}G' \times$$

$$(\nabla \otimes I_T)]\Psi^{-1}J\Omega_*$$

$$= \Omega_* J'\Psi^{-1}[\Psi(I_M \otimes C) - (\nabla \otimes I_T)G(I_M \otimes C)$$

$$+ (\nabla \otimes I_T)G\Psi^{-1}(\nabla \otimes I_T)G(I_M \otimes C)$$

$$- \Psi(I_M \otimes C)\Psi^{-1}G'(\nabla \otimes I_T)$$

$$+ (\nabla \otimes I_T)G(I_M \otimes C)\Psi^{-1}G'(\nabla \otimes I_T)$$

$$+ \Psi(I_M \otimes C)\Psi^{-1}G'(\nabla \otimes I_T)\Psi^{-1}G'(\nabla \otimes I_T)]\Psi^{-1}J\Omega_* + O_p(T^{-g})$$

$$(g \geq 3/2)$$

$$= \Omega_* + \Omega_* J'\Psi^{-1}(\nabla \otimes I_T)G(I_M \otimes C)\Psi^{-1}(\nabla \otimes I_T)\Psi^{-1}J\Omega_* + O_p(T^{-g})$$

$$(g \geq 3/2) \qquad (3.72)$$

where the following results have been used for simplification:

$$G(I_M \otimes C)\Psi^{-1}G' = G(I_M \otimes C)\Psi^{-1}$$

$$G(I_M \otimes C)\Psi^{-1}J = 0$$

$$J'G = 0 .$$

Defining

$$q = h'\Omega_* J'\Psi^{-1}(\tilde{S} \otimes I_T)G(I_M \otimes C)\Psi^{-1}(\tilde{S} \otimes I_T)\Psi^{-1}J\Omega_* h$$

$$= Th'\Omega X'\Psi^{-1}(\tilde{S} \otimes I_T)Q(\tilde{S} \otimes I_T)\Psi^{-1}X\Omega h$$

and using (3.72), one can obtain from (3.71) the following:

$$
\begin{aligned}
\mathrm{CF}(\alpha) &= \mathrm{E}_{\nabla}\left[\exp\left\{-\frac{1}{2}\alpha^2(\theta^2+q)\right\}\right] + O(T^{-g}) \qquad (g \geq 3/2) \\
&= \exp\left\{-\frac{1}{2}\alpha^2\theta^2\right\}\mathrm{E}_{\nabla}\left[\exp\left\{-\frac{1}{2}\alpha^2 q\right\}\right] + O(T^{-g}) \qquad (g \geq 3/2) \\
&= \exp\left\{-\frac{1}{2}\alpha^2\theta^2\right\}\mathrm{E}_{\nabla}\left(1-\frac{1}{2}\alpha^2 q\right) + O(T^{-g}) \qquad (g \geq 3/2) \\
&= \exp\left\{-\frac{1}{2}\alpha^2\theta^2\right\}\left(1-\frac{1}{2}\alpha^2 \mathrm{E}(q)\right) + O(T^{-g}) \qquad (g \geq 3/2)\,. \qquad (3.73)
\end{aligned}
$$

However, from (3.43) we have

$$
\begin{aligned}
\mathrm{E}(q) &= \left(\frac{M}{T}\right)\theta^2 + h'\Omega X'[P - (\Sigma^{-1}\otimes W)]X\Omega h \\
&= \bar{q}\,, \qquad (3.74)
\end{aligned}
$$

say, so that the characteristic function, to order $O(T^{-1})$, of a_h is

$$\left(1-\frac{1}{2}\alpha^2\bar{q}\right)\exp\left\{-\frac{1}{2}\alpha^2\theta^2\right\}. \qquad (3.75)$$

Employing the "inversion theorem," and denoting the univariate standard Normal probability density function as

$$\phi(z) = (2\pi)^{-\frac{1}{2}}\exp\left\{-\frac{1}{2}z^2\right\},$$

the probability density function of a_h is

$$
\begin{aligned}
f_{SU}(a_h) &= \frac{1}{2\pi}\int_{-\infty}^{\infty}\left[\left(1-\frac{1}{2}\sigma^2\bar{q}\right)\exp\{-\sqrt{-1}\,\alpha a_h\}\exp\left\{-\frac{1}{2}\alpha^2\theta^2\right\}\right]d\alpha \\
&= \frac{1}{2\pi}\int_{-\infty}^{\infty}\exp\left\{-\sqrt{-1}\,\alpha a_h - \frac{1}{2}\alpha^2\theta^2\right\}d\alpha \\
&\quad + \frac{\bar{q}}{4\pi}\int_{-\infty}^{\infty}\left[(\sqrt{-1}\,\alpha)^2\exp\left\{-\sqrt{-1}\,\alpha a_h - \frac{1}{2}\alpha^2\theta^2\right\}\right]d\alpha \\
&= \left(\frac{1}{\theta}\right)\phi\left(\frac{a_h}{\theta}\right) - \frac{\bar{q}}{2\theta^2}\left[\frac{a_h^2}{\theta^2}-1\right]\left(\frac{1}{\theta}\right)\phi\left(\frac{a_h}{\theta}\right) \qquad (3.76)
\end{aligned}
$$

up to order $O(T^{-1})$.

From (3.76), one can write the distribution function of a_h, to the same order of approximation, as

$$F_{SU}(m) = \int_{-\infty}^{m} [f_{SU}(a_h)]\, da_h$$

$$= \Phi\left(\frac{m}{\theta}\right) - \frac{m\bar{q}}{2\theta^2}\,\phi\left(\frac{m}{\theta}\right) \tag{3.77}$$

which, following Sargan (1975), can be expressed alternatively as

$$F_{SU}(m) = \Phi\left[\left(1 - \frac{\bar{q}}{2\theta^2}\right)\frac{m}{\theta}\right], \tag{3.78}$$

where $\Phi(\cdot)$ denotes the cumulative distribution function of a standard Normal variate (see Phillips, 1977; pp. 156-157).

It is interesting to note that the first term on the right hand side of (3.77) is the cumulative distribution function of a Normal variate. The second term is of order $O(T^{-1})$, as $\bar{q}$ is of order $O(T^{-1})$ and $\theta^2 = Th'\Omega h$ is of order $O(1)$. The absence of any term of order $O(T^{-\frac{1}{2}})$ implies that for moderate T the sampling distribution of the SUUR estimator may be quite close to the Normal distribution. Further, we have

$$F_{SU}(0) = \Phi(0) = \frac{1}{2}, \tag{3.79}$$

so that the SUUR estimator is median unbiased at least to the order of our approximation (see Phillips, 1977; p. 157).

Finally, note that the asymptotic approximations for the density and distribution functions of a_h, $f_{SU}(\cdot)$ and $F_{SU}(\cdot)$ have their validity in the neighbourhood of the origin, $a_h = 0$, characterized by $\nabla = 0$ and $p = 0$. Relaxing this specification, Kariya and Maekawa (1982) considered a two equation SURE model and worked out the relevant asymptotic approximations. Their principal results are stated below without proof.

If we write

$$(\Sigma^{-\frac{1}{2}} \otimes I_T)X\Omega^{\frac{1}{2}} = \begin{bmatrix} A_1 \\ A_2 \end{bmatrix},$$

$$H = \frac{2(T - K)}{(T - K - 1)}\,[\Omega - \Omega^{\frac{1}{2}}\{(A_1'A_1)^2 + (A_2'A_2)^2\}\Omega^{\frac{1}{2}}], \tag{3.80}$$

$$\bar{q}^* = h'Hh,$$

then the second-order asymptotic approximation for the probability density function of a_h is

$$f^*_{SU}(a_h) = \frac{1}{\theta}\Phi\left(\frac{a_h}{\theta}\right) - \frac{\bar{q}^*}{2\theta^2}\left(\frac{a_h}{\theta^2} - 1\right)\frac{1}{\theta}\,\phi\left(\frac{a_h}{\theta}\right) , \tag{3.81}$$

from which the expression for the second-order asymptotic approximation of the distribution function can be obtained as:

$$F^*_{SU}(m) = \Phi\left(\frac{m}{\theta}\right) - \frac{m\bar{q}^*}{2\theta^2}\,\phi\left(\frac{m}{\theta}\right) . \tag{3.82}$$

Kariya and Maekawa also obtained the asymptotic approximation for the probability density function of $\sqrt{T}(\tilde{\beta}_{SU} - \beta)$. Denoting by $\phi^*(b)$ the multivariate Normal density function of a random vector b with mean vector 0 and variance covariance matrix $(T - K)\Omega^{-1}$, the first-order asymptotic approximation for the exact probability density function of $\sqrt{T}(\tilde{\beta}_{SU} - \beta)$ is $\phi^*(b)$. Similarly, the second-order asymptotic approximation is

$$(1 + \alpha^*)\phi^*(b) , \tag{3.83}$$

where

$$\alpha^* = \frac{1}{(T-K)(T-K-1)}\left[b'\Omega^{-1}b + (T-K-1)\left\{\frac{1}{2}\mathrm{tr}\, H\Omega^{-1} - \frac{1}{(T-K)}b'\Omega^{-1}H\Omega^{-1}b\right\}\right] .$$

Kariya and Maekawa claimed that these results do not require T to be as large as is needed in large-sample asymptotic approximations; merely asymptotic normality is required. These results are valid for all T "not too small."

3.4.2 Comparison with OLS Estimator

We know that the sampling distribution of $(b_0 - \beta)$ is multivariate Normal with mean vector 0 and variance covariance matrix $\Omega_0 = (X'X)^{-1}X'\Psi X(X'X)^{-1}$. So, the probability density function of $a_h = \sqrt{T}\,h'(b_0 - \beta)$ is

$$f_0(a_h) = \frac{1}{\theta_0}\,\phi\left(\frac{a_h}{\theta_0}\right) \tag{3.84}$$

and its cumulative distribution function is

$$F_0(m) = \Phi\left(\frac{m}{\theta_0}\right) \tag{3.85}$$

where $\theta_0 = (Th'\Omega_0 h)^{\frac{1}{2}}$.

Now, in order to compare the SUUR and OLS estimators, we choose the performance criterion to be the concentration probability (CP), popularly known as Pitman's efficiency criterion. This is defined as the probability in an interval symmetric around the true parameter value. From (3.78) and (3.85), the concentration probabilities (to the order of our approximation) associated with the SUUR and OLS estimators are

$$\begin{aligned} CP_{SU} &= \Pr[|T^{\frac{1}{2}}h'(\tilde{\beta}_{SU} - \beta)| \leq m] \\ &= F_{SU}(m) - F_{SU}(-m) \\ &= \Phi\left[\left(1 - \frac{\bar{q}}{2\theta^2}\right)\frac{m}{\theta}\right] - \Phi\left[-\left(1 - \frac{\bar{q}}{2\theta^2}\right)\frac{m}{\theta}\right], \end{aligned} \tag{3.86}$$

$$\begin{aligned} CP_0 &= \Pr[|T^{\frac{1}{2}}h'(b_0 - \beta)| \leq m] \\ &= F_0(m) - F_0(-m) \\ &= \Phi\left(\frac{m}{\theta_0}\right) - \Phi\left(-\frac{m}{\theta_0}\right). \end{aligned} \tag{3.87}$$

Thus, we have

$$(CP_{SU} - CP_0) = 2\left[\Phi\left(\left(1 - \frac{\bar{q}}{2\theta^2}\right)\frac{m}{\theta}\right) - \Phi\left(\frac{m}{\theta_0}\right)\right],$$

from which it follows that the OLS estimator will be more concentrated around β than will the SUUR estimator if

$$\left(1 - \frac{\bar{q}}{2\theta^2}\right) \leq \left(\frac{\theta}{\theta_0}\right). \tag{3.88}$$

Let us examine this condition for a two equation model. If the regressors across the equations are orthogonal (i.e., $X_1'X_2 = 0$), it is seen from (3.56) and (3.57) that

$$h'\Omega h = (1-\rho_{12}^2)\sum_i^2 \sigma_{ii}h_i'(X_i'X_i)^{-1}h_i$$

$$h'\Omega_0 h = \sum_i^2 \sigma_{ii}h_i'(X_i'X_i)^{-1}h_i$$

$$\bar{q} = (1-\rho_{12}^2)\sum_i^2 \sigma_{ii}h_i'(X_i'X_i)^{-1}h_i$$

where $h' = (h_1 h_2)'$, with the vectors h_1 and h_2 being of orders $(K_1 \times 1)$ and $(K_2 \times 1)$ respectively.

Using these results, the condition (3.88) reduces to

$$\left(1-\frac{1}{2T}\right) \leq (1-\rho_{12}^2)^{\frac{1}{2}} ,$$

from which we have

$$\rho_{12}^2 \leq \frac{1}{T}\left(1-\frac{1}{4T}\right) \quad \text{or} \quad T \leq \frac{1+(1-\rho_{12}^2)^{\frac{1}{2}}}{2\rho_{12}^2} , \tag{3.89}$$

which is similar to the condition (3.58) obtained for the dominance of the OLS estimator over the SUUR and SURR estimators with the variance covariance matrix to order $O(T^{-2})$ as the performance criterion. Thus, the OLS estimator will have higher concentration around β than the SUUR estimator for a low correlation between the disturbances across the two equations. The reverse is true when this correlation is moderate to high.

Next, let us consider a two equation SURE model with subset regressors; i.e., X_2 is a submatrix of X_1. As the SUUR and OLS estimators for the coefficient vector in the second equation are identical, we restrict our attention to the coefficient vector in the first equation of the model. For this purpose, we set $h_2 = 0$. It is then easy to see from (3.59) and (3.61) that

$$h'\Omega h = \sigma_{11}h_1'(X_1'X_1)^{-1}X_1'[I_T - \rho_{12}^2\bar{P}_{X_2}]X_1(X_1'X_1)^{-1}h_1$$

$$h'\Omega_0 h = \sigma_{11}h_1'(X_1'X_1)^{-1}h_1$$

$$\bar{q} = \sigma_{11}(1-\rho_{12}^2)h_1'(X_1'X_1)^{-1}X_1'\bar{P}_{X_2}X_1(X_1'X_1)^{-1}h_1$$

from which the condition (3.88) becomes

$$\left[1-\frac{(1-\rho_{12}^2)}{2T}\left(\frac{\mu}{1-\mu\rho_{12}^2}\right)\right] \leq (1-\mu\rho_{12}^2)^{\frac{1}{2}} , \tag{3.90}$$

where μ is given by

$$0 \leq \mu = \frac{h_1'(X_1'X_1)^{-1}X_1'\bar{P}_{X_2}X_1(X_1'X_1)^{-1}h_1}{h_1'(X_1'X_1)^{-1}h_1} \leq 1 .$$

From (3.90), we have

$$T \leq \left(\frac{1-\rho_{12}^2}{\rho_{12}^2}\right)\left[\frac{1+(1-\mu\rho_{12}^2)^{\frac{1}{2}}}{2(1-\mu\rho_{12}^2)}\right], \tag{3.91}$$

which provides the condition for the dominance of the OLS estimator over the SUUR estimator with respect to the concentration probability to the order of our approximation.

3.5 MONTE CARLO EVIDENCE

One approach which may be used to improve upon asymptotic results, and to study the performance of estimators in finite samples is the Monte Carlo technique. Given the relatively intractable nature of the sampling distributions of the FGLS estimators, Kmenta and Gilbert (1968) carried out an interesting Monte Carlo study of the OLS and SURR estimators, and some other estimators to be described in the later chapters of this book. For the moment we confine our attention to the OLS and SURR estimators.

Kmenta and Gilbert discussed four experiments, each dealing with various specifications of the SURE model. The experiments differed in terms of the values given to the explanatory variables, while for a given experiment the models used differed in terms of the properties of the random disturbance terms. Some attention was paid to certain types of misspecification. For instance, although the estimators studied were based on specification (3.1), situations involving heteroscedastic or autocorrelated errors were examined, and the effect of including a one-period lagged value of a dependent variable as a regressor was also studied.

Three of the four experiments were based on a two equation SURE system. They differed according to the degree of correlation between the variables in the two equations, and the presence of the lagged dependent variable as a regressor in one case. The fourth experiment involved a four equation system. Nine different models were used in all, but the different experiments used different sub-sets of these models. Four of the models satisfied the basic assumptions in (3.1), the disturbances in two of them exhibited heteroscedasticity, while the disturbances in the remaining three models followed autoregressive processes. Samples of size 10, 20 and 100 were considered; normally distributed pseudo-random disturbances were used; and a limited range of values was adopted for the parameters.

The analysis undertaken by Kmenta and Gilbert provided a number of illuminating results (keeping in mind the time at which they were reported). The investigation revealed the unbiasedness of the estimators in all of the cases except one, and this involved only one parameter in one of the models considered. Further comparisons were therefore based on the standard deviations of the empirical sampling distributions, rather than on root mean squared errors. The superiority of the SURR estimator over the OLS estimator with respect to asymptotic properties was observed to hold in finite samples too. For instance, the property that the asymptotic variance of the SURR estimator is smaller for a higher correlation between the disturbances of different equations, and for a lower correlation between regressors across the equations, carried over to small samples, even in the presence of specification errors of the types outlined above. The result that an increase in the number of equations leads to a reduction of asymptotic variance was verified for $T = 100$, but no clear trend emerged in smaller samples.

The comparisons made by Kmenta and Gilbert revealed that the SURR estimator is more efficient than the OLS estimator, except in one situation. This situation was associated with cases in which the disturbances across the equations were in fact uncorrelated, and this fact was not taken into account in the SURR estimator. Further, the SURR estimator was found to be robust to the specification errors involving lagged variables and heteroscedastic or autocorrelated disturbances, and the magnitudes of the variances of the disturbances were found to have no influence on the relative performances of the estimators. The SURR estimator was found to be only a little worse (in terms of efficiency, etc.) than the OLS estimator in the case of uncorrelated disturbances across the equations while it was considerably better than the OLS estimator in the majority of other cases. It was therefore recommended by Kmenta and Gilbert that the SURR estimator should be used in preference to the OLS estimator in the estimation of SURE models.

Breusch (1980) pointed out that the numerical values assigned to coefficient vectors in the study conducted by Kmenta and Gilbert are immaterial for the generation of the data. Findings based on certain prespecified values of the coefficient vectors remain unaltered even if we replace them by null vectors. This is due to the invariance of the bias vectors with respect to such changes in the values assigned to the coefficients. Similarly, the relative efficiency of the OLS estimator with respect to the SURR estimator is invariant to these values. In fact, if the exact sampling distribution under study is not affected by changes in certain parameters, the numerical values chosen for empirical investigation obviously will not matter at all. Benefits from such invariance properties, Breusch pointed out, can be derived in designing informative Monte Carlo experiments.

3.6 SOME REMARKS ON LARGE-SAMPLE APPROXIMATION METHODOLOGY AND THE MONTE CARLO TECHNIQUE

The methodology which underlies large-sample asymptotic approximations is as follows. First, employing a Taylor's series expansion we can express the estimation error, ξ, for some estimator of β as an infinite series:

$$\begin{aligned}\xi &= \sum_{j=1}^{\infty} \xi_{-j/2} \\ &= \sum_{j=1}^{g} \xi_{-j/2} + \sum_{j=g+1}^{\infty} \xi_{-j/2} \\ &= \xi(g) + O_p(T^{-j}) \ ; \quad (j > g)\end{aligned} \tag{3.92}$$

where the elements of $\xi_{-j/2}$ are of order $O_p(T^{-j/2})$. We can then approximate the estimation error by the first few leading terms and so approximate the random vector ξ by another random vector, say, $\xi(g)$. The properties of this approximating random vector $\xi(g)$ are referred to as large-sample asymptotic properties. This process raises some interesting issues (see Srinivasan, 1970). Firstly, the expansion (3.92) may not converge so the validity of approximating it by a few leading terms should be checked. Following Taylor (1977; Appendix), it can be demonstrated that the remainder term $(\xi - \xi(g))$ is at most of order $O_p(T^{-(g+1)/2})$ (see also Sargan, 1974). Secondly, the term by term expectations of the series should be checked. In other words, $E(\xi)$ should exist and the expectation of the remainder term should be at most of order $O(T^{-(g+1)/2})$. A similar statement applies to higher-order moments of ξ.

The existence of moments of ξ is considered in the next chapter (see also Taylor, 1977). For the SUUR estimator in a two equation model, Kariya and Maekawa (1982) established the convergence of the remainder term while evaluating the variance covariance matrix to order $O(T^{-2})$ and $O(T^{-3})$. For instance, if R_2 and R_3 denote the differences between the exact variance covariance matrix of the SURR estimator in a two equation model and its approximations to order $O(T^{-2})$ and $O(T^{-3})$ respectively, Kariya and Maekawa (1982; Theorem 4.1) demonstrated that

$$\begin{aligned}(\text{tr } R_2R_2')^{\frac{1}{2}} &\le \frac{3.65}{(T-K+1)(T-K+3)} (\text{tr } \Omega^2)^{\frac{1}{2}} \\ (\text{tr } R_3R_3')^{\frac{1}{2}} &\le \frac{79}{(T-K+1)(T-K+3)(T-K+5)} (\text{tr } \Omega^2)^{\frac{1}{2}} ,\end{aligned} \tag{3.93}$$

provided (T - K) does not fall below eight in value.

If we consider asymptotic approximations for the probability distribution function of $a_h = \sqrt{T}\,h'(\tilde{\beta}_{SU} - \beta)$, the upper bound for the absolute difference between the exact distribution function and its first-order asymptotic approximation, $\Phi(m/\theta)$, is

$$\left(\frac{2}{\pi}\right)^{\frac{1}{2}}\left(\frac{1}{T - K - 3}\right) \tag{3.94}$$

while a similar bound in the context of the second-order asymptotic approximation (3.82) is

$$\frac{4}{\pi}\left[\frac{1}{(T - K - 3)^2} + \frac{2}{(T - K + 1)(T - K + 3)}\right]. \tag{3.95}$$

Similar bounds were obtained by Kariya and Maekawa for the probability density function of $\sqrt{T}(\tilde{\beta}_{SU} - \beta)$. The upper bound turns out to be

$$\frac{K(T - K)^{\frac{1}{2}}}{(2\pi)^{K/2}(T - K - 3)|\Omega|^{\frac{1}{2}}} \tag{3.96}$$

for the first-order asymptotic approximation using the multivariate Normal density function with mean vector 0 and variance covariance matrix $(T - K)\Omega^{-1}$. For the second-order asymptotic approximation (3.83), this bound is

$$\frac{K(T - K)^{\frac{1}{2}}}{2^{K/2+3}\pi^{K}|\Omega|^{\frac{1}{2}}}\left[\frac{4(K + 2)}{(T - K - 3)^2} + \frac{15}{(T - K + 1)(T - K + 3)}\right]. \tag{3.97}$$

It should be pointed out that no general claims can be made that the moments of $\xi(g)$ validly approximate the moments of ξ. The distinction must be made between moments of the approximating random vector, $\xi(g)$, and the approximate moments of the original random vector. In other words, the moments of the limiting distribution and the limiting forms of the moments of the exact distribution cannot be equated, and generally no explicit relationship connecting them can be obtained.

Situations do arise in which the exact moments may be infinite while the moments of some approximating distribution may be finite. For example, consider the first equation of the SURE model and assume that there is merely one regressor in this equation:

$$y_{t1} = \beta_{11}x_{t11} + u_{t1}\,; \quad u_{t1} \sim N(0, \sigma_{11})$$

which, simplifying the subscripts for convenience, may be written as

$$y_t = \beta x_t + u_t \; ; \quad u_t \sim N(0, \sigma^2) \quad . \tag{3.98}$$

Let $\hat{\beta}$ denote the OLS estimator of β. It is then easy to verify that $\hat{\beta}$ has a Normal distribution, $N\left(\beta, \frac{\sigma^2}{\Sigma x_t^2}\right)$. If we estimate $(1/\beta)$ by $(1/\hat{\beta})$, we have

$$\left(\frac{1}{\hat{\beta}} - \frac{1}{\beta}\right) = \sum_{j=1}^{\infty} (-1)^j \left(\frac{\epsilon^j}{\beta^{j+1}}\right) \tag{3.99}$$

where $\epsilon = (\hat{\beta} - \beta) \sim N(0, \sigma^2 / \Sigma x_t^2)$. Now it is easy to verify that $E(1/\hat{\beta} - 1/\beta)$ does not exist. On the other hand, if we approximate $(1/\hat{\beta} - 1/\beta)$ by a finite number of leading terms on the right hand side of (3.99), the expectation of the approximating sum is always finite.

Matching the exact moments of the original random vector with the moments of an approximating random vector is indeed not an important issue. The series approximations are direct stochastic approximations of the estimators in a probability sense and hence in a distributional sense. Their moments contain useful information about the distribution of the original random vector in terms of central tendency, variability, etc., even in situations where the exact moments of the original random vector do not exist. They are subject to the same qualifications and limitations as the standard asymptotic theory in statistical inference. The real issue in judging the merits of any asymptotic approximation procedure lies in its ability to approximate complicated sampling distributions and to provide results that are sufficiently simple and interpretable, while at the same time reasonably accurate under conditions frequently observed in practice. In fact, there may be situations where large-sample asymptotic approximations may be more meaningful than exact results. To illustrate this point, consider again the estimation of the reciprocal of β in the model (3.98). Suppose that practical considerations indicate that both β and its OLS estimator $\hat{\beta}$ are necessarily positive. Normality of the disturbances obviously contravenes the strict positivity of $\hat{\beta}$, although it may be conjectured that the probability associated with non-positive values of $\hat{\beta}$ is very small. The nonexistence of the mean of $(1/\hat{\beta})$ may therefore be a spurious phenomenon, as Theil (1971; p. 377) indicated, caused by a seemingly innocent specification error pertaining to the normality of the disturbances. The presence of such slight specification errors may sometimes tend to invalidate the reliability placed on exact results in any given application and may make asymptotic approximations more meaningful, due to their relative insensitivity to specification error.

The Monte Carlo technique for studying the properties of intractable sampling distributions of estimators essentially comprises, first, the selection of a suitable model that may reveal the salient features of the estimation procedures under study. Next, an appropriate distribution of the disturbances is specified and the disturbances are generated from it through some random device. Values of the explanatory variables are then chosen and numerical values are given to the unknown parameters in the model. Using the values of disturbances, explanatory variables and parameters, observations on the variables to be explained are obtained. The thus derived artificial data set is then used to estimate the parameters by the procedures under study, ignoring the fact that the true values of the parameters are known. This process is repeated a large number of times so as to yield a large number of estimates of any particular parameter. The frequency distribution of these estimates thus provides an empirical estimate of the true sampling distribution of the estimator. This is just like estimating the probability of getting a head in a single toss of a coin by the proportion of heads observed in a large number of tosses. The various features of such empirical sampling distributions may then shed some light on the sampling properties of the estimator of interest. It is obvious that the findings based on an exercise of this type will be conditional upon the specification of the model used, so that generalizations may often be hazardous. The reliability, validity and generality of the results depend upon the design of the Monte Carlo experiment and the analysis adopted. It is therefore quite difficult, if not impossible, to arrive at general but definite conclusions. Moreover, the conclusions are derived from an empirical estimate of the sampling distribution rather than from the true sampling distribution, and thus the technique itself is subject to a sampling error, the seriousness of which cannot always be assessed or easily controlled.

As a result, any investigation of the bias, mean squared error, etc., of an estimator may sometimes lead to erroneous conclusions. The situation becomes exacerbated when the population moments are not finite. For instance, if a particular estimator does not possess a second moment, the average of squares of deviations of estimates from the true value of parameter obtained in Monte Carlo experiments may exhibit considerable variations due to the infinite nature of the true mean squared error, and therefore may provide false conclusions. However, this is not to suggest that Monte Carlo experiments are worthless; they do provide illuminating information, useful insights and important clues for further investigation. To summarize, we believe that Monte Carlo studies should not be treated as a substitute or a competitor for theoretical investigations. They should be taken in the spirit of stimulating and complementing theoretical investigations.

EXERCISES

3.1 Carefully distinguish between the matrices $[X'(\Sigma^{-1} \otimes I_T)X]^{-1}$ and $[\lim_{T\to\infty} \frac{1}{T} X'(\Sigma^{-1} \otimes I_T)X]^{-1}$ in the context of the variance covariance matrix of a consistent FGLS estimator.

3.2 Consider the consistent estimators of the variances and covariances of the disturbance vector for the equations of a SURE model based on restricted and unrestricted OLS residuals and compare their biases and mean squared errors assuming normality of the disturbances.

3.3 Using the residuals implied by the SUUR and SURR estimators of β in a SURE model with normally distributed disturbances, construct consistent estimators for the variance of the disturbance term in the first equation. Evaluate and compare the large-sample approximations for the bias and mean squared error in each case. [Hint: See Srivastava and Tiwari (1976).]

3.4 If S is any unbiased estimator of the variance covariance matrix Σ and $E(S^{-1})$ exists, show that $E(S^{-1} - \Sigma^{-1})$ is a nonnegative definite matrix. [Hint: See Srivastava (1970a).]

3.5 Consider a two equation SURE model

$$y_1 = \beta_1 x_1 + u_1$$

$$y_2 = \beta_2 x_2 + u_2$$

with the following specifications:

$$x_1'x_1 = x_2'x_2 = T\ , \qquad \Sigma = \begin{bmatrix} 1 & \rho_{12} \\ \rho_{12} & 1 \end{bmatrix} .$$

Ignoring the specific form of Σ and taking it to be unrestricted, the SUUR estimators of β_1 and β_2 are found. Assuming that the disturbances follow a bivariate Normal probability law, obtain the large-sample approximation (to order $O(T^{-2})$) for the variance covariance matrix of the SUUR estimators of β_1 and β_2. Examine their efficiency relative to the OLS estimators.

3.6 Establish the equivalence of the expressions (3.77) and (3.78).

3.7 In a two equation SURE model with orthogonal regressors and normally distributed disturbances, suppose that the efficiency of the OLS estimator relative to the SUUR/SURR estimators is measured by the ratio of the determinants of the variance covariance matrices to order $O(T^{-2})$. Indicate the quantities (characterizing the model, such as the values of regressors, variances and covariances of the disturbances, etc.) that have no bearing on this efficiency measure. How will your answer change if instead of orthogonal regressors we have subset regressors?

3.8 Derive the first-order and second-order asymptotic approximations for the probability density function of $\sqrt{T}(\tilde{\beta}_{SU} - \beta)$ in a two equation SURE model with normally distributed disturbances. [Hint: See Kariya and Maekawa (1982).]

3.9 Using the two equation SURE model described in Exercise 3.5, simplify the condition (3.88) for the OLS estimator to be superior to the SUUR/SURR estimator with respect to the criterion of concentration probability to the specified order of approximation.

3.10 Design a Monte Carlo study with the objective of determining situations in which the OLS estimator may have smaller standard errors than the SUUR and SURR estimators.

REFERENCES

Binkley, J. K. (1982). "The effect of variable correlation on the efficiency of seemingly unrelated regression in a two-equation model," Journal of the American Statistical Association 77, 890-895.

Breusch, T. S. (1980). "Useful invariance results for generalized regression models," Journal of Econometrics 13, 327-340.

Kariya, T. and K. Maekawa (1982). "A method for approximations to the pdf's and cdf's of GLSE's and its application to the seemingly unrelated regression model," Annals of the Institute of Statistical Mathematics 34, 281-297.

Kmenta, J. and R. F. Gilbert (1968). "Small sample properties of alternative estimators of seemingly unrelated regressions," Journal of the American Statistical Association 63, 1180-1200.

Phillips, P. C. B. (1977). "An approximation to the finite sample distribution of Zellner's seemingly unrelated regression estimator," Journal of Econometrics 6, 147-164.

Sargan, J. D. (1974). "The validity of Nagar's expansion for the moments of econometric estimators," Econometrica 42, 169-176.

Sargan, J. D. (1975). "Gram-Charlier approximations applied to t-ratios of k-class estimators," Econometrica 43, 327-346.

Srinivasan, T. N. (1970). "Approximations to finite sample moments of estimators whose exact sampling distributions are unknown," Econometrica 38, 533-541.

Srivastava, V. K. (1970). "The efficiency of estimating seemingly unrelated regression equations," Annals of the Institute of Statistical Mathematics 22, 483-493.

Srivastava, V. K. (1970a). "On the expectation of the inverse of a matrix," Sankhyā A 32, 236.

Srivastava, V. K. and R. Tiwari (1976). "Evaluation of expectations of products of stochastic matrices," Scandinavian Journal of Statistics 3, 135-138.

Srivastava, V. K. and S. Upadhyaya (1978). "Large-sample approximations in seemingly unrelated regression equations," Annals of the Institute of Statistical Mathematics 30, 89-96.

Taylor, W. E. (1977). "Small sample properties of a class of two stage Aitken estimators," Econometrica 45, 497-508.

Theil, H. (1971). Principles of Econometrics (Wiley, New York).

Zellner, A. and D. S. Huang (1962). "Further properties of efficient estimators for seemingly unrelated regression equations," International Economic Review 3, 300-313.

4

Exact Finite-Sample Properties of Feasible Generalized Least Squares Estimators

4.1 INTRODUCTION

The last chapter was devoted to the properties of the OLS, SUUR and SURR estimators. Some of the available Monte Carlo evidence concerning the sampling properties of these estimators was considered, and the limitations of this technique were discussed. Large-sample asymptotics were used to explore some limiting properties of the estimators of interest. It was observed that both the SUUR and SURR estimators are unbiased to order $O(T^{-1})$ and that their variance covariance matrices are identical to order $O(T^{-2})$. Thus, these large-sample results fail to yield a statistical basis for choosing between the SUUR and SURR estimators. Of course, working within this large-sample framework has its own general limitations. In particular, properties which are derived under the assumption that T is very large may not hold in finite samples, and so these asymptotic results may be extremely misleading in the context of samples of the size usually encountered in practice.

In applications involving time-series data, sample sizes often are quite limited, despite the increasing availability of data collected on a quarterly or monthly basis. In many cases, sample dates are chosen to ensure that the assumption, that the model's parameters are constant, is a reasonable one. That is, in an empirical application an economist may deliberately limit the sample period (and hence the sample size), and so cast doubt on the relevance of any known asymptotic properties of his inferential procedures. The requirement that the model whose parameters are being estimated be fixed in terms of its specification may lead the researcher to discard time-series observations as they become increasingly "dated." All of this suggests the need to study in detail the finite-sample properties of estimators and test procedures in general, and in the context of the present discussion, the need to consider the features of the finite-sample distributions of various estimators of the SURE model. This topic is addressed in some detail in this chapter.

4.2 UNBIASEDNESS

Consider the SURE model

$$\begin{aligned} y &= X\beta + u, \\ E(u) &= 0, \quad E(uu') = (\Sigma \otimes I_T) = \Psi, \end{aligned} \tag{4.1}$$

for which the FGLS estimator of β is

$$b_F = [X'(S^{-1} \otimes I_T)X]^{-1}X'(S^{-1} \otimes I_T)y . \tag{4.2}$$

We assume that S is a consistent estimator of Σ and that its elements are even functions of the disturbances. The choices $S = \tilde{S}$ and $S = \hat{S}$, which lead to the SUUR and SURR estimators respectively, obviously satisfy this specification. Substituting (4.1) in (4.2), we get

$$(b_F - \beta) = [X'(S^{-1} \otimes I_T)X]^{-1}X'(S^{-1} \otimes I_T)u . \tag{4.3}$$

The expression on the right hand side of (4.3) is an odd function of the disturbances. If the disturbances follow a symmetric probability distribution, their odd-order moments will be 0 and hence the mean of $(b_F - \beta)$ will be a null vector, provided that $E(b_F)$ exists (see Kakwani, 1967). Now, to consider the existence of the mean of b_F we follow the approach used by Srivastava and Raj (1979) and take a linear function $a'X(b_F - \beta)$ in which a is any $(MT \times 1)$ vector with real and nonstochastic elements. As S is nonsingular, we can find a nonsingular Γ such that $S = \Gamma\Gamma'$. Using the Cauchy-Schwarz inequality, we have

$$|a'X(b_F - \beta)| = |a'(\Gamma \otimes I_T)(\Gamma^{-1} \otimes I_T)X(b_F - \beta)|$$

$$\leq [(b_F - \beta)'X'(\Gamma'^{-1} \otimes I_T)(\Gamma^{-1} \otimes I_T)X(b_F - \beta)]^{\frac{1}{2}}$$

$$\cdot [a'(\Gamma \otimes I_T)(\Gamma' \otimes I_T)a]^{\frac{1}{2}}$$

or,

$$|a'X(b_F - \beta)| \leq [(b_F - \beta)'X'(S^{-1} \otimes I_T)X(b_F - \beta)]^{\frac{1}{2}}[a'(S \otimes I_T)a]^{\frac{1}{2}}$$

or equivalently, using (4.3),

$$|a'X(b_F - \beta)| \leq [u'(S^{-1} \otimes I_T)X[X'(S^{-1} \otimes I_T)X]^{-1}X'(S^{-1} \otimes I_T)u]^{\frac{1}{2}}[a'(S \otimes I_T)a]^{\frac{1}{2}}. \tag{4.4}$$

If we write

$$D = (\Gamma^{-1} \otimes I_T)X[X'(S^{-1} \otimes I_T)X]^{-1}X'(\Gamma'^{-1} \otimes I_T),$$

it is easy to see that D is an idempotent matrix of order (MT × MT) having K* unit characteristic roots and (MT - K*) null roots, where $K^* = \Sigma_i K_i$. Further, let P be an orthogonal matrix which diagonalizes D so that

$$PDP' = \begin{bmatrix} I_{K^*} & 0 \\ 0 & 0 \end{bmatrix}$$

and therefore the matrix

$$P[(\text{tr } D)I_{MT} - D]P' = \begin{bmatrix} (K^* - 1)I_{K^*} & 0 \\ 0 & K^*I_{MT-K^*} \end{bmatrix}$$

is positive definite, which implies that $[(\text{tr } D)I_{MT} - D]$ is positive definite. Using this, we observe that

$$u'(S^{-1} \otimes I_T)X[X'(S^{-1} \otimes I_T)X]^{-1}X'(S^{-1} \otimes I_T)u$$

$$= u'(\Gamma'^{-1} \otimes I_T)(\Gamma^{-1} \otimes I_T)X[X'(S^{-1} \otimes I_T)X]^{-1}X'(\Gamma'^{-1} \otimes I_T)(\Gamma^{-1} \otimes I_T)u \tag{4.5}$$

$$= u'(\Gamma'^{-1} \otimes I_T)D(\Gamma^{-1} \otimes I_T)u$$

$$< (\text{tr } D)u'(\Gamma'^{-1} \otimes I_T)(\Gamma^{-1} \otimes I_T)u = K^* u'(S^{-1} \otimes I_T)u .$$

Next, from Rao (1973; p. 60), we have

$$\begin{aligned} u'(S^{-1} \otimes I_T)u &\leq \left(\frac{1}{\lambda_{min}}\right) u'u \\ a'(S \otimes I_T)a &\leq \lambda_{max} a'a , \end{aligned} \tag{4.6}$$

where λ_{min} and λ_{max} denote the smallest and largest characteristic roots of $(S \otimes I_T)$.

Using (4.5) and (4.6), it follows from (4.4) that

$$|a'X(b_F - \beta)| \leq \left[K^*\left(\frac{\lambda_{max}}{\lambda_{min}}\right)(u'u)(a'a)\right]^{\frac{1}{2}} . \tag{4.7}$$

So, using Hölder's inequality and Jensen's inequality (Rao, 1973; p. 149),

$$E[|a'X(b_F - \beta)|] \leq K^{*\frac{1}{2}}(a'a)^{\frac{1}{2}}\left[E\left(\frac{1}{\lambda_{min}}\right)\right]^{\frac{1}{2}}[E(\lambda_{max}u'u)]^{\frac{1}{2}} , \tag{4.8}$$

provided that the expectations on the right hand side of (4.8) exist. The requirement for this is that the disturbances should have finite moments of order four, and that $E(S^{-1})$ is finite (see Fuller and Battese, 1973; pp. 628-629). As finiteness of the expectation of the absolute value of the linear function $a'X(b_F - \beta)$, for any arbitrary nonstochastic real vector a, implies the existence of the mean vector of the FGLS estimator, we observe that the FGLS estimator is unbiased if the disturbances follow a symmetric distribution having finite moments of order four and if the expectation of S^{-1} is finite. In particular, if the disturbances are normally distributed, $T\tilde{S}$ follows a Wishart distribution and $E(\tilde{S}^{-1})$ exists when (T - K - M) exceeds unity (see results (A.2.ii) of the Appendix). Thus, from the comments following (4.3), the SUUR estimator of β is unbiased when the disturbances are normal and (T - K - M) is greater than unity.

Alternative sufficient conditions for the existence of the mean of the FGLS estimator can be obtained from (4.7). As in (4.5), using the Cauchy-Schwarz inequality, we have

$$E[|a'X(b_F - \beta)|] \leq K^{*\frac{1}{2}}(a'a)^{\frac{1}{2}} E\left[\frac{\lambda_{max}}{\lambda_{min}}(u'u)\right]^{\frac{1}{2}} . \tag{4.9}$$

Following Don and Magnus (1980), we find that

$$E\left[\frac{\lambda_{max}}{\lambda_{min}}(u'u)\right]^{\frac{1}{2}} \leq \left[E\left(\frac{\lambda_{max}}{\lambda_{min}}\right)\right]^{\frac{1}{2}} [E(u'u)]^{\frac{1}{2}} \tag{4.10}$$

$$E\left[\frac{\lambda_{max}}{\lambda_{min}}(u'u)\right]^{\frac{1}{2}} \leq [E(\lambda_{max})]^{\frac{1}{2}} [E(\lambda_{min}^2)]^{\frac{1}{4}} [E(u'u)^2]^{\frac{1}{4}} \tag{4.11}$$

$$E\left[\frac{\lambda_{max}}{\lambda_{min}}(u'u)\right]^{\frac{1}{2}} \leq [E(\lambda_{min})]^{\frac{1}{2}} [E(\lambda_{max}^2)]^{\frac{1}{4}} [E(u'u)^2]^{\frac{1}{4}} . \tag{4.12}$$

Using these inequalities in turn, we get from (4.9) the following sufficient conditions for the existence of the mean of the FGLS estimator:

(i) $E[(\text{tr } S)(\text{tr } S^{-1})]$ is finite and the disturbances have second-order moments;
(ii) $E(S)$ and $E(S^{-2})$ are finite and the disturbances have fourth-order moments;
(iii) $E(S^2)$ and $E(S^{-1})$ are finite and the disturbances have fourth-order moments.

4.3 THE TWO EQUATION MODEL

In order to discuss the derivation of the variance covariance matrix of the FGLS estimator we shall consider a two equation model in which the disturbances are normally distributed. To obtain explicit expressions for the FGLS estimators $b_{(1)F}$ and $b_{(2)F}$ of β_1 and β_2 respectively in this case, we introduce the following notation:

$$P_{x_i} = X_i(X_i'X_i)^{-1}X_i' , \qquad \bar{P}_{x_i} = I_T - P_{x_i} ,$$

$$R_{ij} = [X_i'(I_T - r^2 P_{x_j})X_i]^{-1} , \qquad r = \frac{s_{12}}{(s_{11}s_{22})^{\frac{1}{2}}} ; \; i, j = 1, 2 .$$

It is easy to see that

$$X'(S^{-1} \otimes I_T)X = \begin{bmatrix} X_1' & 0 \\ 0 & X_2' \end{bmatrix} \left(\begin{bmatrix} s_{11} & s_{12} \\ s_{12} & s_{22} \end{bmatrix}^{-1} \otimes I_T \right) \begin{bmatrix} X_1 & 0 \\ 0 & X_2 \end{bmatrix}$$

$$= \frac{1}{(s_{11}s_{22} - s_{12}^2)} \begin{bmatrix} X_1' & 0 \\ 0 & X_2' \end{bmatrix} \begin{bmatrix} s_{22}I_T & -s_{12}I_T \\ -s_{12}I_T & s_{11}I_T \end{bmatrix} \begin{bmatrix} X_1 & 0 \\ 0 & X_2 \end{bmatrix}$$

$$= \frac{1}{(s_{11}s_{22} - s_{12}^2)} \begin{bmatrix} s_{22}X_1'X_1 & -s_{12}X_1'X_2 \\ -s_{12}X_2'X_1 & s_{11}X_2'X_2 \end{bmatrix}$$

so that

$$[X'(S^{-1} \otimes I_T)X]^{-1} =$$

$$s_{11}s_{22}(1 - r^2) \begin{bmatrix} \frac{1}{s_{22}} R_{12} & \frac{r^2}{s_{12}} R_{12}X_1'X_2(X_2'X_2)^{-1} \\ \frac{r^2}{s_{12}}(X_2'X_2)^{-1}X_2'X_1R_{12} & \frac{1}{s_{11}}(X_2'X_2)^{-1} + \frac{r^2}{s_{11}}(X_2'X_2)^{-1}X_2'X_1R_{12}X_1'X_2(X_2'X_2)^{-1} \end{bmatrix}. \tag{4.13}$$

Similarly, we find

$$X'(S^{-1} \otimes I_T)y = \frac{1}{s_{11}s_{22}(1 - r^2)} \begin{bmatrix} s_{22}X_1'y_1 - s_{12}X_1'y_2 \\ -s_{12}X_2'y_1 + s_{11}X_2'y_2 \end{bmatrix}. \tag{4.14}$$

Recalling that the FGLS estimators of β_1 and β_2 are given by

$$\begin{bmatrix} b_{(1)F} \\ b_{(2)F} \end{bmatrix} = [X'(S^{-1} \otimes I_T)X]^{-1}X'(S^{-1} \otimes I_T)y$$

and using (4.13) and (4.14), we obtain

$$\begin{aligned} b_{(1)F} &= R_{12}\left[X_1'(I_T - r^2 P_{x_2})y_1 - \frac{s_{12}}{s_{22}} X_1'\bar{P}_{x_2} y_2\right] \\ b_{(2)F} &= (X_2'X_2)^{-1}X_2'y_2 - r^2(X_2'X_2)^{-1}X_2'X_1R_{12}X_1'\bar{P}_{x_2}y_2 \\ &\quad - \left(\frac{s_{12}}{s_{11}}\right)(X_2'X_2)^{-1}X_2'[I_T - X_1R_{12}X_1'(I_T - r^2P_{x_2})]y_1 \, . \end{aligned} \tag{4.15}$$

Interchanging the subscripts (1) and (2) in (4.15) yields alternative expressions for the estimators:

$$\begin{aligned} b_{(1)F} &= (X_1'X_1)^{-1}X_1'y_1 - r^2(X_1'X_1)^{-1}X_1'X_2R_{21}X_2'\bar{P}_{x_1}y_1 \\ &\quad - \frac{s_{12}}{s_{22}}(X_1'X_1)^{-1}X_1'\{I_T - X_2R_{21}X_2'(I_T - r^2P_{x_1})\}y_2 \\ b_{(2)F} &= R_{21}\left[X_2'(I_T - r^2P_{x_1})y_2 - \frac{s_{12}}{s_{11}} X_2'\bar{P}_{x_1}y_1\right] . \end{aligned} \tag{4.16}$$

From the first equation in (4.15) and the second equation in (4.16), the FGLS estimator can be interpreted as the ordinary least squares estimator purged by removing terms for the correlation between the disturbances of the two equations. These expressions form the basis for our discussion of the efficiency of certain FGLS estimators in the next three sections.

4.4 EFFICIENCY PROPERTIES UNDER ORTHOGONAL REGRESSORS

First, we discuss some results obtained by Zellner (1963; 1972), assuming that the regressors in the two equations are orthogonal, i.e.,

$$X_1'X_2 = 0 \, . \tag{4.17}$$

Using this condition in (4.15) or (4.16), we get

$$b_{(1)F} = (X_1'X_1)^{-1}X_1'y_1 - \frac{s_{12}}{s_{22}}(X_1'X_1)^{-1}X_1'y_2$$

$$b_{(2)F} = (X_2'X_2)^{-1}X_2'y_2 - \frac{s_{12}}{s_{11}}(X_2'X_2)^{-1}X_2'y_1 .$$

(4.18)

Now we restrict our attention to two special cases, viz., the SUUR and SURR estimators.

4.4.1 The Variance Covariance Matrix

Let us first consider the SUUR estimator of β_1. It is easy to see that

$$(\tilde{\beta}_{(1)SU} - \beta_1) = (X_1'X_1)^{-1}X_1'u_1 - \frac{\tilde{s}_{12}}{\tilde{s}_{22}}(X_1'X_1)^{-1}X_1'u_2$$

(4.19)

so that its variance covariance matrix is

$$V(\tilde{\beta}_{(1)SU}) = E(\tilde{\beta}_{(1)SU} - \beta_1)(\tilde{\beta}_{(1)SU} - \beta_1)'$$

$$= (X_1'X_1)^{-1}E\left[X_1'u_1u_1'X_1 - \frac{\tilde{s}_{12}}{\tilde{s}_{22}}X_1'(u_1u_2' + u_2u_1')X_1 \right.$$

$$\left. + \left[\frac{\tilde{s}_{12}}{\tilde{s}_{22}}\right]^2 X_1'u_2u_2'X_1\right](X_1'X_1)^{-1} .$$

(4.20)

Notice that here Z is just the $[T \times (K_1 + K_2)]$ matrix $(X_1 \quad X_2)$ so that

$$Z(Z'Z)^{-1}Z' = (X_1 \quad X_2)\begin{bmatrix} X_1'X_1 & 0 \\ 0 & X_2'X_2 \end{bmatrix}^{-1}\begin{pmatrix} X_1' \\ X_2' \end{pmatrix}$$

$$= X_1(X_1'X_1)^{-1}X_1' + X_2(X_2'X_2)^{-1}X_2'$$

$$= (P_{x_1} + P_{x_2}) ,$$

by the orthogonality of the regressors. Thus we have

$$\tilde{s}_{ij} = \frac{1}{T} y_i' [I_T - Z(Z'Z)^{-1} Z'] y_j$$

$$= \frac{1}{T} (\beta_i' X_i' + u_i')[I_T - Z(Z'Z)^{-1} Z'](X_j \beta_j + u_j)$$

$$= \frac{1}{T} u_i' [I_T - Z(Z'Z)^{-1} Z'] u_j$$

$$= \frac{1}{T} u_i' \bar{P}_z u_j \qquad (i, j = 1, 2) \qquad (4.21)$$

so that the normality of the disturbances implies that the vectors $X_1'u_1$, $X_1'u_2$, $X_2'u_1$ and $X_2'u_2$ are stochastically independent of $\tilde{s}_{11}$, $\tilde{s}_{22}$ and $\tilde{s}_{12}$, since $X_i'\bar{P}_z = 0$ for $i = 1, 2$. Using this result in (4.20), we get

$$V(\tilde{\beta}_{(1)SU}) = \left[\sigma_{11} - 2\rho_{12}(\sigma_{11}\sigma_{22})^{\frac{1}{2}} E\left[\frac{\tilde{s}_{12}}{\tilde{s}_{22}}\right] + \sigma_{22} E\left[\frac{\tilde{s}_{12}}{\tilde{s}_{22}}\right]^2\right](X_1'X_1)^{-1}. \qquad (4.22)$$

To evaluate the two expectations in (4.22), write

$$u_1 = \delta u_2 + v, \qquad (4.23)$$

where u_2 and v are independent random vectors and $\delta = (\sigma_{12}/\sigma_{22}) = \rho_{12}\sqrt{(\sigma_{11}/\sigma_{22})}$. It is easy to check that v follows a multivariate Normal distribution with mean vector 0 and variance covariance matrix $\sigma_{11}(1-\rho_{12}^2)I_T$.

Using (4.21) and (4.23), we have

$$\frac{\tilde{s}_{12}}{\tilde{s}_{22}} = \delta + \frac{v'\bar{P}_z u_2}{u_2'\bar{P}_z u_2},$$

so that

$$E\left[\frac{\tilde{s}_{12}}{\tilde{s}_{22}}\right] = \delta + E\left[\frac{E(v')\bar{P}_z u_2}{u_2'\bar{P}_z u_2}\right]$$

$$= \delta \qquad (4.24)$$

$$E\left[\frac{\tilde{s}_{12}}{\tilde{s}_{22}}\right]^2 = \delta^2 + 2\delta E\left[\frac{E(v')\bar{P}_z u_2}{u_2'\bar{P}_z u_2}\right] + E\left[\frac{u_2'\bar{P}_z E(vv')\bar{P}_z u_2}{(u_2'\bar{P}_z u_2)^2}\right]$$

$$= \delta^2 + \sigma_{11}(1 - \rho_{12}^2) E\left[\frac{1}{u_2'\bar{P}_z u_2}\right]$$

$$= \frac{\sigma_{11}}{\sigma_{22}}\left(\rho_{12}^2 + \frac{1-\rho_{12}^2}{T-K-2}\right), \tag{4.25}$$

because $u_2'\bar{P}_Z u_2$ follows a χ^2 distribution with (T - K) degrees of freedom.

Substituting (4.24) and (4.25) in (4.22) gives us the result obtained by Zellner (1963; 1972):

$$\begin{aligned} V(\tilde{\beta}_{(1)SU}) &= \sigma_{11}\left[1 - 2\rho_{12}^2 + \rho_{12}^2 + \frac{1-\rho_{12}^2}{T-K-2}\right](X_1'X_1)^{-1} \\ &= \sigma_{11}(1-\rho_{12}^2)\left[1 + \frac{1}{T-K-2}\right](X_1'X_1)^{-1} \quad (T-K) > 2 \end{aligned} \tag{4.26}$$

Analogously, the variance covariance matrix of the SUUR estimator of β_2 is

$$V(\tilde{\beta}_{(2)SU}) = \sigma_{22}(1-\rho_{12}^2)\left[1 + \frac{1}{T-K-2}\right](X_2'X_2)^{-1} \quad (T-K) > 2\,. \tag{4.27}$$

Similarly, the expression for the cross-moments matrix of the SUUR estimators of β_1 and β_2 is

$$\begin{aligned} &E(\tilde{\beta}_{(1)SU} - \beta_1)(\tilde{\beta}_{(2)SU} - \beta_2)' \\ &= (X_1'X_1)^{-1}E\left[X_1'u_1u_2'X_2 - \frac{\tilde{s}_{12}}{\tilde{s}_{22}}X_1'u_2u_2'X_2 - \frac{\tilde{s}_{12}}{\tilde{s}_{11}}X_1'u_1u_1'X_2 \right. \\ &\qquad \left. + \frac{\tilde{s}_{12}^2}{\tilde{s}_{11}\tilde{s}_{22}}X_1'u_2u_1'X_2\right](X_2'X_2)^{-1} \\ &= (X_1'X_1)^{-1}\left[\sigma_{12}X_1'X_2 - \sigma_{22}E\left[\frac{\tilde{s}_{12}}{\tilde{s}_{22}}\right]X_1'X_2 - \sigma_{11}E\left[\frac{\tilde{s}_{12}}{\tilde{s}_{11}}\right]X_1'X_2 \right. \\ &\qquad \left. + \sigma_{12}E\left[\frac{\tilde{s}_{12}^2}{\tilde{s}_{11}\tilde{s}_{22}}\right]X_1'X_2\right](X_2'X_2)^{-1} \\ &= 0 \end{aligned} \tag{4.28}$$

by virtue of the orthogonality of X_1 and X_2.

Thus, the complete variance covariance matrix of the SUUR estimator of β is, from (4.26), (4.27) and (4.28),

$$V(\tilde{\beta}_{SU}) = (1-\rho_{12}^2)\left[1+\frac{1}{T-K-2}\right]\begin{bmatrix}\sigma_{11}(X_1'X_1)^{-1} & 0 \\ 0 & \sigma_{22}(X_2'X_2)^{-1}\end{bmatrix}. \tag{4.29}$$

Proceeding as for the SUUR estimator, the variance covariance matrix of the SURR estimator of β_1 is

$$\begin{aligned} V(\hat{\beta}_{(1)SR}) &= E(\hat{\beta}_{(1)SR}-\beta_1)(\hat{\beta}_{(1)SR}-\beta_1)' \\ &= (X_1'X_1)^{-1}E\left[X_1'u_1u_1'X_1 - \frac{\hat{s}_{12}}{\hat{s}_{22}}X_1'(u_1u_2'+u_2u_1')X_1\right. \\ &\quad \left. + \left[\frac{\hat{s}_{12}}{\hat{s}_{22}}\right]^2 X_1'u_2u_2'X_1\right](X_1'X_1)^{-1} \\ &= \sigma_{11}(X_1'X_1)^{-1} - (X_1'X_1)^{-1}X_1'E\left[\frac{\hat{s}_{12}}{\hat{s}_{22}}(u_1u_2'+u_2u_1')\right]X_1(X_1'X_1)^{-1} \\ &\quad + (X_1'X_1)^{-1}X_1'E\left[\left[\frac{\hat{s}_{12}}{\hat{s}_{22}}\right]^2 u_2u_2'\right]X_1(X_1'X_1)^{-1}. \end{aligned} \tag{4.30}$$

Now, we have

$$\bar{P}_{X_1}\bar{P}_{X_2} = \bar{P}_{X_2} - P_{X_1}$$

by virtue of orthogonality of X_1 and X_2, so by using (4.23) we find

$$\begin{aligned} E&\left[\frac{\hat{s}_{12}}{\hat{s}_{22}}u_1u_2'\right] \\ &= E\left[\frac{u_1'\bar{P}_{X_1}\bar{P}_{X_2}u_2}{u_2'\bar{P}_{X_2}u_2}u_1u_2'\right] \\ &= E\left[\frac{u_1'(\bar{P}_{X_2}-P_{X_1})u_2}{u_2'\bar{P}_{X_2}u_2}u_1u_2'\right] \\ &= \delta^2E\left[\frac{u_2'(\bar{P}_{X_2}-P_{X_1})u_2}{u_2'\bar{P}_{X_2}u_2}u_2u_2'\right] - \delta E\left[\frac{E(v')(\bar{P}_{X_2}-P_{X_1})u_2}{u_2'\bar{P}_{X_2}u_2}u_2u_2'\right] \end{aligned} \tag{4.31}$$

$$- \delta E\left[\frac{u_2'(\bar{P}_{x_2} - P_{x_1})u_1}{u_2'\bar{P}_{x_2}u_2} E(v)u_2'\right] + E\left[\frac{1}{u_2'\bar{P}_{x_2}u_2} E(vv')(\bar{P}_{x_2} - P_{x_1})u_2u_2'\right]$$

$$= \delta^2 E\left[\left(1 - \frac{u_2'P_{x_1}u_2}{u_2'\bar{P}_{x_2}u_2}\right) u_2u_2'\right] + \sigma_{11}(1-\rho_{12}^2)(\bar{P}_{x_2} - P_{x_1})E\left[\frac{1}{u_2'\bar{P}_{x_2}u_2} u_2u_2'\right]$$

$$= \delta^2\sigma_{22}I_T - \delta^2 E\left[\frac{u_2'P_{x_1}u_2}{u_2'\bar{P}_{x_2}u_2}u_2u_2'\right] + \sigma_{11}(1-\rho_{12}^2)(\bar{P}_{x_2} - P_{x_1})E\left[\frac{1}{u_2'\bar{P}_{x_2}u_2}u_2u_2'\right] \tag{4.31}$$

$$E\left[\left[\frac{\hat{s}_{12}}{\hat{s}_{22}}\right]^2 u_2u_2'\right]$$

$$= E\left[\left[\frac{u_1'(\bar{P}_{x_2} - P_{x_1})u_2}{u_2'\bar{P}_{x_2}u_2}\right]^2 u_2u_2'\right]$$

$$= \delta^2 E\left[\left[\frac{u_2'(\bar{P}_{x_2} - P_{x_1})u_2}{u_2'\bar{P}_{x_2}u_2}\right]^2 u_2u_2'\right] + 2\delta E\left[\frac{E(v')\bar{P}_{x_2} - P_{x_1})u_2u_2'(\bar{P}_{x_2} - P_{x_1})u_2}{(u_2'\bar{P}_{x_2}u_2)^2}u_2u_2'\right]$$

$$+ E\left[\frac{u_2'(\bar{P}_{x_2} - P_{x_1})E(vv')(\bar{P}_{x_2} - P_{x_1})u_2}{(u_2'\bar{P}_{x_2}u_2)^2}\right]u_2u_2'$$

$$= \delta^2 E\left[\left[1 - \frac{u_2'P_{x_1}u_2}{u_2'\bar{P}_{x_2}u_2}\right]^2 u_2u_2'\right] + \sigma_{11}(1-\rho_{12}^2)E\left[\frac{1}{u_2'\bar{P}_{x_2}u_2}\left(1 - \frac{u_2'P_{x_1}u_2}{u_2'\bar{P}_{x_2}u_2}\right)u_2u_2'\right]$$

$$= \delta^2\sigma_{22}I_T - 2\delta^2 E\left[\frac{u_2'P_{x_1}u_2}{u_2'\bar{P}_{x_2}u_2} u_2u_2'\right] + \delta^2 E\left[\left[\frac{u_2'P_{x_1}u_2}{u_2'\bar{P}_{x_2}u_2}\right]^2 u_2u_2'\right]$$

$$+ \sigma_{11}(1 - \rho_{12}^2)E\left[\frac{1}{u_2'\bar{P}_{x_2}u_2} u_2u_2'\right] - \sigma_{11}(1 - \rho_{12}^2)E\left[\frac{u_2'P_{x_1}u_2}{(u_2'\bar{P}_{x_2}u_2)^2} u_2u_2'\right]. \tag{4.32}$$

Substituting (4.31) and (4.32) in (4.30), we get

$$V(\hat{\beta}_{(1)SR})$$

$$= (\sigma_{11} - \delta^2\sigma_{22})(X_1'X_1)^{-1} + \delta^2 E\left[\left[\frac{u_2'P_{x_1}u_2}{u_2'\bar{P}_{x_2}u_2}\right]^2 (X_1'X_1)^{-1}X_1'u_2u_2'X_1(X_1'X_1)^{-1}\right]$$

$$+ \sigma_{11}(1-\rho_{12}^2)E\left[\left\{\frac{1}{u_2'\bar{P}_{x_2}u_2} - \left[\frac{1}{u_2'\bar{P}_{x_2}u_2}\right]^2\right\}(X_1'X_1)^{-1}X_1'u_2u_2'X_1(X_1'X_1)^{-1}\right]. \tag{4.33}$$

Defining

$$w = (X_1'X_1)^{-\frac{1}{2}}X_1'u_2\,, \quad \chi^2 = \left(\frac{1}{\sigma_{22}}\right)u_2'(\bar{P}_{x_2} - P_{x_1})u_2\,,$$

we observe that w has a multivariate Normal distribution with mean vector 0 and variance covariance matrix I_{K_1}, while χ^2 has a χ^2 distribution with (T - K) degrees of freedom. Further, from result (A.2.iii) of the Appendix, w and χ^2 are stochastically independent. Now, using results (B.2) of the Appendix, we obtain

$$V(\hat{\beta}_{(1)SR})$$

$$= \sigma_{11}(1-\rho_{12}^2)(X_1'X_1)^{-1} + \rho_{12}^2\sigma_{11}(X_1'X_1)^{-\frac{1}{2}}E\left[\left(\frac{w'w}{\chi^2 + w'w}\right)^2 ww'\right](X_1'X_1)^{-\frac{1}{2}}$$

$$+ \sigma_{11}(1-\rho_{12}^2)(X_1'X_1)^{-\frac{1}{2}}E\left[\frac{1}{\chi^2 + w'w}ww' - \frac{w'w}{(\chi^2 + w'w)^2}ww'\right](X_1'X_1)^{-\frac{1}{2}}$$

$$= \sigma_{11}(1-\rho_{12}^2)(X_1'X_1)^{-1} + \rho_{12}^2\sigma_{11}\frac{(K_1+2)(K_1+4)}{(T-K_2+2)(T-K_2+4)}(X_1'X_1)^{-1}$$

$$+ \sigma_{11}(1-\rho_{12}^2)\left[\frac{1}{T-K_2} - \frac{K_1+2}{(T-K_2)(T-K_2+2)}\right](X_1'X_1)^{-1}$$

$$= \sigma_{11}(1-\rho_{12}^2)\left[1 + \frac{(T-K)}{(T-K_2)(T-K_2+2)} + \frac{(K_1+2)(K_1+4)}{(T-K_2+2)(T-K_2+4)}\left(\frac{\rho_{12}^2}{1-\rho_{12}^2}\right)\right](X_1'X_1)^{-1}. \tag{4.34}$$

This result was obtained by Revankar (1976; p. 185).

Similarly, the expression for the variance covariance matrix of the SURR estimator of β_2 can be written as

$$V(\hat{\beta}_{(2)SR})$$

$$= \sigma_{22}(1-\rho_{12}^2)\left[1+\frac{(T-K)}{(T-K_1)(T-K_1+2)}+\frac{(K_2+2)(K_2+4)}{(T-K_1+2)(T-K_1+4)}\left(\frac{\rho_{12}^2}{1-\rho_{12}^2}\right)\right](X_2'X_2)^{-1}. \tag{4.35}$$

For the cross-moments, it is easy to verify that

$$E(\hat{\beta}_{(1)SR}-\beta_1)(\hat{\beta}_{(2)SR}-\beta_2)'$$

$$= (X_1'X_1)^{-1}X_1'E\left[u_1u_2' - \frac{\hat{s}_{12}}{\hat{s}_{22}}u_2u_2' - \frac{\hat{s}_{12}}{\hat{s}_{11}}u_1u_1' + \frac{\hat{s}_{12}^2}{\hat{s}_{11}\hat{s}_{22}}u_2u_1'\right]X_2(X_2'X_2)^{-1}$$

$$= 0\,. \tag{4.36}$$

The expressions (4.34), (4.35) and (4.36) provide us with the complete variance covariance matrix of the SURR estimator of β.

4.4.2 Efficiency Comparisons

Revankar (1976) compared the efficiencies of the SUUR, SURR and OLS estimators. Recalling that the OLS estimator of β_1 is unbiased with variance covariance matrix

$$V(b_{(1)0}) = \sigma_{11}(X_1'X_1)^{-1}, \tag{4.37}$$

it may be observed from (4.26) that

$$V(b_{(1)0}) - V(\tilde{\beta}_{(1)SU}) = \sigma_{11}\left[\rho_{12}^2 - \frac{1-\rho_{12}^2}{T-K-2}\right](X_1'X_1)^{-1} \tag{4.38}$$

is positive definite when

$$\rho_{12}^2 > \frac{1}{T-K-1} \qquad (T-K) > 2 \tag{4.39}$$

so that the SUUR estimator dominates the OLS estimator with the variance covariance matrix as the performance criterion if ρ_{12}^2 exceeds $1/(T-K-1)$.

On the other hand, the OLS estimator dominates the SUUR estimator for values of ρ_{12}^2 falling below $1/(T - K - 1)$.

Comparing (4.37) with (4.34), we see that the SURR estimator dominates the OLS estimator if (Revankar, 1976; p. 186)

$$\rho_{12}^2 > \frac{1}{1 + (T - K_2)\left(1 + \frac{K_1 + K_2 + 2}{T - K_2 + 4}\right)} \tag{4.40}$$

while the opposite is the case when inequality (4.40) is reversed. Further, it may be seen that the range of ρ_{12}^2 values for which the SURR estimator is "better" than the OLS estimator gets wider as T and/or K_1 increases.

Similarly, comparing (4.26) and (4.34), we obtain (Revankar, 1976; p. 186):

$$V(\hat{\beta}_{(1)SR}) - V(\tilde{\beta}_{(1)SU}) = \sigma_{11} h (X_1' X_1)^{-1} \tag{4.41}$$

where

$$h = \left(\frac{K_1 + 2}{T - K_2 + 2}\right)\left[\frac{(K_1 + 4)\rho_{12}^2}{(T - K_2 + 4)} - \frac{2\left(T - K + \frac{1}{2}K_1\right)(1 - \rho_{12}^2)}{(T - K_2)(T - K - 2)}\right] . \tag{4.42}$$

So, we see that the SUUR estimator is superior to the SURR estimator, in the sense of positive definiteness of the difference matrix (4.41), when h is positive. This in turn requires that

$$\rho_{12}^2 > \frac{1}{1 + \left(2 + \frac{1}{2}K_1\right)\left(\frac{T - K_2}{T - K_2 + 4}\right)\left(\frac{T - K - 2}{T - K + \frac{1}{2}K_1}\right)} \tag{4.43}$$

from which it follows that the range of ρ_{12}^2 values over which the SUUR estimator dominates the SURR estimator becomes greater as T and/or K_1 increases. Further, we see that inequality (4.43) holds as long as ρ_{12}^2 is greater than $1/(3 + \frac{1}{2}K_1)$. Revankar (1976) presented values of h for various combinations of ρ_{12}, $(T - K)$ and K_1, and observed that the SUUR estimator is "better" than the SURR estimator when $|\rho_{12}|$, $(T - K)$ and K_1 are moderately large. For moderately small values, the reverse is true. For instance, the SUUR estimator is superior to the SURR estimator at least as long as $|\rho_{12}|$ and $(T - K)$ do not exceed 0.5 and 10 respectively.

Now let us examine the loss in efficiency arising from the replacement of Σ in the GLS estimator by its consistent estimators, $\tilde{S}$ and $\hat{S}$. For this purpose, we observe that the variance covariance matrix of the GLS estimator of β_1 is

$$\Omega_{11} = \sigma_{11}(1 - \rho_{12}^2)(X_1'X_1)^{-1}, \tag{4.44}$$

so that

$$V(\tilde{\beta}_{(1)SU}) = \left[1 + \frac{1}{T-K-2}\right]\Omega_{11}, \tag{4.45}$$

implying that the variance covariance matrix is inflated by $(T-K-1)/(T-K-2)$ as a result of this replacement.

Similarly, it is easy to verify that the inflation factor in the case of the SURR estimator of β_1 is

$$\left[1 + \frac{(T-K)}{(T-K_2)(T-K_2+2)} + \frac{(K_1+2)(K_1+4)}{(T-K_2+2)(T-K_2+4)}\left(\frac{\rho_{12}^2}{1-\rho_{12}^2}\right)\right], \tag{4.46}$$

which is larger than the inflation factor in the case of the SUUR estimator. This implies that the use of unrestricted residuals to form a consistent estimator of Σ leads to a smaller loss of precision (in comparison with the use of the restricted residuals) in all cases, despite the fact that the unrestricted residuals fail to use the information regarding the omission of K_2 regressors from the first equation and K_1 regressors from the second equation.

Finally, consider the case in which the disturbances in the two equations are uncorrelated ($\rho_{12} = 0$) but this information is ignored in the GLS estimation of the coefficients. Then, (4.26) and (4.34) reduce to

$$\begin{aligned} V(\tilde{\beta}_{(1)SU}) &= \sigma_{11}\left[1 + \frac{1}{T-K-2}\right](X_1'X_1)^{-1} \\ V(\hat{\beta}_{(1)SR}) &= \sigma_{11}\left[1 + \frac{(T-K)}{(T-K_2)(T-K_2+2)}\right](X_1'X_1)^{-1}. \end{aligned} \tag{4.47}$$

Comparing these expressions with (4.37), it is obvious that there is a loss in the efficiency of the SUUR and SURR estimators with respect to the OLS estimator when we ignore the information that the disturbances are uncorrelated when estimating the coefficients.

4.4.3 Large-Sample Asymptotic Approximations

In order to study the adequacy of large-sample asymptotic approximations to the variance covariance matrices of some of these estimators we write

$$\begin{aligned} \frac{1}{T-K-2} &= \frac{1}{T}\left[1 - \frac{K+2}{T}\right]^{-1} \\ &= \frac{1}{T} + \frac{K+2}{T^2} + \cdots \end{aligned}$$

provided T is larger than $(K+2)$.

Thus, considering the SUUR estimator of β_1 in the special two equation model, from (4.26) we have

$$V(\tilde{\beta}_{(1)SU}) = \sigma_{11}(1 - \rho_{12}^2)\left[1 + \frac{1}{T-K-2}\right](X_1'X_1)^{-1}$$

$$= \sigma_{11}(1 - \rho_{12}^2)\left[1 + \frac{1}{T} + \frac{K+2}{T^2} + \cdots\right](X_1'X_1)^{-1} \tag{4.48}$$

from which we see that the variance covariance matrix, to order $O(T^{-1})$, of this estimator is

$$V_1 = \sigma_{11}(1 - \rho_{12}^2)(X_1'X_1)^{-1}, \tag{4.49}$$

and to order $O(T^{-2})$ it is

$$V_2 = \sigma_{11}(1 - \rho_{12}^2)\left[1 + \frac{1}{T}\right](X_1'X_1)^{-1}. \tag{4.50}$$

These results match the first-order and second-order large-sample asymptotic approximations obtained in Section 3.3 (see Kakwani, 1974, and Srivastava and Upadhyaya, 1978, for further details).

Now, assuming that T is greater than K_2 and using

$$\frac{(T-K)}{(T-K_2)(T-K_2+2)} = \frac{1}{T}\left(1 - \frac{K}{T}\right)\left[1 - \frac{2(K_2-1)}{T} + \frac{K_2(K_2-2)}{T^2}\right]^{-1}$$

$$= \frac{1}{T} - \frac{K_1 - K_2 + 2}{T^2} + \cdots$$

$$\frac{1}{(T-K_2+2)(T-K_2+4)} = \frac{1}{T^2}\left[1 - \frac{2(K_2-3)}{T} + \frac{(K_2-2)(K_2-4)}{T^2}\right]^{-1}$$

$$= \frac{1}{T^2} + \cdots$$

it is easy to verify from the exact expression for the variance covariance matrix of the SURR estimator that the approximations to order $O(T^{-1})$ and $O(T^{-2})$ are V_1 and V_2 respectively and therefore they agree with the first-order and second-order large-sample asymptotic approximations given in Section 3.3. It is thus seen that the exact variance covariance matrix of the approximate sampling distribution coincides with the approximate variance covariance matrix of the exact sampling distribution, at least as far as first-order and second-order asymptotic approximations are concerned. If we retain terms up to order $O(T^{-3})$, we get

$$V(\tilde{\beta}_{(1)SU}) = V_2 + \frac{\sigma_{11}(1-\rho_{12}^2)}{T^2}(K+2)(X_1'X_1)^{-1}$$
$$V(\hat{\beta}_{(1)SR}) = V_2 + \frac{\sigma_{11}(1-\rho_{12}^2)}{T^2}\left[-(K_1 - K_2 + 2) + (K_1+2)(K_1+4)\left(\frac{\rho_{12}^2}{1-\rho_{12}^2}\right)\right](X_1'X_1)^{-1} \quad (4.51)$$

from which it follows that the SUUR estimator is superior to the SURR estimator if

$$\rho_{12}^2 > \frac{2}{K_1 + 6}\ , \quad (4.52)$$

implying that the SUUR estimator will definitely perform better than the SURR estimator, according to this approximation, if ρ_{12}^2 exceeds 1/3.

It may be observed that the accuracy of these large-sample asymptotic approximations can be studied by comparing (4.49), (4.50) and (4.51) with the exact variance covariance matrix expressions for the SUUR and SURR estimators. To illustrate, let us consider the SUUR estimator of β. Suppose that R_g denotes a matrix by which the exact variance covariance exceeds its approximation to order $O(T^{-g})$. It is then easy to see that

$$R_g = \ell_g(1 - \rho_{12}^2)\begin{bmatrix} \sigma_{11}(X_1'X_1)^{-1} & 0 \\ 0 & \sigma_{22}(X_2'X_2)^{-1} \end{bmatrix} \quad (4.53)$$

where ℓ_g is a scalar quantity measuring the accuracy of the approximation (to order $O(T^{-g})$) relative to the exact result.

From (4.29), we have

$$\ell_1 = \frac{1}{(T - K - 2)}$$
$$\ell_2 = \frac{(K+2)}{T(T - K - 2)} \quad (4.54)$$
$$\ell_3 = \frac{(K+2)^2}{T^2(T - K - 2)}\ ,$$

provided that T exceeds (K + 2).

Now, let us consider the upper bounds, as obtained by Kariya and Maekawa (1982), for $\sqrt{(\mathrm{tr}\ R_g R_g')}$ with g = 2, 3. From (3.93), we have

$$(\text{tr } R_2R_2')^{\frac{1}{2}} \le \frac{3.65(1-\rho_{12}^2)}{(T-K+1)(T-K+3)}[\sigma_{11}^2 \text{ tr}(X_1'X_1)^{-2} + \sigma_{22}^2 \text{ tr}(X_2'X_2)^{-2}]^{\frac{1}{2}}$$

$$(\text{tr } R_3R_3')^{\frac{1}{2}} \le \frac{79(1-\rho_{12}^2)}{(T-K+1)(T-K+3)(T-K+5)}[\sigma_{11}^2 \text{ tr}(X_1'X_1)^{-2} + \sigma_{22}^2 \text{ tr}(X_2'X_2)^{-2}]^{\frac{1}{2}} \tag{4.55}$$

but from (4.53) we have

$$(\text{tr } R_gR_g')^{\frac{1}{2}} = \ell_g(1-\rho_{12}^2)[\sigma_{11}^2 \text{ tr}(X_1'X_1)^{-2} + \sigma_{22}^2 \text{ tr}(X_2'X_2)^{-2}]^{\frac{1}{2}}$$

so that the inequalities (4.55) assume simple forms:

$$\ell_2 \le \frac{3.65}{(T-K+1)(T-K+3)}$$

$$\ell_3 \le \frac{79}{(T-K+1)(T-K+3)(T-K+5)}\,. \tag{4.56}$$

A comparison of the bounds with the exact values from (4.54) may shed some light on the sharpness of these bounds.

4.4.4 Sampling Distribution of the SUUR Estimator

Now let us consider the sampling distribution of the SUUR estimator of β_1, which was derived by Zellner (1963; 1972). If we write

$$w_i = (X_1'X_1)^{-1}X_1'u_i \qquad (i = 1,\ 2)$$

we observe that the vector $(w_1' \quad w_2')$ follows a multivariate Normal distribution with mean vector 0 and variance covariance matrix $[\Sigma \otimes (X_1'X_1)^{-1}]$. Further, w_1 and w_2 are stochastically independent of $\ell = (\tilde{s}_{12}/\tilde{s}_{22})$. Now, to derive the probability density function of ℓ, we follow Zellner (1963) and observe that

$$\begin{bmatrix} a_{11} & a_{12} \\ a_{12} & a_{22} \end{bmatrix}$$

follows a Wishart distribution $W_2(n, \Sigma)$ where $a_{ij} = T\tilde{s}_{ij}$ and $n = (T - K)$. Thus, the joint probability density function is given by

$$p(a_{11}, a_{22}, a_{12})$$

$$= \frac{(a_{11}a_{22} - a_{12}^2)^{(n-3)/2}}{2^n|\Sigma|^{n/2}\Gamma\left(\frac{1}{2}\right)\Gamma\left(\frac{n}{2}\right)\Gamma\left(\frac{n-1}{2}\right)} \exp\left\{-\frac{1}{2(1-\rho_{12}^2)}\left[\frac{a_{11}}{\sigma_{11}} + \frac{a_{22}}{\sigma_{22}} - \frac{2\rho_{12}a_{12}}{\sqrt{(\sigma_{11}\sigma_{22})}}\right]\right\}. \tag{4.57}$$

Applying the transformation

$$a_{11} = a_{11}, \quad a_{22} = a_{22}, \quad \ell = \frac{a_{12}}{a_{22}}$$

and observing that the Jacobian of this transformation is a_{22}, we find the joint probability density function of a_{11}, a_{22} and ℓ as follows:

$$p(a_{11}, a_{22}, \ell) = f(\ell)\left[\frac{(a_{11} - \ell^2 a_{22})^{(n-3)/2} \exp\left\{-\frac{a_{11} - \ell^2 a_{22}}{2\sigma_{11}(1-\rho_{12}^2)}\right\}}{[2\sigma_{11}(1-\rho_{12}^2)]^{(n-2)/2}\Gamma\left(\frac{n-1}{2}\right)}\right] \cdot \left[\frac{a_{22}^{(n-1)/2} \exp\left\{-\frac{a_{22}}{2\sigma_{11}(1-\rho_{12}^2)}\left(\ell^2 - 2\ell\delta + \frac{\sigma_{11}}{\sigma_{22}}\right)\right\}}{[2\sigma_{11}(1-\rho_{12}^2)]^{(n+1)/2}\Gamma\left(\frac{n+1}{2}\right)\left(\ell^2 - 2\ell\delta + \frac{\sigma_{11}}{\sigma_{22}}\right)^{-(n+1)/2}}\right] \tag{4.58}$$

where

$$f(\ell) = \left[\frac{\sigma_{11}}{\sigma_{22}}(1-\rho_{12}^2)\right]^{n/2} \cdot \frac{\Gamma\left(\frac{n+1}{2}\right)}{\Gamma\left(\frac{1}{2}\right)\Gamma\left(\frac{n}{2}\right)} \cdot \frac{1}{\left(\ell^2 - 2\ell\delta + \frac{\sigma_{11}}{\sigma_{22}}\right)^{(n+1)/2}}, \tag{4.59}$$

and $\delta = \rho_{12}\sqrt{\sigma_{11}/\sigma_{22}}$.

Integrating $p(a_{11}, a_{22}, \ell)$ with respect to a_{11} and a_{22}, we obtain the probability density function of ℓ as $f(\ell)$, which is Student-t in form. Thus the joint probability density function of w_1, w_2 and ℓ is

$$p(w_1, w_2, \ell) = f(w_1, w_2)f(\ell) \tag{4.60}$$

where

$$f(w_1, w_2) = \frac{1}{(2\pi)^{K_1}|\Sigma \otimes (X_1'X_1)^{-1}|^{\frac{1}{2}}} \exp\left\{-\frac{1}{2}(w_1' \quad w_2')(\Sigma^{-1} \otimes X_1'X_1)\begin{pmatrix} w_1 \\ w_2 \end{pmatrix}\right\}$$

$$= \frac{1}{(2\pi)^{K_1}[\sigma_{11}\sigma_{22}(1-\rho_{12}^2)]^{K_1/2}|X_1'X_1|^{-1}}$$

$$\cdot \exp\left\{-\frac{1}{2\sigma_{11}(1-\rho_{12}^2)}\left[w_1'X_1'X_1w_1 - 2\delta w_1'X_1'X_1w_2 + \frac{\sigma_{11}}{\sigma_{22}}w_2'X_1'X_1w_2\right]\right\}. \tag{4.61}$$

From (4.19), we have

$$(\tilde{\beta}_{(1)SU} - \beta_1) = (w_1 - \ell w_2) .$$

Employing the transformation

$$\xi_1 = (w_1 - \ell w_2), \quad w_2 = w_2, \quad \ell = \ell$$

and observing that the Jacobian of this transformation is 1, we find from (4.60) that the joint probability density function of ξ_1, w_2 and ℓ is

$$\frac{c\exp\left\{-\frac{1}{2\sigma_{11}(1-\rho_{12}^2)}\left[\xi_1'X_1'X_1\xi_1 - 2(\delta-\ell)\xi_1'X_1'X_1w_2 + \left(\ell^2 - 2\ell\delta + \frac{\sigma_{11}}{\sigma_{22}}\right)w_2'X_1'X_1w_2\right]\right\}}{\left(\ell^2 - 2\ell\delta + \frac{\sigma_{11}}{\sigma_{22}}\right)^{(n+1)/2}} \tag{4.62}$$

where

$$c = \frac{\left[\frac{\sigma_{11}}{\sigma_{22}}(1-\rho_{12}^2)\right]^{n/2}}{(2\pi)^{K_1}[\sigma_{11}\sigma_{22}(1-\rho_{12}^2)]^{K_1/2}|X_1'X_1|^{-1}} \cdot \frac{\Gamma\left(\frac{n+1}{2}\right)}{\Gamma\left(\frac{1}{2}\right)\Gamma\left(\frac{n}{2}\right)} .$$

If we write

$$c^* = \frac{\left[\frac{\sigma_{11}}{\sigma_{22}}(1-\rho_{12}^2)\right]^{n/2}}{(2\pi\sigma_{22})^{\frac{1}{2}K_1}|X_1'X_1|^{-\frac{1}{2}}} \cdot \frac{\Gamma\left(\frac{n+1}{2}\right)}{\Gamma\left(\frac{1}{2}\right)\Gamma\left(\frac{n}{2}\right)}$$

$$f^*(\xi_1, \ell) = \frac{\exp\left\{-\frac{\xi_1'X_1'X_1\xi_1}{2\sigma_{22}\left(\ell^2 - 2\ell\delta + \frac{\sigma_{11}}{\sigma_{22}}\right)}\right\}}{\left(\ell^2 - 2\ell\delta + \frac{\sigma_{11}}{\sigma_{22}}\right)^{(n+K_1+1)/2}}$$

then (4.62) becomes

$$c^*f^*(\xi_1, \ell)\left[\frac{\ell^2 - 2\ell\delta + \frac{\sigma_{11}}{\sigma_{22}}}{2\pi\sigma_{11}(1-\rho_{12}^2)}\right]^{K_1/2} |X_1'X_1|^{\frac{1}{2}}$$

$$\cdot \exp\left\{-\frac{\left(\ell^2 - 2\ell\delta + \frac{\sigma_{11}}{\sigma_{22}}\right)}{2\sigma_{11}(1-\rho_{12}^2)}\left[w_2' - \frac{(\delta-\ell)}{\left(\ell^2 - 2\ell\delta + \frac{\sigma_{11}}{\sigma_{22}}\right)}\xi_1'\right](X_1'X_1)\right.$$

$$\left.\cdot\left[w_2 - \frac{(\delta-\ell)}{\left(\ell^2 - 2\ell\delta + \frac{\sigma_{11}}{\sigma_{22}}\right)}\xi_1'\right]\right\} .$$

Integrating out w_2, we obtain the joint probability density function of $\xi_1 = (\tilde{\beta}_{(1)SU} - \beta_1)$ and ℓ as $c^*f^*(\xi_1, \ell)$, which can be expressed as

$$c^* \sum_{\alpha=0}^{\infty} \frac{(-1)^\alpha}{\alpha!}\left[\frac{\xi_1'X_1'X_1\xi_1}{2\sigma_{22}}\right]^\alpha \cdot \frac{1}{\left(\ell^2 - 2\ell\delta + \frac{\sigma_{11}}{\sigma_{22}}\right)^{\left(\frac{n+K_1+1}{2}+\alpha\right)}} . \quad (4.63)$$

Now, because for $m \geq 1$

$$\int_{-\infty}^{\infty} \frac{d\ell}{\left(\ell^2 - 2\ell\delta + \frac{\sigma_{11}}{\sigma_{22}}\right)^m} = \left[\frac{\sigma_{22}}{\sigma_{11}(1-\rho_{12}^2)}\right]^{m-1/2} \int_0^1 (1-t)^{-1/2} t^{m-3/2}\, dt$$

$$= \left[\frac{\sigma_{22}}{\sigma_{11}(1-\rho_{12}^2)}\right]^{m-1/2} \cdot \frac{\Gamma\left(\frac{1}{2}\right)\Gamma\left(m-\frac{1}{2}\right)}{\Gamma(m)} \quad (4.64)$$

we see that integrating the expression (4.63) with respect to ℓ gives the probability density function of $\xi_1 = (\tilde{\beta}_{(1)SU} - \beta_1)$ as

$$g(\xi_1) = c^*\left[\frac{\sigma_{22}}{\sigma_{11}(1-\rho_{12}^2)}\right]^{(n+K_1)/2} \sum_{\alpha=0}^{\infty} \frac{(-1)^\alpha}{\alpha!}\left[\frac{\xi_1'X_1'X_1\xi_1}{2\sigma_{11}(1-\rho_{12}^2)}\right]^\alpha \frac{\Gamma\left(\frac{1}{2}\right)\Gamma\left(\frac{n+K_1}{2}+\alpha\right)}{\Gamma\left(\frac{n+K_1+1}{2}+\alpha\right)}$$

$$= \frac{1}{[2\pi\sigma_{11}(1-\rho_{12}^2)]^{\frac{1}{2}K_1}|X_1'X_1|^{-\frac{1}{2}}} \sum_{\alpha=0}^{\infty} \frac{(-1)^\alpha}{\alpha!}\left[\frac{\sigma_{22}\xi_1'X_1'X_1\xi_1}{\sigma_{11}(1-\rho_{12}^2)}\right]^\alpha \frac{\Gamma\left(\frac{n+1}{2}\right)\Gamma\left(\frac{n+K_1}{2}+\alpha\right)}{\Gamma\left(\frac{n}{2}\right)\Gamma\left(\frac{n+K_1+1}{2}+\alpha\right)} .$$

(4.65)

Zellner (1963) derived the expression (4.65) under the assumption that $(X_1'X_1)$ is an identity matrix, and observed that the distribution of ξ_1 tends to Normal as n grows large. Such an observation is more clearly visible if, instead of expressing $c^*f^*(\xi_1, \ell)$ in the form (4.63), we write it as

$$c^* \exp\left\{-\frac{\xi_1'X_1'X_1\xi_1}{2\sigma_{11}(1-\rho_{12}^2)}\right\} \sum_{\alpha=0}^{\infty} \frac{1}{\alpha!}\left[\frac{\xi_1'X_1'X_1\xi_1}{2\sigma_{11}(1-\rho_{12}^2)}\right]^{\alpha} \frac{(\delta-\ell)^{2\alpha}}{\left(\ell^2-2\ell\delta+\frac{\sigma_{11}}{\sigma_{22}}\right)^{\left(\frac{n+K_1+1}{2}+\alpha\right)}} \cdot \tag{4.66}$$

Now, as in the case of (4.64), we observe that

$$\int_{-\infty}^{\infty} \frac{(\delta-\ell)^{2\alpha}}{\left(\ell^2-2\ell\delta+\frac{\sigma_{11}}{\sigma_{22}}\right)^{m}} d\ell = \left[\frac{\sigma_{22}}{\sigma_{11}(1-\rho_{12}^2)}\right]^{m-\alpha-\frac{1}{2}} \frac{\Gamma\left(\alpha+\frac{1}{2}\right)\Gamma\left(m-\alpha-\frac{1}{2}\right)}{\Gamma(m)} \tag{4.67}$$

so that integrating out ℓ in (4.66), we get an alternative expression for the probability density function of $\xi_1 = (\tilde{\beta}_{(1)SU} - \beta_1)$:

$$g(\xi_1) = \phi(\cdot) \sum_{\alpha=0}^{\infty} \frac{1}{\alpha!}\left[\frac{\xi_1'X_1'X_1\xi_1}{2\sigma_{11}(1-\rho_{12}^2)}\right]^{\alpha} \frac{\Gamma\left(\frac{n+1}{2}\right)\Gamma\left(\alpha+\frac{1}{2}\right)\Gamma\left(\frac{n+K_1}{2}\right)}{\Gamma\left(\frac{n}{2}\right)\Gamma\left(\frac{1}{2}\right)\Gamma\left(\frac{n+K_1+1}{2}+\alpha\right)} \tag{4.68}$$

where $\phi(\cdot)$ denotes the probability density function of a multivariate Normal distribution with mean vector 0 and variance covariance matrix $\sigma_{11}(1-\rho_{12}^2)(X_1'X_1)^{-1}$.

From (4.65) and (4.68), it is obvious that the sampling distribution of the SUUR estimator is symmetric. To examine the kurtosis of this distribution, we may express the k'th element of ξ_1 as

$$\xi_{1k} = \sum_{t=1}^{T} g_{kt}(u_{t1} - \ell u_{t2}) \tag{4.69}$$

where g_{kt} denotes the (k, t)'th element of $(X_1'X_1)^{-1}X_1'$.

From the properties of Student's t distribution, the first four moments of ℓ are

$$
\begin{aligned}
E(\ell) &= \left(\frac{\sigma_{11}}{\sigma_{22}}\right)^{\frac{1}{2}} \rho_{12} = \delta , \\
E(\ell^2) &= \frac{\sigma_{11}}{\sigma_{22}}\left[\rho_{12}^2 + \frac{1-\rho_{12}^2}{n-2}\right] , \\
E(\ell^3) &= \left(\frac{\sigma_{11}}{\sigma_{22}}\right)^{\frac{3}{2}}\left[\rho_{12}^3 + \frac{3\rho_{12}(1-\rho_{12}^2)}{n-2}\right] , \\
E(\ell^4) &= \left(\frac{\sigma_{11}}{\sigma_{22}}\right)^{2}\left[\rho_{12}^4 + \frac{3(1-\rho_{12}^2)}{(n-2)}\left(2\rho_{12}^2 + \frac{1-\rho_{12}^2}{n-4}\right)\right] .
\end{aligned}
\tag{4.70}
$$

Now, from (4.69), all odd-order moments of ξ_{1k} are 0, and

$$
E(\xi_{1k}^2) = \sigma_{11}(1-\rho_{12}^2)\left(\frac{n-1}{n-2}\right)\left(\sum_{t=1}^{T} g_{kt}^2\right) \; ; \; n > 2 \tag{4.71}
$$

$$
E(\xi_{1k}^4) = 3\sigma_{11}^2(1-\rho_{12}^2)^2\left[\frac{(n-1)(n-3)}{(n-2)(n-4)}\right]\left(\sum_{t=1}^{T} g_{kt}^2\right)^2 \; ; \; n > 4 \tag{4.72}
$$

from which Pearson's measure for excess of kurtosis is given by

$$
\frac{E(\xi_{1k}^4)}{[E(\xi_{1k}^2)]^2} - 3 = \frac{6}{(n-1)(n-4)} \; ; \quad n > 4 \tag{4.73}
$$

implying that the sampling distribution of the SUUR estimator is leptokurtic and its kurtosis measure does not depend upon the correlation between the disturbances of different equations.

4.5 EFFICIENCY PROPERTIES UNDER SUBSET REGRESSORS

Next we consider a two equation SURE model in which the regressors in the second equation are a proper subset of the regressors in the first equation, i.e., X_2 is a submatrix of X_1. Notice that Z is equal to X_1 here. To simplify the exposition, let us assume that X_1 is of the form

$$
X_1 = (X_2 \quad X_0) , \tag{4.74}
$$

where X_0 is a $[T \times (K_1 - K_2)]$ matrix of T observations on $(K_1 - K_2)$ explanatory variables deleted from the second equation.

Under the specification (4.74), it is easy to see that $X_2'\bar{P}_{x_1}$ is a null matrix, while $X_1'\bar{P}_{x_2}$ is not null. Using these results in (4.16), we obtain

$$
\begin{aligned}
b_{(1)F} &= (X_1'X_1)^{-1}X_1'y_1 - \left(\frac{s_{12}}{s_{22}}\right)(X_1'X_1)^{-1}X_1'\bar{P}_{x_2}y_2 , \\
b_{(2)F} &= (X_2'X_2)^{-1}X_2'y_2 .
\end{aligned}
\tag{4.75}
$$

Thus, we see that the FGLS estimator of β_2 is just the OLS estimator in this case. Revankar (1974; 1976) considered the efficiency of $b_{(1)F}$.

4.5.1 The Variance Covariance Matrix

First, we shall set $S = \tilde{S}$. The SUUR estimator of β_1 is

$$
\tilde{\beta}_{(1)SU} = (X_1'X_1)^{-1}X_1'y_1 - \frac{\tilde{s}_{12}}{\tilde{s}_{22}}(X_1'X_1)^{-1}X_1'\bar{P}_{x_2}y_2
$$

where $\tilde{s}_{ij}$ (i, j = 1, 2) is the same as in (4.21) with $Z = X_1$. The variance covariance matrix of the SUUR estimator is

$$
\begin{aligned}
V(\tilde{\beta}_{(1)SU}) &= (X_1'X_1)^{-1}E\left[X_1'u_1u_1'X_1 - \frac{\tilde{s}_{12}}{\tilde{s}_{22}}X_1'(u_1u_2'\bar{P}_{x_2} + \bar{P}_{x_2}u_2u_1')X_1\right. \\
&\quad \left. + \left[\frac{\tilde{s}_{12}}{\tilde{s}_{22}}\right]^2 X_1'\bar{P}_{x_2}u_2u_2'\bar{P}_{x_2}X_1\right](X_1'X_1)^{-1} \\
&= \sigma_{11}(X_1'X_1)^{-1} - \left[2\rho_{12}(\sigma_{11}\sigma_{22})^{\frac{1}{2}}E\left[\frac{\tilde{s}_{12}}{\tilde{s}_{22}}\right] - \sigma_{22}E\left[\frac{\tilde{s}_{12}}{\tilde{s}_{22}}\right]^2\right](X_1'X_1)^{-1}X_1' \\
&\quad \cdot \bar{P}_{x_2}X_1(X_1'X_1)^{-1}
\end{aligned}
\tag{4.76}
$$

because $\bar{P}_{x_1}u_1$ and $\bar{P}_{x_2}u_2$ are stochastically independent of $X_1'u_1$ and $X_1'\bar{P}_{x_2}u_2$.

Substituting X_1 for Z in (4.24) and (4.25) we have

$$
E\left[\frac{\tilde{s}_{12}}{\tilde{s}_{22}}\right] = \delta \tag{4.77}
$$

and

$$E\left[\frac{\tilde{s}_{12}}{\tilde{s}_{22}}\right]^2 = \left(\frac{\sigma_{11}}{\sigma_{22}}\right)\left(\rho_{12}^2 + \frac{1-\rho_{12}^2}{T-K_1-2}\right) \; ; \; (T-K_1) > 2 \, . \tag{4.78}$$

Substituting (4.77) and (4.78) in (4.76), we find the result obtained by Revankar (1974; p. 189):

$$V(\tilde{\beta}_{(1)SU}) = \sigma_{11}(X_1'X_1)^{-1} - \sigma_{11}\left(\rho_{12}^2 - \frac{1-\rho_{12}^2}{T-K_1-2}\right)(X_1'X_1)^{-1}X_1'\bar{P}_{x_2}X_1(X_1'X_1)^{-1} \, . \tag{4.79}$$

The SUUR estimator of β_2, being identical to the OLS estimator, has the variance covariance matrix

$$V(\tilde{\beta}_{(2)SU}) = \sigma_{22}(X_2'X_2)^{-1} \, . \tag{4.80}$$

Finally, we have

$$\begin{aligned} E(\tilde{\beta}_{(1)SU} - \beta_1)(\tilde{\beta}_{(2)SU} - \beta_2)' &= (X_1'X_1)^{-1}E\left[X_1'u_1u_2'X_2 - \frac{\tilde{s}_{12}}{\tilde{s}_{22}}X_1'\bar{P}_{x_2}u_2u_2'X_2\right](X_2'X_2)^{-1} \\ &= (X_1'X_1)^{-1}\left[\sigma_{12}X_1'X_2 - \sigma_{22}E\left[\frac{\tilde{s}_{12}}{\tilde{s}_{22}}\right]X_1'\bar{P}_{x_2}X_2\right](X_2'X_2)^{-1} \\ &= \sigma_{12}(X_1'X_1)^{-1}X_1'X_2(X_2'X_2)^{-1} \end{aligned} \tag{4.81}$$

where use has been made of the result $\bar{P}_{x_2}X_2 = 0$.

Combining (4.79), (4.80) and (4.81), the complete variance covariance matrix of the SUUR estimator of β is

$$\begin{bmatrix} \sigma_{11}(X_1'X_1)^{-1} - \sigma_{11}\left(\rho_{12}^2 - \frac{1-\rho_{12}^2}{T-K_1-2}\right) \cdot (X_1'X_1)^{-1}X_1'\bar{P}_{x_2}X_1(X_1'X_1)^{-1} & \sigma_{12}(X_1'X_1)^{-1}X_1'X_2(X_2'X_2)^{-1} \\ \sigma_{12}(X_2'X_2)^{-1}X_2'X_1(X_1'X_1)^{-1} & \sigma_{22}(X_2'X_2)^{-1} \end{bmatrix} \, . \tag{4.82}$$

Now, let us set $S = \hat{S}$ in the FGLS estimator to obtain the SURR estimator. To obtain its variance covariance matrix, we observe from (4.23) that

$$E\left[\frac{\hat{s}_{12}}{\hat{s}_{22}} u_1 u_2'\right] = E\left[\frac{u_1'\bar{P}_{X_1}\bar{P}_{X_2}u_2}{u_2'\bar{P}_{X_2}u_2} u_1 u_2'\right]$$

$$= E\left[\frac{u_1'\bar{P}_{X_1}u_2}{u_2'\bar{P}_{X_2}u_2} u_1 u_2'\right]$$

$$= \delta^2 E\left[\frac{u_2'\bar{P}_{X_1}u_2}{u_2'\bar{P}_{X_2}u_2} u_2 u_2'\right] + \delta E\left[\frac{E(v')\bar{P}_{X_1}u_2}{u_2'\bar{P}_{X_2}u_2} u_2 u_2'\right]$$

$$+ \delta E\left[\frac{u_2'\bar{P}_{X_1}u_2}{u_2'\bar{P}_{X_2}u_2} E(v) u_2'\right] + E\left[\frac{1}{u_2'\bar{P}_{X_2}u_2} E(vv')\bar{P}_{X_1} u_2 u_2'\right]$$

$$= \delta^2 E\left[\frac{u_2'\bar{P}_{X_1}u_2}{u_2'\bar{P}_{X_2}u_2} u_2 u_2'\right] + \sigma_{11}(1-\rho_{12}^2)\bar{P}_{X_1} E\left[\frac{1}{u_2'\bar{P}_{X_2}u_2} u_2 u_2'\right] \quad (4.83)$$

$$E\left[\left[\frac{\hat{s}_{12}}{\hat{s}_{22}}\right]^2 u_2 u_2'\right] = E\left[\left[\frac{u_1'\bar{P}_{X_1}u_2}{u_2'\bar{P}_{X_2}u_2}\right]^2 u_2 u_2'\right]$$

$$= \delta^2 E\left[\left[\frac{u_2'\bar{P}_{X_1}u_2}{u_2'\bar{P}_{X_2}u_2}\right]^2 u_2 u_2'\right] + 2\delta E\left[\frac{u_2'\bar{P}_{X_1}u_2 E(v')\bar{P}_{X_1}u_2}{(u_2'\bar{P}_{X_2}u_2)^2} u_2 u_2'\right]$$

$$+ E\left[\frac{u_2'\bar{P}_{X_1}E(vv')\bar{P}_{X_1}u_2}{(u_2'\bar{P}_{X_2}u_2)^2} u_2 u_2'\right]$$

$$= \delta^2 E\left[\left[\frac{u_2'\bar{P}_{X_1}u_2}{u_2'\bar{P}_{X_2}u_2}\right]^2 u_2 u_2'\right] + \sigma_{11}(1-\rho_{12}^2) E\left[\frac{u_3'\bar{P}_{X_1}u_2}{(u_2'\bar{P}_{X_2}u_2)^2} u_2 u_2'\right] . \quad (4.84)$$

Using (4.83) and (4.84), we find

$$
\begin{aligned}
V(\hat{\beta}_{(1)SR}) & \\
&= (X_1'X_1)^{-1}X_1'E\left[u_1u_1' - \frac{\hat{s}_{12}}{\hat{s}_{22}}(u_1u_2'\bar{P}_{X_2} + \bar{P}_{X_2}u_2u_1') \right.\\
&\quad \left. + \left[\frac{\hat{s}_{12}}{\hat{s}_{22}}\right]^2 \bar{P}_{X_2}u_2u_2'\bar{P}_{X_2}\right] X_1(X_1'X_1)^{-1} \\
&= \sigma_{11}(X_1'X_1)^{-1} - \delta^2(X_1'X_1)^{-1}X_1'E\left[\left[\frac{u_2'\bar{P}_{X_1}u_2}{u_2'\bar{P}_{X_2}u_2}\right](u_2u_2'\bar{P}_{X_2} + \bar{P}_{X_2}u_2u_2')\right]X_1(X_1'X_1)^{-1} \\
&\quad + (X_1'X_1)^{-1}X_1'\bar{P}_{X_2}E\left[\frac{u_2'\bar{P}_{X_1}u_2(\delta^2u_2'\bar{P}_{X_1}u_2 + \sigma_{11}(1-\rho_{12}^2))}{(u_2'\bar{P}_{X_2}u_2)^2}u_2u_2'\right]\bar{P}_{X_2}X_1(X_1'X_1)^{-1}\,.
\end{aligned} \tag{4.85}
$$

Observing that $P_{X_2}u_2$ is stochastically independent of $\bar{P}_{X_2}u_2$ and $\bar{P}_{X_1}u_2$, we can write

$$
\begin{aligned}
E&\left[\frac{u_2'\bar{P}_{X_1}u_2}{u_2'\bar{P}_{X_2}u_2}u_2u_2'\bar{P}_{X_2}\right] \\
&= E\left[\frac{u_2'\bar{P}_{X_1}u_2}{u_2'\bar{P}_{X_2}u_2}(P_{X_2}u_2 + \bar{P}_{X_2}u_2)u_2'\bar{P}_{X_2}\right] \\
&= E\left[\frac{u_2'\bar{P}_{X_1}u_2}{u_2'\bar{P}_{X_2}u_2}E(P_{X_2}u_2)u_2'\bar{P}_{X_2}\right] + E\left[\frac{u_2'\bar{P}_{X_1}u_2}{u_2'\bar{P}_{X_2}u_2}\bar{P}_{X_2}u_2u_2'\bar{P}_{X_2}\right] \\
&= \bar{P}_{X_2}E\left[\frac{u_2'\bar{P}_{X_1}u_2}{u_2'\bar{P}_{X_2}u_2}u_2u_2'\;\bar{P}_{X_2}\right].
\end{aligned} \tag{4.86}
$$

Further, we have

$$
X_1'\bar{P}_{X_2}u_2u_2'\bar{P}_{X_2}X_1 = \underset{\qquad (K_2) \qquad\qquad (K_1-K_2)}{\begin{bmatrix} 0 & 0 \\ 0 & X_0'\bar{P}_{X_2}u_2u_2'\bar{P}_{X_2}X_0 \end{bmatrix}} \begin{matrix} (K_2) \\ (K_1-K_2) \end{matrix}\,. \tag{4.87}
$$

Let Γ be a nonsingular matrix such that $\Gamma' X_0'\bar{P}_{X_2}X_0\Gamma = I_{(K_1-K_2)}$. Defining

$$\chi^2 = \frac{1}{\sigma_{22}} u_2' \bar{P}_{X_1} u_2$$
$$w = \sigma_{22}^{-\frac{1}{2}} \Gamma' X_0' \bar{P}_{X_2} u_2 \qquad (4.88)$$

so that

$$\begin{aligned} w'w &= \frac{1}{\sigma_{22}} u_2' \bar{P}_{X_2} X_0 \Gamma \Gamma' X_0' \bar{P}_{X_2} u_2 \\ &= \frac{1}{\sigma_{22}} u_2' \bar{P}_{X_2} X_0 (X_0' \bar{P}_{X_2} X_0)^{-1} X_0' \bar{P}_{X_2} u_2 \\ &= \frac{1}{\sigma_{22}} u_2' (\bar{P}_{X_2} - \bar{P}_{X_1}) u_2 \ , \end{aligned}$$

it is seen that

$$\begin{aligned} & E\left[\frac{u_2' \bar{P}_{X_1} u_2}{u_2' \bar{P}_{X_2} u_2}\left[2 - \frac{u_2' \bar{P}_{X_1} u_2}{u_2' \bar{P}_{X_2} u_2}\right] X_0' \bar{P}_{X_2} u_2 u_2' \bar{P}_{X_2} X_0\right] \\ &= \sigma_{22} \Gamma'^{-1} \left[E(ww') - E\left(\frac{w'w}{\chi^2 + w'w}\right)^2 ww'\right] \Gamma^{-1} \\ &= \sigma_{22} \left[1 - \frac{(K_1 - K_2 + 2)(K_1 - K_2 + 4)}{(T - K_2 + 2)(T - K_2 + 4)}\right] (\Gamma\Gamma')^{-1} \\ &= \sigma_{22} \left[1 - \frac{(K_1 - K_2 + 2)(K_1 - K_2 + 4)}{(T - K_2 + 2)(T - K_2 + 4)}\right] X_0' \bar{P}_{X_2} X_0 \end{aligned} \qquad (4.89)$$

$$\begin{aligned} & E\left[\frac{u_2' \bar{P}_{X_1} u_2}{u_2' \bar{P}_{X_2} u_2} X_0' \bar{P}_{X_2} u_2 u_2' \bar{P}_{X_2} X_0\right] \\ &= \Gamma'^{-1} E\left[\frac{\chi^2}{(\chi^2 + w'w)^2} ww'\right] \Gamma^{-1} \\ &= \Gamma'^{-1} E\left[\frac{1}{\chi^2 + w'w} ww'\right] \Gamma^{-1} - \Gamma'^{-1} E\left[\frac{w'w}{(\chi^2 + w'w)^2} ww'\right] \Gamma^{-1} \\ &= \left[\frac{1}{(T - K_2)} - \frac{(K_1 - K_2 + 2)}{(T - K_2)(T - K_2 + 2)}\right] (\Gamma\Gamma')^{-1} \\ &= \frac{(T - K_1)}{(T - K_2)(T - K_2 + 2)} X_0' \bar{P}_{X_2} X_0 \end{aligned} \qquad (4.90)$$

where use has been made of the results (B.2) of the Appendix.

Using (4.86), (4.87), (4.89) and (4.90), we obtain from (4.85) the following result reported by Revankar (1976; p. 185):

$$
\begin{aligned}
V(\hat{\beta}_{(1)SR}) &= \sigma_{11}(X_1'X_1)^{-1} - \sigma_{11}\rho_{12}^2(X_1'X_1)^{-1}X_1'\bar{P}_{X_2}X_1(X_1'X_1)^{-1} \\
&+ \sigma_{11}\left[\frac{(K_1-K_2+2)(K_1-K_2+4)\rho_{12}^2}{(T-K_2+2)(T-K_2+4)} + \frac{(T-K_1)(1-\rho_{12}^2)}{(T-K_2)(T-K_2+2)}\right] \\
&\cdot (X_1'X_1)^{-1}X_1'\bar{P}_{X_2}X_1(X_1'X_1)^{-1} . \qquad (4.91)
\end{aligned}
$$

Similarly, it can be shown that

$$
\begin{aligned}
E(\hat{\beta}_{(1)SR} - \beta_1)(\hat{\beta}_{(2)SR} - \beta_2)' &= (X_1'X_1)^{-1}X_1'E\left[u_1u_2' - \frac{\hat{s}_{12}}{\hat{s}_{22}}\bar{P}_{X_2}u_2u_2'\right]X_2(X_2'X_2)^{-1} \\
&= \sigma_{12}(X_1'X_1)^{-1}X_1'X_2(X_2'X_2)^{-1} . \qquad (4.92)
\end{aligned}
$$

The complete variance covariance matrix of the SURR estimator of β can be obtained from (4.91), (4.92) and

$$V(\hat{\beta}_{(2)SR}) = \sigma_{22}(X_2'X_2)^{-1} .$$

4.5.2 Efficiency Comparisons

Using (4.37) and (4.79), we may compare the performances of the OLS and SUUR estimators of β_1 in the two equation SURE model with subset regressors as follows:

$$V(b_{(1)0}) - V(\tilde{\beta}_{(1)SU}) = \sigma_{11}\left[\rho_{12}^2 - \frac{1-\rho_{12}^2}{T-K_1-2}\right](X_1'X_1)^{-1}X_1'\bar{P}_{X_2}X_1(X_1'X_1)^{-1} \qquad (4.93)$$

which is at least positive semidefinite if

$$\rho_{12}^2 > \frac{1}{(T-K_1-1)} . \qquad (4.94)$$

Thus, the SUUR estimator dominates the OLS estimator when ρ_{12}^2 exceeds $1/(T-K_1-1)$. The converse is true when inequality (4.94) is reversed. Similarly, using (4.37) and (4.91), the SURR estimator dominates the OLS estimator if

$$\rho_{12}^2 > \frac{1}{1 + (T - K_2)\left(1 + \frac{K_1 - K_2 + 2}{T - K_2 + 4}\right)} \tag{4.95}$$

while the SURR estimator is dominated by the OLS estimator when inequality (4.95) is reversed. Notice that the interval of ρ_{12}^2 for which the SURR estimator dominates the OLS estimator is bigger for larger values of T and K_1. Further, the lower bound specified by (4.95) is larger than the corresponding bound in (4.40) obtained under the assumption of orthogonal regressors.

Comparing the expressions (4.79) and (4.91), we have

$$V(\hat{\beta}_{(1)SR}) - V(\tilde{\beta}_{(1)SU}) = \sigma_{11} h (X_1'X_1)^{-1} X_1' \bar{P}_{x_2} X_1 (X_1'X_1)^{-1} \tag{4.96}$$

where

$$h = \left(\frac{K_1 - K_2 + 2}{T - K_2 + 2}\right)\left[\frac{(K_1 - K_2 + 4)\rho_{12}^2}{T - K_2 + 4} - \frac{2\left(T - \frac{K_1 + K_2}{2}\right)(1 - \rho_{12}^2)}{(T - K_2)(T - K_1 - 2)}\right]. \tag{4.97}$$

Revankar (1976) computed the values of h for a few selected values of $|\rho_{12}|$, $(T - K_1)$ and $(K_1 - K_2)$ and observed that for moderately small values of these quantities, the SURR estimator is superior to the SUUR estimator, while the SUUR estimator is superior for moderately large values. (Recall the earlier results with orthogonal regressors.) Revankar also noted that the gain in efficiency arising from the choice of one of these estimators over the other is not substantial in a wide range of practical situations, so essentially there is little to choose between them. From the positivity of h, as defined by (4.97), it is seen that (Revankar, 1976; p. 185)

$$\rho_{12}^2 > \frac{1}{1 + \left(2 + \frac{K_1 - K_2}{2}\right)\left(\frac{T - K_2}{T - K_2 + 4}\right)\left(\frac{T - K_1 - 2}{T - \frac{K_1 + K_2}{2}}\right)} \tag{4.98}$$

from which we observe that the range of ρ_{12}^2 values over which the SUUR estimator performs better than the SURR estimator increases with an increase in T and $(K_1 - K_2)$. These results may be compared with (4.43).

Now, for the SURE model with subset regressors, let us analyze the loss in efficiency attributable to the replacement of Σ by its consistent estimators $\tilde{S}$ and $\hat{S}$ when obtaining the FGLS estimators. In this case, no loss occurs so far as the estimation of β_2 is concerned. For β_1, we observe that the variance covariance matrix of the GLS estimator is

$$\Omega_{11} = \sigma_{11}(X_1'X_1)^{-1} - \sigma_{11}\rho_{12}^2 (X_1'X_1)^{-1} X_1' \bar{P}_{X_2} X_1 (X_1'X_1)^{-1}, \tag{4.99}$$

and so

$$[V(\tilde{\beta}_{(1)SU}) - \Omega_{11}]$$

$$= \sigma_{11}(1-\rho_{12}^2)\left[\frac{1}{T-K_1-2}\right](X_1'X_1)^{-1}X_1'\bar{P}_{X_2}X_1(X_1'X_1)^{-1}$$

$$[V(\hat{\beta}_{(1)SR}) - \Omega_{11}]$$

$$= \sigma_{11}(1-\rho_{12}^2)\left[\frac{(T-K_1)}{(T-K_2)(T-K_2+2)} + \frac{(K_1-K_2+2)(K_1-K_2+4)}{(T-K_2+2)(T-K_2+4)}\left(\frac{\rho_{12}^2}{1-\rho_{12}^2}\right)\right](X_1'X_1)^{-1}X_1'\bar{P}_{X_2}X_1(X_1'X_1)^{-1}. \quad (4.100)$$

The above expressions measure the increase in the variability of the GLS estimator of β_1 arising from the replacement of Σ by its consistent estimators $\tilde{S}$ and $\hat{S}$. This increase is smaller for the SUUR estimator. This suggests that, in the context of subset regressors, when developing a consistent estimator of Σ the use of unrestricted residuals may be more fruitful than the use of restricted residuals, despite the fact that unrestricted residuals ignore the information related to the omission of $(K_1 - K_2)$ regressors in the second equation of our two equation model.

Finally, consider the consequences of unnecessarily applying the FGLS estimator when the disturbances between the two equations are uncorrelated. Substituting $\rho_{12} = 0$ in (4.79) and (4.91) yields

$$V(\tilde{\beta}_{(1)SU})$$

$$= \sigma_{11}(X_1'X_1)^{-1} + \sigma_{11}\left[\frac{1}{(T-K_1-2)}\right](X_1'X_1)^{-1}X_1'\bar{P}_{X_2}X_1(X_1'X_1)^{-1}$$

$$V(\hat{\beta}_{(1)SR}) \quad (4.101)$$

$$= \sigma_{11}(X_1'X_1)^{-1} + \sigma_{11}\left[\frac{(T-K_1)}{(T-K_2)(T-K_2+2)}\right](X_1'X_1)^{-1}X_1'\bar{P}_{X_2}X_1(X_1'X_1)^{-1}.$$

A comparison of these expressions with the variance covariance matrix of the OLS estimator clearly reveals the loss in efficiency due to ignoring the uncorrelated nature of the disturbances across the equations. The loss in efficiency may be substantial when K_1 and K_2 are small and T is large, at least in this two equation model with subset regressors.

4.5.3 Large-Sample Asymptotic Approximations

Retaining this same special model, when T is large, we can write

$$V(\tilde{\beta}_{(1)SU})$$
$$= \sigma_{11}(X_1'X_1)^{-1} - \sigma_{11}\left[\rho_{12}^2 - \frac{1-\rho_{12}^2}{T}\left(1 - \frac{K_1+2}{T}\right)^{-1}\right](X_1'X_1)^{-1}X_1'\bar{P}_{X_2}X_1(X_1'X_1)^{-1}$$
$$= \sigma_{11}(X_1'X_1)^{-1} - \sigma_{11}\left[\rho_{12}^2 - (1-\rho_{12}^2)\left(\frac{1}{T} + \frac{K_1+2}{T^2} + \cdots\right)\right](X_1'X_1)^{-1}X_1'\bar{P}_{X_2}X_1(X_1'X_1)^{-1} \tag{4.102}$$

provided that T is larger than $(K_1 + 2)$.

If V_1, V_2 and $\tilde{V}_3$ denote the variance covariance matrix of the SUUR estimator of β_1 to order $O(T^{-1})$, $O(T^{-2})$ and $O(T^{-3})$ respectively, we have

$$V_1 = \sigma_{11}(X_1'X_1)^{-1} - \sigma_{11}\rho_{12}^2(X_1'X_1)^{-1}X_1'\bar{P}_{X_2}X_1(X_1'X_1)^{-1}$$
$$V_2 = V_1 + \sigma_{11}\frac{1-\rho_{12}^2}{T}(X_1'X_1)^{-1}X_1'\bar{P}_{X_2}X_1(X_1'X_1)^{-1} \tag{4.103}$$
$$\tilde{V}_3 = V_2 + \sigma_{11}(1-\rho_{12}^2)\left(\frac{K_1+2}{T^2}\right)(X_1'X_1)^{-1}X_1'\bar{P}_{X_2}X_1(X_1'X_1)^{-1}.$$

Similarly, assuming that $T > K_2$ and using

$$\frac{1}{(T-K_2+2)(T-K_2+4)} = \frac{1}{T^2}\left[1 - \frac{2(K_2-3)}{T} + \frac{(K_2-2)(K_2-4)}{T^2}\right]^{-1}$$
$$= \frac{1}{T^2} + \cdots$$

$$\frac{T-K_1}{(T-K_2)(T-K_2+2)} = \frac{1}{T}\left(1 - \frac{K_1}{T}\right)\left[1 - \frac{2(K_2-1)}{T} + \frac{K_2(K_2-2)}{T^2}\right]^{-1}$$
$$= \frac{1}{T} + \frac{2(K_2-1)-K_1}{T^2} + \cdots$$

we find that the variance covariance matrix of the SURR estimator of β_1 is equal to V_1 to order $O(T^{-1})$ and V_2 to order $O(T^{-2})$ (see Srivastava and Upadhyaya, 1978). To order $O(T^{-3})$, this matrix is

$$\hat{V}_3 = V_2 + \sigma_{11}\frac{1-\rho_{12}^2}{T^2}\left[2K_2 - 2 - K_1 + (K_1-K_2+2)(K_1-K_2+4)\left(\frac{\rho_{12}^2}{1-\rho_{12}^2}\right)\right]$$
$$\cdot (X_1'X_1)^{-1}X_1'\bar{P}_{X_2}X_1(X_1'X_1)^{-1}. \tag{4.104}$$

Thus we observe that the variance covariance matrices of the SUUR and SURR estimators of β_1 are identical up to order $O(T^{-2})$, which coincides with the result (3.59) of Chapter 3. Comparing $\hat{V}_3$ and $\tilde{V}_3$, we find that the

SUUR estimator is more efficient than the SURR estimator with respect to the variance covariance matrix to order $O(T^{-3})$ if

$$\rho_{12}^2 > \frac{2}{(K_1 - K_2 + 6)} \tag{4.105}$$

which implies that the SUUR estimator will certainly be better than the SURR estimator, to the order of our approximation, if ρ_{12}^2 is known to exceed 1/3.

The appropriateness of large-sample asymptotic approximations can be judged by considering the difference between the approximating matrix and the exact variance covariance matrix. For instance, if R_g denotes the matrix by which the exact variance covariance matrix of the SUUR estimators of β exceeds its approximating matrix to order $O(T^{-g})$, we have

$$R_g = \sigma_{11}(1-\rho_{12}^2)\ell_g \begin{bmatrix} (X_1'X_1)^{-1}X_1'\bar{P}_{X_2}X_1(X_1'X_1)^{-1} & 0 \\ 0 & 0 \end{bmatrix} \tag{4.106}$$

where

$$\ell_1 = \frac{1}{(T-K_1-2)}, \quad \ell_2 = \frac{(K_1+2)}{T(T-K_1-2)}, \quad \ell_3 = \frac{(K_1+2)^2}{T^2(T-K_1-2)} .$$

Thus the values of ℓ_g provide some idea about the accuracy of our asymptotic approximations.

4.5.4 Sampling Distribution of the SUUR Estimator

It is extremely difficult to derive the sampling distribution of the SUUR estimator, even for our special SURE model. However, the features of this distribution can be studied by means of its moments, and these are relatively easy to obtain. This was shown by Ullah and Rafiquzzman (1977): define

$$\xi_1 = (\tilde{\beta}_{(1)SU} - \beta_1)$$

and look at the moments of its k'th element

$$\xi_{1k} = \sum_{t=1}^{T} \left[g_{kt}u_{t1} - \frac{\tilde{s}_{12}}{\tilde{s}_{22}} g_{kt}^* u_{t2} \right] , \tag{4.107}$$

where g_{kt} and g_{kt}^* denote the (k, t)'th element of the matrices $(X_1'X_1)^{-1}X_1'$ and $(X_1'X_1)^{-1}X_1'\bar{P}_{X_2}$ respectively.

From (4.107), we observe that all odd order moments of ξ_{1k} are zero so that the sampling distribution of the SUUR estimator is symmetric. Further, using (4.70), it is easy to derive from (4.107) the following results:

$$E(\xi_{1k}^2) = \sigma_{11}\left[\frac{\sum_{t=1}^{T} g_{tk}^2}{\sum_{t=1}^{T} g_{tk}^{*2}} - \rho_{12}^2 + \frac{1-\rho_{12}^2}{(n-2)}\right]\left(\sum_{t=1}^{T} g_{tk}^{*2}\right) ; \quad n > 2 \tag{4.108}$$

$$E(\xi_{1k}^4) = 3\sigma_{11}^2\left[\left\{\frac{\sum_{t=1}^{T} g_{tk}^2}{\sum_{t=1}^{T} g_{tk}^{*2}} - \rho_{12}^2 + \frac{1-\rho_{12}^2}{(n-2)}\right\}^2 + \frac{2(n-1)(1-\rho_{12}^2)^2}{(n-2)^2(n-4)}\right]\left(\sum_{t=1}^{T} g_{tk}^{*2}\right)^2 ; \quad n > 4 \tag{4.109}$$

from which Pearson's measure for excess of kurtosis is obtained as

$$\frac{E(\xi_{1k}^4)}{[E(\xi_{1k}^2)]^2} - 3 = \left[\frac{6(n-1)(1-\rho_{12}^2)^2}{(n-2)^2(n-4)\left[\frac{\sum_{t=1}^{T} g_{tk}^2}{\sum_{t=1}^{T} g_{tk}^{*2}} - \rho_{12}^2 + \frac{1-\rho_{12}^2}{(n-2)}\right]^2}\right] \tag{4.110}$$

which indicates that the sampling distribution of the SUUR estimator in this case is leptokurtic. For further details see Ullah and Rafiquzzaman (1977).

4.5.5 Similarity with Covariance Analysis

Conniffe (1982a) has pointed out the close correspondence between the approaches adopted in the SURE model and the analysis of covariance model in experimental designs. Corresponding to $X_1 = (X_2 \quad X_0)$, let us partition the vector β_1 as $\beta_1' = (\gamma_1' \quad \gamma_0')$, so that our two equation SURE model with subset regressors can be written as

$$\begin{aligned} y_1 &= X_2\gamma_1 + X_0\gamma_0 + u_1 \\ y_2 &= X_2\beta_2 + u_2 \,. \end{aligned} \tag{4.111}$$

This defines the analysis of covariance model, for example in a Latin square design, if we take X_2 as the full column rank matrix of covariates and X_0 as the full column rank incidence matrix.

Using (4.23), it is easy to see that the maximum likelihood estimators (denoted by an asterisk * over the parameters) of δ, γ_1, γ_0 and β_2 are given by

$$\delta^* = \frac{\tilde{s}_{12}}{\tilde{s}_{22}}, \quad \begin{bmatrix} \gamma_1^* \\ \gamma_0^* \end{bmatrix} = \tilde{\beta}_{(1)SU}, \quad \beta_2^* = \tilde{\beta}_{(2)SU} = b_{(2)0}.$$

Thus, the solution of model (4.111) for γ_0 is identical with the standard least squares solution of the model

$$y_1 = X_2\gamma_1 + X_0\gamma_0 + \delta y_2 + v, \tag{4.112}$$

which is the form of the model for the analysis of covariance (see Rao, 1973 p. 289).

4.6 EFFICIENCY PROPERTIES UNDER UNCONSTRAINED REGRESSORS

Now we consider a two equation SURE model in which no constraints, such as $X_1'X_2 = 0$ or $X_1 = (X_2 \quad X_0)$, are placed on the regressor matrices and we limit our discussion to the efficiency of the SUUR estimator.

4.6.1 The Variance Covariance Matrix

Considering the SUUR estimator of β_1 we have, from the first equation of (4.15),

$$(\tilde{\beta}_{(1)SU} - \beta_1) = R_{12}\left[X_1'(I_T - r^2 P_{x_2})u_1 - \frac{\tilde{s}_{12}}{\tilde{s}_{22}} X_1'\bar{P}_{x_2}u_2\right]. \tag{4.113}$$

As X_1 and X_2 are submatrices of Z, it is easy to see that $X_i'\bar{P}_z$ is a null matrix (i = 1, 2). Using this result, together with the normality of the disturbances, we observe from (4.21) that the vectors $X_1'u_1$, $X_2'u_1$, $X_1'u_2$ and $X_2'u_2$ are stochastically independent of $\tilde{s}_{11}$, $\tilde{s}_{22}$ and $\tilde{s}_{12}$. Thus, from (4.113) the variance covariance matrix of $\tilde{\beta}_{(1)SU}$ is

$$V(\tilde{\beta}_{(1)SU}) = E(\tilde{\beta}_{(1)SU} - \beta_1)(\tilde{\beta}_{(1)SU} - \beta_1)'$$

$$= E\left[\sigma_{11} R_{12} X_1'(I_T - r^2 P_{X_2})^2 X_1 R_{12} - 2\sigma_{12}\left[\frac{\tilde{s}_{12}}{\tilde{s}_{22}}\right] R_{12} X_1' \bar{P}_{X_2} X_1 R_{12}\right.$$

$$\left. + \sigma_{22}\left[\frac{\tilde{s}_{12}}{\tilde{s}_{22}}\right]^2 R_{12} X_1' \bar{P}_{X_2} X_1 R_{12}\right]$$

$$= E\left[\sigma_{11} R_{12} X_1'(I_T - r^2 P_{X_2})^2 X_1 R_{12} - 2\rho_{12} r(\sigma_{11}\sigma_{22})^{\frac{1}{2}}\left[\frac{\tilde{s}_{11}}{\tilde{s}_{22}}\right]^{\frac{1}{2}} R_{12} X_1' \bar{P}_{X_2} X_1 R_{12}\right.$$

$$\left. + \sigma_{22} r^2\left[\frac{\tilde{s}_{11}}{\tilde{s}_{22}}\right] R_{12} X_1' \bar{P}_{X_2} X_1 R_{12}\right] \tag{4.114}$$

where $r = \tilde{s}_{12}/(\tilde{s}_{11}\tilde{s}_{22})^{\frac{1}{2}}$. To evaluate (4.114) we adopt what is essentially the approach taken by Hall (1977). Assuming K_1 to be greater than K_2 (without any loss of generality), denote the roots of the equation

$$|X_1' P_{X_2} X_1 - \theta X_1' X_1| = 0$$

by $\theta_1, \theta_2, \cdots, \theta_{K_1}$. These are the squared canonical correlation coefficients between the sets of explanatory variables in the two equations.

Notice that all θ_i's lie between 0 and 1. When X_1 and X_2 are orthogonal all of the θ_i's are zero. Similarly, if X_2 is a submatrix of X_1, say, $X_1 = (X_2 \quad X_0)$, we have

$$(X_1' P_{X_2} X_1 - \theta X_1' X_1) = \begin{bmatrix} (1-\theta)X_2'X_2 & (1-\theta)X_2'X_0 \\ (1-\theta)X_0'X_2 & X_0'(P_{X_2} - \theta I_T)X_0 \end{bmatrix}$$

and therefore the determinant

$$\begin{aligned} |X_1' P_{X_2} X_1 - \theta X_1' X_1| &= |(1-\theta)X_2'X_2| \cdot |X_0'(P_{X_2} - \theta I_T)X_0 - (1-\theta)X_0' P_{X_2} X_0| \\ &= (1-\theta)^{K_2}(-\theta)^{K_1-K_2} |X_2'X_2| \cdot |X_0' \bar{P}_{X_2} X_0| , \end{aligned}$$

which when set equal to 0 reveals that K_2 roots among $\theta_1, \theta_2, \cdots, \theta_{K_1}$ are one and the remaining $(K_1 - K_2)$ roots are zero.

Now, we can find a nonsingular matrix C^{-1} such that

$$\begin{aligned} X_1'X_1 &= C'C \quad \text{or} \quad C'^{-1} X_1' X_1 C^{-1} = I_{K_1} \\ X_1' P_{X_2} X_1 &= C'\Theta C \quad \text{or} \quad C'^{-1} X_1' P_{X_2} X_1 C^{-1} = \Theta \end{aligned} \tag{4.115}$$

where Θ is a $(K_1 \times K_1)$ diagonal matrix with $\theta_1, \theta_2, \cdots, \theta_{K_1}$ as the diagonal elements.

Observing that

$$\begin{aligned} R_{12} &= [X_1'(I_T - r^2 P_{X_2})X_1]^{-1} \\ &= [CC' - r^2 C'\Theta C]^{-1} \\ &= C^{-1}GC'^{-1} \end{aligned}$$

with $G = (I_{K_1} - r^2\Theta)^{-1}$, we can write (4.114) as

$$V(\tilde{\beta}_{(1)SU}) = \sigma_{11}C^{-1}HC'^{-1} \tag{4.116}$$

where H is a $(K_1 \times K_1)$ diagonal matrix defined as

$$H = E\left[G(I_{K_1} - r^2(2-r^2)\Theta)G - 2\rho_{12}r\left(\frac{\sigma_{22}\tilde{s}_{11}}{\sigma_{11}\tilde{s}_{22}}\right)^{\frac{1}{2}}G(I_{K_1} - \Theta)G + r^2\left(\frac{\sigma_{22}\tilde{s}_{11}}{\sigma_{11}\tilde{s}_{22}}\right)G(I_{K_1} - \Theta)G\right]. \tag{4.117}$$

We now need to consider the evaluation of the expectation in (4.117). If h_k denotes the k'th diagonal element of the matrix H, we have

$$h_k = E\left[\frac{1 - r^2(2-r^2)\theta_k - 2\rho_{12}r\left(\frac{\sigma_{22}\tilde{s}_{11}}{\sigma_{11}\tilde{s}_{22}}\right)^{\frac{1}{2}}(1-\theta_k) + r^2\frac{\sigma_{22}\tilde{s}_{11}}{\sigma_{11}\tilde{s}_{22}}(1-\theta_k)}{(1 - r^2\theta_k)^2}\right]. \tag{4.118}$$

In order to evaluate this expectation, we define

$$\phi(d; m) = \left[E\ \frac{r^d}{(1 - r^2\theta_k)^2}\left(\frac{\sigma_{22}\tilde{s}_{11}}{\sigma_{11}\tilde{s}_{22}}\right)^{m/2}\right] \tag{4.119}$$

where $d = 0, 1, 2, 4$ and $m = 0, 1, 2$.

Writing $a_{ij} = T\tilde{s}_{ij}$ for simplicity in notation, we observe that

$$T\tilde{S} = \begin{bmatrix} a_{11} & a_{12} \\ a_{12} & a_{22} \end{bmatrix}$$

follows a Wishart distribution, $W_2(n, \Sigma)$, with $n = (T - K)$ so that the joint probability density function of a_{11}, a_{22} and a_{12} is

$$p(a_{11}, a_{22}, a_{12}) = \frac{(a_{11}a_{22} - a_{12}^2)^{(n-3)/2} \exp\left\{-\frac{1}{2(1-\rho_{12}^2)}\left[\frac{a_{11}}{\sigma_{11}} - \frac{2\rho_{12}a_{12}}{\sqrt{\sigma_{11}\sigma_{22}}} + \frac{a_{22}}{\sigma_{22}}\right]\right\}}{2^n |\Sigma|^{n/2}\Gamma\left(\frac{1}{2}\right)\Gamma\left(\frac{n}{2}\right)\Gamma\left(\frac{n-1}{2}\right)} . \tag{4.120}$$

Applying the transformation

$$a_{11} = a_{11}, \quad a_{22} = a_{22}, \quad a_{12} = r(a_{11}a_{22})^{\frac{1}{2}}$$

we get the joint probability density function of a_{11}, a_{22} and r as follows:

$$\begin{aligned} p(a_{11}, a_{22}, r) &= \overset{*}{C}\left(\frac{a_{11}a_{22}}{\sigma_{11}\sigma_{22}}\right)^{(n/2)-1}(1-r^2)^{(n-3)/2}\exp\left\{-\frac{1}{2(1-\rho_{12}^2)}\left[\frac{a_{11}}{\sigma_{11}} - 2r\rho_{12}\left(\frac{a_{11}a_{22}}{\sigma_{11}\sigma_{22}}\right)^{\frac{1}{2}} + \frac{a_{22}}{\sigma_{22}}\right]\right\} \\ &= \overset{*}{C}(1-r^2)^{(n-3)/2}\exp\left\{-\frac{1}{2(1-\rho_{12}^2)}\left[\frac{a_{11}}{\sigma_{11}} + \frac{a_{22}}{\sigma_{22}}\right]\right\}\sum_{\alpha=0}^{\infty}\left(\frac{a_{11}a_{22}}{\sigma_{11}\sigma_{22}}\right)^{[(n+\alpha)/2]-1}\left[\frac{r\rho_{12}}{1-\rho_{12}^2}\right]^{\alpha}\left(\frac{1}{\alpha!}\right) \end{aligned}$$

where

$$\overset{*}{C} = \left[2^n\sigma_{11}\sigma_{22}(1-\rho_{12}^2)^{n/2}\Gamma\left(\frac{1}{2}\right)\Gamma\left(\frac{n}{2}\right)\Gamma\left(\frac{n-1}{2}\right)\right]^{-1} .$$

Thus from (4.119) we have

$$\phi(d;m) = \overset{*}{C}\sum_{\alpha=0}^{\infty} f_\alpha(1)\, f_\alpha(2)\left[\frac{\rho_{12}}{1-\rho_{12}^2}\right]^{\alpha}\left(\frac{1}{\alpha!}\right)\int_{-1}^{1}\frac{r^{d+\alpha}(1-r^2)^{(n-3)/2}}{(1-r^2\theta_k^2)^2}\,dr \tag{4.121}$$

where

$$f_\alpha(1) = \int_0^\infty \left(\frac{a_{11}}{\sigma_{11}}\right)^{[(n+m+\alpha)/2]-1}\exp\left\{-\frac{a_{11}}{2\sigma_{11}(1-\rho_{12}^2)}\right\}da_{11}$$

$$f_\alpha(2) = \int_0^\infty \left(\frac{a_{22}}{\sigma_{22}}\right)^{[(n-m+\alpha)/2]-1}\exp\left\{-\frac{a_{22}}{2\sigma_{22}(1-\rho_{12}^2)}\right\}da_{22} .$$

Using the transformation

$$t = \frac{a_{11}}{2\sigma_{11}(1-\rho_{12}^2)} ,$$

we find

$$f_\alpha(1) = \sigma_{11}[2(1-\rho_{12}^2)]^{(n+m+\alpha)/2} \int_0^\infty t^{[(n+m+\alpha)/2]-1} \exp\{-t\}\, dt$$

$$= \sigma_{11}[2(1-\rho_{12}^2)]^{(n+m+\alpha)/2} \Gamma\left(\frac{n+m+\alpha}{2}\right) . \qquad (4.122)$$

Similarly, we have

$$f_\alpha(2) = \sigma_{22}[2(1-\rho_{12}^2)]^{(n-m+\alpha)/2} \Gamma\left(\frac{n-m+\alpha}{2}\right) , \qquad (4.123)$$

provided that $(n-m+\alpha)$ is nonnegative, which in turn holds as long as n exceeds 2, because m is at most 2.

Substituting (4.122) and (4.123) in (4.121), we get

$$\phi(d;m) = \frac{(1-\rho_{12}^2)^{n/2}}{\Gamma\left(\frac{1}{2}\right)\Gamma\left(\frac{n}{2}\right)\Gamma\left(\frac{n-1}{2}\right)} \sum_{\alpha=0}^{\infty} \frac{(2\rho_{12})^\alpha}{\alpha!} \Gamma\left(\frac{n+m+\alpha}{2}\right)\Gamma\left(\frac{n-m+\alpha}{2}\right)$$

$$\cdot \int_{-1}^{1} \frac{r^{d+\alpha}(1-r^2)^{(n-3)/2}}{(1-r^2\theta_k)^2}\, dr . \qquad (4.124)$$

When d is an even integer, it is easy to see that the integral on the right hand side of (4.124) vanishes for odd values of α. Dropping such terms, we set $\alpha = 2i$ so that the expression (4.124) becomes

$$\frac{(1-\rho_{12}^2)^{n/2}}{\Gamma\left(\frac{1}{2}\right)\Gamma\left(\frac{n}{2}\right)\Gamma\left(\frac{n-1}{2}\right)} \sum_{i=0}^{\infty} \frac{(2\rho_{12})^{2i}}{(2i)!} \Gamma\left(\frac{n+m}{2}+i\right)\Gamma\left(\frac{n-m}{2}+i\right) \int_{-1}^{1} \frac{r^{d+2i}(1-r^2)^{(n-3)/2}}{(1-r^2\theta_k)^2}\, dr$$

$$= \frac{(1-\rho_{12}^2)^{n/2}}{\sqrt{\pi}\,\Gamma\left(\frac{n}{2}\right)\Gamma\left(\frac{n-1}{2}\right)} \sum_{i=0}^{\infty} \frac{(2\rho_{12})^{2i}}{(2i)!} \Gamma\left(\frac{n+m}{2}+i\right)\Gamma\left(\frac{n-m}{2}+i\right) \int_{0}^{1} \frac{(r^2)^{i+(d-1)/2}(1-r^2)^{(n-3)/2}}{(1-r^2\theta_k)^2}\, dr^2$$

$$(4.125)$$

Using the duplication formula

$$\sqrt{\pi}\,(2i)! = 2^{2i}\, i!\,\Gamma\left(i+\frac{1}{2}\right),$$

and the result

$$\int_0^1 (r^2)^{i+(d-1)/2}(1-r^2)^{(n-3)/2}(1-r^2\theta_k)^{-2}\,dr^2$$

$$= \sum_{j=0}^{\infty}(j+1)\theta_k^j \int_0^1 (r^2)^{i+j+(d-1)/2}(1-r^2)^{(n-3)/2}\,dr^2$$

$$= \sum_{j=0}^{\infty}(j+1)\theta_k^j \frac{\Gamma\left(\frac{d+1}{2}+i+j\right)\Gamma\left(\frac{n-1}{2}\right)}{\Gamma\left(\frac{n+d}{2}+i+j\right)},$$

we get

$$\phi(d;m) = (1-\rho_{12}^2)^{n/2}\sum_{i=0}^{\infty}\sum_{j=0}^{\infty}(j+1)\theta_k^j \frac{(\rho_{12}^2)^i}{i!}\,\frac{\Gamma\left(\frac{d+1}{2}+i+j\right)\Gamma\left(\frac{n+m}{2}+i\right)\Gamma\left(\frac{n-m}{2}+i\right)}{\Gamma\left(\frac{n}{2}\right)\Gamma\left(\frac{n+d}{2}+i+j\right)\Gamma\left(i+\frac{1}{2}\right)} \tag{4.126}$$

provided that d is an even integer.

When d is an odd integer, the integral in (4.124) vanishes for even values of α. Dropping such terms, using the duplication formula

$$\sqrt{\pi}\,(2i+1)! = 2^{2i+1}\, i!\,\Gamma\left(i+\frac{3}{2}\right),$$

and then evaluating the integral in the same manner as indicated for (4.126), we find:

$$\phi(d;m) = \rho_{12}(1-\rho_{12}^2)^{n/2}\sum_{i=0}^{\infty}\sum_{j=0}^{\infty}(j+1)\theta_k^j \frac{(\rho_{12}^2)^i}{i!}\,\frac{\Gamma\left(\frac{d}{2}+i+j+1\right)\Gamma\left(\frac{n+m+1}{2}+i\right)\Gamma\left(\frac{n-m+1}{2}+i\right)}{\Gamma\left(\frac{n}{2}\right)\Gamma\left(\frac{n+d+1}{2}+i+j\right)\Gamma\left(i+\frac{3}{2}\right)} \tag{4.127}$$

provided that d is odd.

Employing (4.126) and (4.127) in (4.118), we get

$$h_k = \phi(0;0) - [2\phi(2;0) - \phi(4;0)]\theta_k - [2\rho_{12}\phi(1;1) - \phi(2;2)](1-\theta_k) , \qquad (4.128)$$

from which a diagonal matrix with diagonal element h_k can be formed, and the expression for the variance covariance matrix of the SUUR estimator of β_1 can be obtained from (4.116). Noting the symmetry of (4.15) and (4.16), the expression for the variance covariance matrix of the SUUR estimator of β_2 can be easily written down, and the expression for the cross-moments of the SUUR estimators of β_1 and β_2 can be derived in a similar manner.

Using an alternative approach to analyse this problem, Mehta and Swamy (1976) started with the Wishart distribution and from it obtained the joint probability density function of $\sqrt{(\sigma_{22}a_{11}/\sigma_{11}a_{22})}$, r and a_{22} through a transformation. Integrating this density with respect to a_{22}, they found the joint density of r and $t^* = \sqrt{(\sigma_{22}a_{11}/\sigma_{11}a_{22})}$ as follows:

$$p(r, t^*) = \frac{2(1-\rho_{12}^2)^{n/2}\Gamma(n)}{\sqrt{\pi}\,\Gamma\left(\frac{n-1}{2}\right)\Gamma\left(\frac{n}{2}\right)} \cdot \frac{(1-r^2)^{(n-3)/2} t^{*n-1}}{(t^{*2} - 2r\rho_{12}t^* + 1)^n}$$

from which we have

$$h_k = \int_0^{\infty}\int_{-1}^{1} \frac{1 - r^2(2-r^2)\theta_k - 2r\rho_{12}t^*(1-\theta_k) + r^2t^{*2}(1-\theta_k)}{(1-r^2\theta_k)^2} p(r, t^*)\, dr\, dt^*$$

$$= 1 + (1-\theta_k)\int_0^{\infty}\int_{-1}^{1} \frac{r^4\theta_k - 2r\rho_{12}t^* + r^2t^{*2}}{(1-r^2\theta_k)^2} p(r, t^*)\, dr\, dt^* .$$

Applying the transformation

$$t = \frac{t^*}{1+t^*} ,$$

Mehta and Swamy established the convergence of the above double integral and used bivariate integration methods to evaluate $(h_k - 1)$ numerically. Hall (1977) obtained an alternative expression for h_k and produced a method for the numerical approximation of the expression (4.125) for computing h_k. Finally, defining

$$\overset{*}{Y} = (y_1 \quad y_2)$$

$$\overset{*}{X} = (X_1 \quad X_2)$$

and denoting the generalized inverse of $(\overset{*}{X}'\overset{*}{X})$ by $(\overset{*}{X},\overset{*}{X})^{+}$, so that n is now T minus the rank of $\overset{*}{X}$, Kunitomo (1977) considered the following form for the a_{ij}'s:

$$\begin{bmatrix} a_{11} & a_{12} \\ a_{12} & a_{22} \end{bmatrix} = \overset{*}{Y}'[I_T - \overset{*}{X}(\overset{*}{X}'\overset{*}{X})^{+}\overset{*}{X}']\overset{*}{Y} ,$$

and evaluated the expression for h_k.

Now let us consider the special case when $\theta_k = 0$, and determine the expression for h_k. Note that all of the θ_k's are zero if and only if X_1 and X_2 are orthogonal. Substituting this condition in (4.126) and (4.127) we observe that

$$\begin{aligned} \phi(0;0) &= (1-\rho_{12}^2)^{n/2} \sum_{i=0}^{\infty} \frac{(\rho_{12}^2)^i}{i!} \cdot \frac{\Gamma\left(\frac{n}{2}+i\right)}{\Gamma\left(\frac{n}{2}\right)} \\ &= (1-\rho_{12}^2)^{n/2} \cdot \frac{1}{(1-\rho_{12}^2)^{n/2}} \\ &= 1 \end{aligned} \tag{4.129}$$

$$\begin{aligned} \phi(1;1) &= \rho_{12}(1-\rho_{12}^2)^{n/2} \sum_{i=0}^{\infty} \frac{(\rho_{12}^2)^i}{i!} \cdot \frac{\Gamma\left(\frac{n}{2}+i\right)}{\Gamma\left(\frac{n}{2}\right)} \\ &= \rho_{12}(1-\rho_{12}^2)^{n/2} \cdot \frac{1}{(1-\rho_{12}^2)^{n/2}} \\ &= \rho_{12} \end{aligned} \tag{4.130}$$

$$\begin{aligned} \phi(2;2) &= (1-\rho_{12}^2)^{n/2} \sum_{i=0}^{\infty} \frac{(\rho_{12}^2)^i}{i!} \cdot \frac{\Gamma\left(\frac{n}{2}-1+i\right)}{\Gamma\left(\frac{n}{2}\right)} \left(i+\frac{1}{2}\right) \\ &= (1-\rho_{12}^2)^{n/2} \sum_{i=1}^{\infty} \frac{(\rho_{12}^2)^i}{(i-1)!} \cdot \frac{\Gamma\left(\frac{n}{2}-1+i\right)}{\Gamma\left(\frac{n}{2}\right)} + \frac{1}{2}(1-\rho_{12}^2)^{n/2} \sum_{i=0}^{\infty} \frac{(\rho_{12}^2)^i}{i!} \cdot \frac{\Gamma\left(\frac{n}{2}-1+i\right)}{\Gamma\left(\frac{n}{2}\right)} \end{aligned}$$

$$= \rho_{12}^2(1-\rho_{12}^2)^{n/2} \sum_{i=0}^{\infty} \frac{(\rho_{12}^2)^i}{i!} \cdot \frac{\Gamma\left(\frac{n}{2}+i\right)}{\Gamma\left(\frac{n}{2}\right)} + \frac{(1-\rho_{12}^2)^{n/2}}{(n-2)} \sum_{i=0}^{\infty} \frac{(\rho_{12}^2)^i}{i!} \cdot \frac{\Gamma\left(\frac{n}{2}-1+i\right)}{\Gamma\left(\frac{n}{2}-1\right)}$$

$$= \rho_{12}^2(1-\rho_{12}^2)^{n/2} \cdot \frac{1}{(1-\rho_{12}^2)^{n/2}} + \frac{(1-\rho_{12}^2)^{n/2}}{(n-2)} \cdot \frac{1}{(1-\rho_{12}^2)^{(n/2)-1}}$$

$$= \rho_{12}^2 + \frac{(1-\rho_{12}^2)}{(n-2)}, \qquad (4.131)$$

from which we have

$$h_k = 1 - 2\rho_{12}^2 + \rho_{12}^2 + \left(\frac{1-\rho_{12}^2}{n-2}\right)$$

$$= (1-\rho_{12}^2)\left(1+\frac{1}{n-2}\right)$$

or

$$H = (1-\rho_{12}^2)\left(1+\frac{1}{n-2}\right) I_{K_1} .$$

Substituting this in (4.116) replicates the expression (4.26) for the variance covariance matrix of the SUUR estimator of β_1 when the θ_k's are all zero (i.e., when X_1 and X_2 are orthogonal).

As for the expression for the variance covariance matrix of the SURR estimator for the unconstrained regressors case, no results are available in the literature. Actually a difficulty arises from the outset in this case because the distribution of $T\hat{S}$ is not Wishart in form. This is still an open problem for investigation.

4.6.2 Empirical Findings

So far as a comparison of the OLS and SUUR estimators is concerned, we note that the complex nature of the expression for the variance covariance matrix of the SUUR estimator does not permit us to draw any clear conclusions. If we take the variance covariance matrix as the performance criterion, we observe that

$$V(b_{(1)0}) = \sigma_{11}(X_1'X_1)^{-1}$$

$$= \sigma_{11}C^{-1}C'^{-1} \qquad (4.132)$$

so that

$$V(\tilde{\beta}_{(1)SU}) - V(b_{(1)0}) = \sigma_{11} C^{-1}(H - I_{K_1})C'^{-1} \tag{4.133}$$

from which we see that the elements of $(H - I_{K_1})$ will shed some light on the efficiency of the OLS estimator relative to the SUUR estimator. On the other hand, if we choose the generalized variance as the performance criterion, the efficiency of the OLS estimator relative to the SUUR estimator is defined by

$$\frac{|V(b_{(1)0})|}{|V(\tilde{\beta}_{(1)SU})|} = \frac{|\sigma_{11} C^{-1} C'^{-1}|}{|\sigma_{11} C^{-1} H C'^{-1}|}$$

$$= \frac{1}{|H|} = \prod_{k=1}^{K_1} \left(\frac{1}{h_k}\right). \tag{4.134}$$

In view of (4.133), Mehta and Swamy (1976) carried out a numerical evaluation of $(h_k - 1)$ for a few selected values of θ_k, ρ_{12} and n (= (T - K)), viz., $\theta_k = 0\ (0.1)\ 0.9$, $\rho_{12} = -1\ (0.1)\ 1$ and n = 3, 5, 9, 13, 23. Kunimoto (1977) computed h_k for the same θ_k and ρ_{12} values assuming n = 8 and $K_1 = K_2 = 2$. On the other hand, in view of (4.134), Hall (1977) computed the values of h_k for $\theta_k = 0.1\ (0.2)\ 0.9$, $|\rho_{12}| = 0.1\ (0.2)\ 0.9$ and n = 20 (20) 100. Srivastava (1982) prepared some graphs depicting h_k against $|\rho_{12}|$ for a few selected values of θ_k and n. The principal result emerging from these investigations is that a low correlation between the disturbances, a high canonical correlation between the regressors across the equations, and a low value for the excess of the number of observations over the number of distinct regressors in the model, are generally not favourable to the SUUR estimator relative to the OLS estimator. On the other hand, the SUUR estimator is preferable when the disturbances in the different equations are highly correlated, regressors across the equations have low canonical correlations and the excess of observations over the number of distinct regressors in the model is large. More specifically, as long as $n > 20$ and $|\rho_{12}| \geq 0.3$, the SUUR estimator is more efficient than the OLS estimator for all values of θ_k. For n not much below 20, a substantial gain in efficiency may be achieved when $|\rho_{12}| \geq 0.5$ and $\theta_k \leq 0.7$. On the other hand, little loss in relative efficiency arises when $|\rho_{12}| < 0.5$ and $\theta_k > 0.7$. For small n (say, below 13) the SUUR estimator generally loses its advantage of being relatively efficient, even though $|\rho_{12}|$ may be quite close to 1 and θ_k may be quite close to 0.

Next, in the context of the general two equation model, let us examine the loss in efficiency arising from the replacement of the true unobservable disturbance variance covariance matrix in the GLS estimator by $\tilde{S}$, to give the SUUR estimator. For this purpose, let us consider the GLS estimator

$b_{(1)G}$ of β_1. The explicit expression for this estimator can be derived in the same manner as in (4.15). This yields

$$b_{(1)G} = [X_1'(I_T - \rho_{12}^2 P_{x_2})X_1]^{-1}\left[X_1'(I_T - \rho_{12}^2 P_{x_2})y_1 - \rho_{12}\left(\frac{\sigma_{11}}{\sigma_{22}}\right)^{\frac{1}{2}} X_1'\bar{P}_{x_2}y_2\right] \tag{4.135}$$

from which it is easy to obtain its variance covariance matrix as

$$\begin{aligned} V(b_{(1)G}) &= E(b_{(1)G} - \beta_1)(b_{(1)G} - \beta_1)' \\ &= \sigma_{11}(1 - \rho_{12}^2)C^{-1}(I_{K_1} - \rho_{12}^2\Theta)C'^{-1} . \end{aligned} \tag{4.136}$$

Considering the ratio

$$\frac{|V(b_{(1)G})|}{|V(\tilde{\beta}_{(1)SU})|} = \prod_{k=1}^{K_1} \frac{(1 - \rho_{12}^2)}{(1 - \rho_{12}^2\theta_k)h_k} ,$$

Srivastava (1982) carried out a numerical evaluation of the quantity

$$\eta_k = \frac{100(1 - \rho_{12}^2)}{(1 - \rho_{12}^2\theta_k)h_k}$$

for a few selected values of $|\rho_{12}|$, θ_k and n, and graphed η_k against $|\rho_{12}|$. As expected, it was observed that the relative loss in efficiency due to the estimation of Σ by means of unrestricted residuals is not substantial for large degrees of freedom, n. However, just the reverse is true in the case of small n. For instance, this relative loss is approximately seventy-five percent for n as small as 5. Given a prespecified n, the loss declines with an increase in the value of θ_k.

It may be noted that the quantity η_k also throws some light on the appropriateness of an asymptotic approximation to the exact result. It provides a measure of the under-estimation of the true variability when the asymptotic formula is used, and thus it may help in judging the adequacy of this asymptotic approximation. The gap between the true value and the asymptotic approximation for given n and θ_k does not vary much with changes in the values of $|\rho_{12}|$, and for large n the asymptotic approximations are very close to the true values, as expected.

4.7 SOME FURTHER RESULTS

We now present some further results relating to the SUUR estimator in the context of two equation models without any constraints on the regressors.

4.7.1 Existence of Moments

The existence of the moments of the SUUR estimator was discussed by Kariya and Maekawa (1982; pp. 289-290). Recalling that $\Omega = [X'(\Sigma^{-1} \otimes I_T)X]^{-1}$, let

$$\overset{*}{X} = (\Sigma^{-\frac{1}{2}} \otimes I_T)X\Omega^{-\frac{1}{2}}$$

$$V = (\Sigma^{-\frac{1}{2}}\tilde{S}\Sigma^{-\frac{1}{2}} \otimes I_T) \tag{4.137}$$

$$D = n[X'(\tilde{S}^{-1} \otimes I_T)X]^{-1}X'(\tilde{S}^{-\frac{1}{2}}\Sigma\tilde{S}^{-\frac{1}{2}} \otimes I_T)X[X'(\tilde{S}^{-1} \otimes I_T)X]^{-1} ,$$

where Σ and $\tilde{S}$ satisfy $(\Sigma^{\frac{1}{2}})^2 = \Sigma$ and $(\tilde{S}^{\frac{1}{2}})^2 = \tilde{S}$ due to their positive definiteness. Note that $V(\tilde{\beta}_{SU}) = \frac{1}{n}E(D)$.

Now, we observe that

$$D = n\Omega^{-\frac{1}{2}}(\overset{*}{X}'V^{-1}\overset{*}{X})^{-1}\overset{*}{X}'V^{-2}\overset{*}{X}(\overset{*}{X}'V^{-1}\overset{*}{X})^{-1}\Omega^{-\frac{1}{2}} \tag{4.138}$$

from which it follows that D remains invariant if we replace V^{-1} in D by

$$\begin{aligned}\overset{*}{V}{}^{-1} &= \left[\frac{2}{\operatorname{tr} \Sigma^{\frac{1}{2}}\tilde{S}^{-1}\Sigma^{\frac{1}{2}}}\right]V^{-1}\\ &= \left[\frac{2}{\operatorname{tr} \Sigma^{\frac{1}{2}}\tilde{S}^{-1}\Sigma^{\frac{1}{2}}}\Sigma^{\frac{1}{2}}\tilde{S}^{-1}\Sigma^{\frac{1}{2}}\right] \otimes I_T\\ &= \left(\frac{2}{\operatorname{tr} C}C\right) \otimes I_T ,\end{aligned} \tag{4.139}$$

where C is the matrix $(\Sigma^{\frac{1}{2}}\tilde{S}^{-1}\Sigma^{\frac{1}{2}})$.

Let Γ be a (2×2) orthogonal matrix such that

$$\frac{2}{\operatorname{tr} C}\Gamma C\Gamma' = \operatorname{Diag}(\ell_1, \ell_2)$$

where ℓ_1 and ℓ_2 $(\geq \ell_1)$ are the characteristic roots of the matrix $\left(\frac{2}{\operatorname{tr} C}\right)C$.

Taking the trace of both the sides, we have

$$\ell_1 + \ell_2 = 2 , \tag{4.140}$$

implying that $0 \leq \ell_1 \leq 1 \leq \ell_2 \leq 2$ (see Kariya and Maekawa, 1982; p. 290).

Now, we observe that

$$(\overset{*}{X}{}'\overset{*}{X}) = \Omega^{-\frac{1}{2}} X'(\Sigma^{-1} \otimes I_T) X \Omega^{-\frac{1}{2}} = I_K , \tag{4.141}$$

so that $[I_T - \overset{*}{X}(\overset{*}{X}{}'\overset{*}{X})^{-1}\overset{*}{X}{}'] = [I_T - \overset{*}{X}\overset{*}{X}{}']$ is an idempotent matrix of rank n.

If we partition

$$(\Gamma \otimes I_T)\overset{*}{X} = \begin{bmatrix} B_1 \\ B_2 \end{bmatrix} \tag{4.142}$$

with matrices B_1 and B_2 of order $(T \times K)$, we have

$$\overset{*}{X}{}'(\Gamma' \otimes I_T)(\Gamma \otimes I_T)\overset{*}{X} = I_K$$

or

$$(B_1'B_1 + B_2'B_2) = I_K .$$

From (4.139) and (4.142), we find that

$$\begin{aligned}(\overset{*}{X}{}'\overset{*}{V}{}^{-1}\overset{*}{X}) &= \overset{*}{X}{}'(\Gamma' \otimes I_T)(\Gamma \otimes I_T)\left(\frac{2}{\operatorname{tr} C} C \otimes I_T\right)(\Gamma' \otimes I_T)(\Gamma \otimes I_T)\overset{*}{X} \\ &= [B_1' \quad B_2']\left(\frac{2}{\operatorname{tr} C} \Gamma C \Gamma' \otimes I_T\right)\begin{bmatrix} B_1 \\ B_2 \end{bmatrix} \\ &= [B_1' \quad B_2']\begin{bmatrix} \ell_1 I_T & 0 \\ 0 & \ell_2 I_T \end{bmatrix}\begin{bmatrix} B_1 \\ B_2 \end{bmatrix} \\ &= (\ell_1 B_1'B_1 + \ell_2 B_2'B_2) .\end{aligned}$$

Similarly, we have

$$(\overset{*}{X}{}'\overset{*}{V}{}^{-2}\overset{*}{X}) = (\ell_1^2 B_1'B_1 + \ell_2^2 B_2'B_2) .$$

Thus, we get

$$\frac{1}{n}\Omega^{-\frac{1}{2}}D\Omega^{-\frac{1}{2}} = (\overset{*}{X}'V^{-1}\overset{*}{X})^{-1}\overset{*}{X}'V^{-2}\overset{*}{X}(\overset{*}{X}'V^{-1}\overset{*}{X})^{-1}$$
$$= (\overset{*}{X}'\overset{*}{V}^{-1}\overset{*}{X})\overset{*}{X}'\overset{*}{V}^{-2}\overset{*}{X}(\overset{*}{X}'\overset{*}{V}^{-1}\overset{*}{X})^{-1}$$
$$= (\ell_1B_1'B_1 + \ell_2B_2'B_2)^{-1}(\ell_1^2B_1'B_1 + \ell_2^2B_2'B_2)(\ell_1B_1'B_1 + \ell_2B_2'B_2)^{-1} . \tag{4.143}$$

Now, we can find an orthogonal matrix Δ of order $(K\times K)$ such that

$$\Delta B_2'B_2\Delta' = \text{Diag}(\delta_1, \delta_2, \ldots, \delta_K) ,$$

where $\delta_1, \delta_2, \ldots, \delta_K$ are the characteristic roots of $B_2'B_2$ lying between 0 and 1.

From (4.143), we have

$$\frac{1}{n}\Delta\Omega^{-\frac{1}{2}}D\Omega^{-\frac{1}{2}}\Delta' = \text{Diag}(\lambda_1, \lambda_2, \ldots, \lambda_K) , \tag{4.144}$$

where

$$\lambda_k = \frac{(2-\ell_2)^2 + 4(\ell_2-1)\delta_k}{[2-\ell_2+2(\ell_2-1)\delta_k]^2} \qquad (k = 1, 2, \ldots, K).$$

If we maximize λ_k with respect to δ_k, we see that the maximum value of λ_k is $1/\ell_2(2-\ell_2)$, which is attained at $\delta_k = (1-\ell_2/2)$. Thus, we have

$$\delta_k \leq \frac{1}{\ell_2(2-\ell_2)} \qquad \text{for } 1 \leq \ell_2 < 2$$

so that from (4.144) we find that the matrix

$$\frac{1}{\ell_2(2-\ell_2)} I_K - \frac{1}{n}\Delta\Omega^{-\frac{1}{2}}D\Omega^{-\frac{1}{2}}\Delta'$$

is positive semidefinite, and so, using the orthogonality of Δ, the matrix

$$\left[\frac{n}{\ell_2(2-\ell_2)}\Omega^{\frac{1}{2}}\Delta'\Delta\Omega^{\frac{1}{2}} - \Omega^{\frac{1}{2}}\Delta'\Delta\Omega^{-\frac{1}{2}}D\Omega^{-\frac{1}{2}}\Delta'\Delta\Omega^{\frac{1}{2}}\right] = \frac{n}{\ell_2(2-\ell_2)}\Omega - D \tag{4.145}$$

is also positive semidefinite. Using this result, we have

$$t'Dt \leq \frac{n}{\ell_2(2-\ell_2)} t'\Omega t \qquad \text{for all } t$$

where t is any column vector with real elements.

Thus, for any nonnegative integer α, we find that

$$E(t'Dt)^{\alpha} \leq (t'\Omega t)^{\alpha} E\left[\frac{n}{\ell_2(2-\ell_2)}\right]^{\alpha} \qquad \text{for all } t\ . \tag{4.146}$$

Employing the probability density function of ℓ_2 from (A.2) of the Appendix, we observe that

$$\begin{aligned} E\left[\frac{n}{\ell_2(2-\ell_2)}\right]^{\alpha} &= n^{\alpha}(n-1)\int_1^2 (\ell_2-1)[\ell_2(2-\ell_2)]^{(n-3)/2-\alpha}\, d\ell_2 \\ &= \frac{n^{\alpha}(n-1)}{2}\int_0^1 w^{(n-3)/2-\alpha}\, dw \\ &= \frac{n^{\alpha}(n-1)}{n-1-2\alpha}\ ; \quad \left[\frac{1}{2}(n-2)-\alpha\right] \geq 0 \end{aligned} \tag{4.147}$$

where we have used the transformation $w = \ell_2(2-\ell_2)$. Using the above result, it is observed from (4.145) that the α'th moment of t'Dt will be finite if

$$\alpha \leq \frac{1}{2}(n-2)\ . \tag{4.148}$$

As $T\tilde{S}$ follows a Wishart distribution and the elements of $\tilde{S}$ are stochastically independent of $X_i'u_j$ (i, j = 1, 2), the conditional distribution of $\xi = \sqrt{n}(\tilde{\beta}_{SU} - \beta)$, given $\tilde{S}$, is multivariate Normal with mean vector 0 and variance covariance matrix D. In particular, the odd-order moments of a linear function $t'\xi$ are zero, provided that they exist.

Now, consider the characteristic function of ξ:

$$\begin{aligned} CF &= E(\exp\{\sqrt{-1}\, t'\xi\}) \\ &= E[E(\exp\{\sqrt{-1}\, t'\xi\} \mid \tilde{S})] \\ &= E[\exp\{\sqrt{-1}\, t'Dt\}]\ . \end{aligned} \tag{4.149}$$

If the moments of t'Dt are finite up to order α, we can write:

$$CF \simeq \sum_{j=0}^{\alpha} \frac{(-1)^{j/2}}{j!} E(t'Dt)^{j}$$

from which it follows that the existence of the α'th moment of $t'Dt$ ensures the existence of the 2α'th moment of $t'\xi$ which, in turn, is equivalent to the existence of the 2α'th order moments of the elements of ξ or equivalently the SUUR estimator (see Kariya and Maekawa, 1982; p. 290). From (4.148), it is thus seen that the SUUR estimator has finite moments of order equal to double of the largest integer involved in $(n-3)/2$, i.e., of order $(n-2)$ for even n and $(n-3)$ for odd n. Recall that here $n = (T-K)$, the degrees of freedom.

4.7.2 Bounds for the Variance Covariance Matrix

Because of the stochastic independence of the elements of $\tilde{S}$ and $X_i'u_j$ (i, j = 1, 2), we observe that the variance covariance matrix of the SUUR estimator of β is

$$V(\tilde{\beta}_{SU}) = \frac{1}{n}E(D)$$

so that from (4.145) and (4.147) an upper bound for $V(\tilde{\beta}_{SU})$ is

$$E\left[\frac{1}{\ell_2(2-\ell_2)}\right]\Omega = \left(\frac{n-1}{n-3}\right)\Omega ,$$

provided that n exceeds 3.

For a lower bound, we observe from (4.138) and (4.141) that the matrix

$$\begin{aligned}(D-n\Omega) &= n\Omega^{\frac{1}{2}}(\overset{*}{X}{}'V^{-1}\overset{*}{X})^{-1}\overset{*}{X}{}'V^{-2}\overset{*}{X}(\overset{*}{X}{}'V^{-1}\overset{*}{X})^{-1}\Omega^{\frac{1}{2}} - n\Omega \\ &= n\Omega^{\frac{1}{2}}(\overset{*}{X}{}'V^{-1}\overset{*}{X})^{-1}\overset{*}{X}{}'V^{-1}[I - \overset{*}{X}\overset{*}{X}{}']V^{-1}\overset{*}{X}(\overset{*}{X}{}'V^{-1}\overset{*}{X})^{-1}\Omega^{\frac{1}{2}}\end{aligned}$$

is positive semidefinite so that the matrix

$$V(\tilde{\beta}_{SU}) - \Omega = \frac{1}{n}E(D) - \Omega$$

is also positive semidefinite. This supplies the lower bound.

Thus we find that the lower and upper bounds for the variance covariance matrix of the SUUR estimator of β are (Kariya, 1981; pp. 976-977)

$$\left[\Omega, \left(1+\frac{2}{n-3}\right)\Omega\right] \tag{4.150}$$

provided that $n > 3$. Similarly, the bounds for the variance covariance matrix of the OLS estimator may be obtained as

$$\left[\Omega, \frac{(\mu_1 + \mu_2)^2}{4\mu_1\mu_2}\Omega\right], \tag{4.151}$$

where μ_1 and μ_2 are the characteristic roots of the (2×2) matrix Σ.

It is interesting to note from (4.150) and (4.151) that the upper bound in the case of the SUUR estimator depends upon the degrees of freedom n while in the case of the OLS estimator it does not. Considering generalised variances, we have

$$\begin{aligned} 1 &\leq \frac{|V(\tilde{\beta}_{SU})|}{|\Omega|} \leq \left[1 + \frac{2}{n-3}\right]^K \\ 1 &\leq \frac{|V(b_0)|}{|\Omega|} \leq \left[\frac{(\mu_1 + \mu_2)^2}{4\mu_1\mu_2}\right]^K \end{aligned} \tag{4.152}$$

where K is the number of distinct regressors in the model. From this it follows (Kariya, 1981; p. 978) that

$$\left[\frac{4\mu_1\mu_2}{(\mu_1 + \mu_2)^2}\right]^K \leq \frac{|V(\tilde{\beta}_{SU})|}{|V(b_0)|} \leq \left[1 + \frac{2}{n-3}\right]^K \tag{4.153}$$

which provides some broad information on the efficiency of the SUUR estimator relative to OLS.

4.7.3 Momentless Estimators

The contention that an estimator should be highly concentrated around the true value of the associated parameter has led to the use of second-order moments for judging the goodness of estimators and comparing the performance of alternative estimators. The existence of second-order moments is thus an attractive property when considering possible estimators, but if a certain estimator does not possess this property, should we discard it out of hand? Presumably not, because there are a number of other considerations, such as computational simplicity, that might govern the choice of an estimator. Maintaining the criterion of the clustering of estimates around the true parameter value, we can choose the measure of concentration probability as the performance criterion, and can base our analysis on this. As an illustration, let us consider a two equation model with orthogonal regressors. From the expression (4.29) it is clear that the variance covariance matrix of the SUUR estimator of β is infinite if n is 2. So, in this case the variance covariance matrix cannot form the basis for judging the performance of the SUUR estimator. Choosing the concentration probability of the estimator for any linear function $h'\beta$ as an alternative performance criterion, we can derive from (4.65) the expressions for concentration probabilities of the SUUR and OLS estimators:

$$CP_{SU} = P[T^{\frac{1}{2}}|h'(\tilde{\beta}_{SU} - \beta)| \leq m] , \tag{4.154}$$

which may be compared with

$$\begin{aligned} CP_0 &= P[T^{\frac{1}{2}}|h'(b_0 - \beta)| \leq m] \\ &= [\Phi(m) - \Phi(-m)] \end{aligned} \tag{4.155}$$

where $\Phi(\cdot)$ denotes the cumulative distribution function of a Normal variate with mean zero and variance $h'(X'X)^{-1}X'\Sigma X(X'X)^{-1}h$. A comparison of (4.154) and (4.155) reveals whether or not the SUUR estimator is superior to the OLS estimator.

4.7.4 The Telser-Conniffe Estimator

The relationship (4.23) can be exploited to develop an interesting estimator whose properties are easy to analyze (Telser, 1964, Conniffe, 1982b). If we substitute (4.23) in the first equation of the two equation SURE model, we get

$$y_1 = X_1\beta_1 + \delta u_2 + v$$

or

$$(y_1 - \delta u_2) = X_1\beta_1 + v .$$

Replacing u_2 on the left hand side by the OLS residual vector $\hat{u}_2 = \bar{P}_{X_2}y_2$ and assuming δ to be known for a moment, we apply the OLS method to estimate β_1. This gives the following "estimator":

$$(X_1'X_1)^{-1}X_1'y_1 - \delta(X_1'X_1)^{-1}X_1'\bar{P}_{X_2}y_2 . \tag{4.156}$$

Now, as δ is unknown, Telser (1964) and Conniffe (1982b) suggested replacing δ by its estimator, (s_{12}/s_{22}), giving the following estimator of β_1:

$$\overset{*}{b}_{(1)F} = (X_1'X_1)^{-1}X_1'y_1 - \frac{s_{12}}{s_{22}}(X_1'X_1)^{-1}X_1'\bar{P}_{X_2}y_2 . \tag{4.157}$$

If we employ unrestricted residuals, we have

$$\overset{*}{\beta}_{(1)SU} = (X_1'X_1)^{-1}X_1'y_1 - \frac{\tilde{s}_{12}}{\tilde{s}_{22}}(X_1'X_1)^{-1}X_1'\bar{P}_{X_2}y_2 \tag{4.158}$$

It is interesting to note that this estimator has the same form as $\tilde{\beta}_{(1)SU}$ when X_2 is a submatrix of X_1. Telser's iterative extension of (4.158) is discussed in Chapter 5. With Normal disturbances, we see that $\overset{*}{\beta}_{(1)SU}$ is unbiased, and its variance covariance matrix is

$$V(\overset{*}{\beta}_{(1)SU}) = E(\overset{*}{\beta}_{(1)SU} - \beta_1)(\overset{*}{\beta}_{(1)SU} - \beta_1)'$$

$$= (X_1'X_1)^{-1}\left[E(X_1'u_1u_1'X_1) - E\left[\frac{\tilde{s}_{12}}{\tilde{s}_{22}}\right]E(X_1'u_1u_2'\bar{P}_{X_2}X_1 + X_1'\bar{P}_{X_2}u_2u_1'X_1)\right.$$

$$\left. + E\left[\frac{\tilde{s}_{12}}{\tilde{s}_{22}}\right]^2 E(X_1'\bar{P}_{X_2}u_2u_2'\bar{P}_{X_2}X_1)\right](X_1'X_1)^{-1}$$

$$= \sigma_{11}(X_1'X_1)^{-1} - \left[2\sigma_{12}E\left[\frac{\tilde{s}_{12}}{\tilde{s}_{22}}\right] - \sigma_{22}E\left[\frac{\tilde{s}_{12}}{\tilde{s}_{22}}\right]^2\right](X_1'X_1)^{-1}X_1'\bar{P}_{X_2}X_1(X_1'X_1)^{-1}\,. \tag{4.159}$$

Now, we observe that

$$\frac{\tilde{s}_{12}}{\tilde{s}_{22}} = \frac{u_1'\bar{P}_Z u_2}{u_2'\bar{P}_Z u_2}$$

$$= \delta + \frac{v'\bar{P}_Z u_2}{u_2'\bar{P}_Z u_2}$$

$$= \rho_{12}\left(\frac{\sigma_{11}}{\sigma_{22}}\right)^{\frac{1}{2}} + \left(\frac{v'\bar{P}_Z u_2}{u_2'\bar{P}_Z u_2}\right)$$

from which it follows that

$$E\left[\frac{\tilde{s}_{12}}{\tilde{s}_{22}}\right] = \rho_{12}\left(\frac{\sigma_{11}}{\sigma_{22}}\right)^{\frac{1}{2}} + E\left[\frac{E(v')\bar{P}_Z u_2}{u_2'\bar{P}_Z u_2}\right]$$

$$= \rho_{12}\left(\frac{\sigma_{11}}{\sigma_{22}}\right)^{\frac{1}{2}}$$

$$E\left[\frac{\tilde{s}_{12}}{\tilde{s}_{22}}\right]^2 = \rho_{12}^2\left(\frac{\sigma_{11}}{\sigma_{22}}\right) + 2\rho_{12}\left(\frac{\sigma_{11}}{\sigma_{22}}\right)^{\frac{1}{2}} E\left[\frac{E(v')\bar{P}_Z u_2}{(u_2'\bar{P}_Z u_2)}\right] + E\left[\frac{u_2'\bar{P}_Z E(vv')\bar{P}_Z u_2}{(u_2'\bar{P}_Z u_2)^2}\right] \tag{4.160}$$

$$= \rho_{12}^2\left(\frac{\sigma_{11}}{\sigma_{22}}\right) + \sigma_{11}(1-\rho_{12}^2)E\left[\frac{1}{u_2'\bar{P}_Z u_2}\right]$$

$$= \rho_{12}^2\left(\frac{\sigma_{11}}{\sigma_{22}}\right) + \sigma_{11}(1-\rho_{12}^2)\left[\frac{1}{\sigma_{22}(n-2)}\right].$$

Substituting these results in (4.159), we get (Conniffe, 1982b; p. 230)

$$V(\overset{*}{\beta}_{(1)SU}) = \sigma_{11}(X_1'X_1)^{-1} - \sigma_{11}\left[\rho_{12}^2 - \frac{1-\rho_{12}^2}{n-2}\right](X_1'X_1)^{-1}X_1'\bar{P}_{X_2}X_1(X_1'X_1)^{-1} \tag{4.161}$$

which is identical to the expression (4.79) for the variance covariance matrix of the SUUR estimator of β_1 when X_2 is a submatrix of X_1. Recall that $n = (T-K)$. Comparing (4.161) with the variance covariance matrix of the OLS estimator of β_1, it is observed that the OLS estimator will be "better" than $\overset{*}{\beta}_{(1)SU}$ so long as ρ_{12}^2 does not exceed $1/(n-1)$. The reverse is true when ρ_{12}^2 exceeds $1/(n-1)$.

Using (4.115), we can write

$$V(\overset{*}{\beta}_{(1)SU}) = \sigma_{11}C^{-1}\left[\left(\frac{n-1}{n-2}\right)(1-\rho_{12}^2)I_{K_1} + \left(\rho_{12}^2 - \frac{1-\rho_{12}^2}{n-2}\right)\Theta\right]C'^{-1} \tag{4.162}$$

so that comparing this quantity with the variance covariance matrix of the SUUR estimator of β_1, we find

$$V(\tilde{\beta}_{(1)SU}) - V(\overset{*}{\beta}_{(1)SU}) = \sigma_{11}C^{-1}\overset{*}{H}C'^{-1} \tag{4.163}$$

where $\overset{*}{H}$ is a diagonal matrix with k'th diagonal element given as

$$h_k^* = h_k - \left(\frac{n-1}{n-2}\right)(1-\rho_{12}^2) - \left(\rho_{12}^2 - \frac{1-\rho_{12}^2}{n-2}\right)\theta_k \tag{4.164}$$

with h_k defined by (4.128).

Conniffe (1982b; pp. 231-232) computed values of h_k^* for $\theta_k = 0.1\ (0.1)\ (0.9)$, $|\rho_{12}| = 0.4\ (0.1)\ 0.9$ and $n = 5, 9, 13$. His tabulated values indicated that the SUUR estimator is superior to the new proposed estimator when the correlation between the disturbances across the equations is very high ($|\rho_{12}| > 0.8$). When $|\rho_{12}|$ is less than 0.8, the proposed estimator (4.158) is superior to the SUUR estimator for small n, but its superiority declines as n becomes moderately large. However, it may be noted that the proof of unbiasedness and the derivation of the variance covariance matrix of the SUUR estimator require the independence of X_1 with u_2 and X_2 with u_1, in addition to the standard assumptions for the application of OLS to each of the two equations. Conniffe (1982c) noted that this independence may not hold in practice, in which case the usually claimed properties of the SUUR and SURR estimators would lose their validity. Accordingly, he discussed a simple test for this independence. In contrast, the new estimator discussed above does not require this independence condition.

4.8 THE MULTI-EQUATION MODEL

Let us consider the estimation of β_1, the coefficient vector in the first equation of the SURE model containing M equations. To study the properties of the SUUR estimator of β_1, we carry out the following partitioning:

$$y = \begin{bmatrix} y_1 \\ y_{[1]} \end{bmatrix} \begin{matrix} (T) \\ ((M-1)T) \end{matrix}, \quad X = \begin{bmatrix} X_1 & 0 \\ 0 & X_{[1]} \end{bmatrix} \begin{matrix} (T) \\ ((M-1)T) \end{matrix}, \quad u = \begin{bmatrix} u_1 \\ u_{[1]} \end{bmatrix} \begin{matrix} (T) \\ ((M-1)T) \end{matrix},$$

$$\qquad (K_1) \quad \left(\sum_{i=2}^{M} K_i\right)$$

$$\beta = \begin{bmatrix} \beta_1 \\ \beta_{[1]} \end{bmatrix} \begin{matrix} (K_1) \\ \left(\sum_{i=2}^{M} K_i\right) \end{matrix}, \quad \Sigma = \begin{bmatrix} \sigma_{11} & \sigma'_{[1]} \\ \sigma_{[1]} & \Sigma_{[1]} \end{bmatrix} \begin{matrix} (1) \\ (M-1) \end{matrix},$$

$$\qquad (1) \quad (M-1)$$

$$\tilde{S} = \begin{bmatrix} \tilde{s}_{11} & \tilde{s}'_{[1]} \\ \tilde{s}_{[1]} & \tilde{S}_{[1]} \end{bmatrix} \begin{matrix} (1) \\ (M-1) \end{matrix}, \quad \tilde{S}^{-1} = \begin{bmatrix} \tilde{s}^{11} & \tilde{s}^{[1]'} \\ \tilde{s}^{[1]} & \tilde{S}^{[1]} \end{bmatrix} \begin{matrix} (1) \\ (M-1) \end{matrix}$$

$$(1) \quad (M-1) \qquad\qquad (1) \quad (M-1)$$

so that

$$\tilde{s}^{11} = \frac{1}{(\tilde{s}_{11} - \tilde{s}'_{[1]}\tilde{S}^{-1}_{[1]}\tilde{s}_{[1]})} = \frac{1}{(\tilde{s}_{11} - d'\tilde{S}_{[1]}d)}$$

and

$$\tilde{s}^{[1]'} = -\frac{1}{(\tilde{s}_{11} - \tilde{s}'_{[1]}\tilde{S}^{-1}_{[1]}\tilde{s}_{[1]})}\tilde{s}'_{[1]}\tilde{S}^{-1}_{[1]} = -\frac{1}{(\tilde{s}_{11} - d'\tilde{S}_{[1]}d)}d',$$

where $d = \tilde{S}^{-1}_{[1]}\tilde{s}_{[1]}$.

Thus, separating the first equation from the rest, the model can be written as

$$\begin{aligned} y_1 &= X_1\beta_1 + u_1 \\ y_{[1]} &= X_{[1]}\beta_{[1]} + u_{[1]} \ . \end{aligned} \tag{4.165}$$

Assuming normality of the disturbances and defining U_1 as a $[T \times (M-1)]$ matrix with column vectors $u_2, u_3, \ldots, u_M$, we can write

$$\begin{aligned} u_1 &= U_1\delta + v \\ &= (\delta' \otimes I_T)u_{[1]} + v \ , \end{aligned} \tag{4.166}$$

where $\delta = \Sigma^{-1}_{[1]}\sigma_{[1]}$ is now an $[(M-1) \times 1]$ vector and v is a random vector stochastically independent of U_1, or equivalently $u_{[1]}$. Further, v follows a multivariate Normal distribution with mean vector 0 and variance covariance matrix $\sigma_{11.[1]}I_T$, with

$$\begin{aligned} \sigma_{11.[1]} &= (\sigma_{11} - \delta'\Sigma_{[1]}\delta) \\ &= (\sigma_{11} - \sigma'_{[1]}\Sigma^{-1}_{[1]}\sigma_{[1]}) \\ &= \sigma_{11}(1 - R^2_{1\cdot[1]}) \ , \end{aligned}$$

where $R_{1\cdot[1]}$ denotes the multiple correlation coefficient between the disturbance of the first equation and the disturbances in the remaining equations.

Next, we observe that the matrix $T\tilde{S}$ follows a Wishart distribution, $W_M(n, \Sigma)$. Further, the elements of $\tilde{S}$ are stochastically independent of $(X'_i u_j)$ (i, j = 1, 2, ..., M). From results (A.2) in the Appendix, it therefore follows that the conditional distribution of $d' = \tilde{s}'_{[1]}\tilde{S}^{-1}_{[1]}$ given $\tilde{S}^{-1}_{[1]}$ or U_1, is multivariate Normal with mean vector δ' and variance covariance matrix $(\sigma_{11} - \sigma'_{[1]}\Sigma^{-1}_{[1]}\sigma_{[1]})\tilde{S}^{-1}_{[1]} = \sigma_{11.[1]}\tilde{S}^{-1}_{[1]}$.

4.8.1 Pairwise Orthogonal Regressors

Suppose that the regressor matrices in the M equations of the model are pairwise orthogonal, i.e.,

$$X_i'X_j = 0 \qquad (i,\ j = 1,\ 2,\ \ldots,\ M;\ i \neq j) .$$

In this case we see that the SUUR estimator of β_1 is given by

$$\begin{aligned}\tilde{\beta}_{(1)SU} &= (X_1'X_1)^{-1}X_1'y_1 - (X_1'X_1)^{-1}X_1'(\tilde{s}_{[1]}'\tilde{S}_{[1]}^{-1} \otimes I_T)y_{[1]} \\ &= (X_1'X_1)^{-1}X_1'y_1 - (X_1'X_1)^{-1}X_1'(d' \otimes I_T)y_{[1]} ,\end{aligned} \tag{4.167}$$

from which we find

$$\begin{aligned}(\tilde{\beta}_{(1)SU} - \beta_1) &= (X_1'X_1)^{-1}X_1'u_1 - (X_1'X_1)^{-1}X_1'(d' \otimes I_T)u_{[1]} \\ &= (X_1'X_1)^{-1}X_1'v - (X_1'X_1)^{-1}X_1'[(d' - \delta') \otimes I_T]u_{[1]} \\ &= (X_1'X_1)^{-1}X_1'v - (X_1'X_1)^{-1}X_1'(d - \delta) ,\end{aligned} \tag{4.168}$$

where we have used (4.165).

By the normality of the disturbances, we see from (4.168) that $\tilde{\beta}_{(1)SU}$ is an unbiased estimator of β_1. Its variance covariance matrix is

$$\begin{aligned}V(\tilde{\beta}_{(1)SU}) &= E(\tilde{\beta}_{(1)SU} - \beta_1)(\tilde{\beta}_{(1)SU} - \beta_1)' \\ &= (X_1'X_1)^{-1}E[X_1'vv'X_1 - X_1'U_1(d-\delta)v'X_1 - X_1'v(d-\delta)'U_1'X_1 \\ &\quad + X_1'U_1(d-\delta)(d-\delta)'U_1X_1](X_1'X_1)^{-1} \\ &= \sigma_{11\cdot[1]}(X_1'X_1)^{-1} - (X_1'X_1)^{-1}E[X_1'U_1(d-\delta)v'X_1 \\ &\quad + X_1'v(d-\delta)'U_1'X_1 - X_1'U_1(d-\delta)(d-\delta)'U_1X_1](X_1'X_1)^{-1} .\end{aligned} \tag{4.169}$$

It is easy to see that

$$\begin{aligned}E[X_1'U_1(d-\delta)v'X_1] &= E[X_1'U_1(d-\delta)E(v')X_1] \\ &= 0\end{aligned} \tag{4.170}$$

$$E[X_1'U_1(d-\delta)(d-\delta)'U_1'X_1] = E[X_1'U_1E\{(d-\delta)(d-\delta)'|\,U_1\}U_1'X_1]$$

$$= \sigma_{11\cdot[1]} E[X_1' U_1 \tilde{S}_{[1]}^{-1} U_1' X_1]$$

$$= \sigma_{11\cdot[1]} E[X_1' U_1 E(\tilde{S}_{[1]}^{-1}) U_1' X_1]$$

$$= \frac{\sigma_{11\cdot[1]}}{(n-M)} E[X_1' U_1 \Sigma_{[1]}^{-1} U_1' X_1]$$

$$= \frac{\sigma_{11\cdot[1]}}{(n-M)} (\mathrm{tr}\ \Sigma_{[1]} \Sigma_{[1]}^{-1})(X_1' X_1)$$

$$= \sigma_{11\cdot[1]} \left(\frac{M-1}{n-M}\right)(X_1' X_1) , \tag{4.171}$$

provided that $n > M$.

Using (4.170) and (4.171), we get from (4.169) the expression for the variance covariance matrix of the SUUR estimator of β_1:

$$V(\tilde{\beta}_{(1)SU}) = \sigma_{11\cdot[1]}\left[1 + \left(\frac{M-1}{n-M}\right)\right](X_1' X_1)^{-1}$$

$$= \sigma_{11}(1 - R_{1\cdot[1]}^2)\left[1 + \left(\frac{M-1}{n-M}\right)\right](X_1' X_1)^{-1} , \tag{4.172}$$

provided that n exceeds M. This result was derived by Kataoka (1974) and is consistent with Zellner's (1963) results when $M = 2$. Comparing (4.172) with the variance covariance matrix of the OLS estimator of β_1, we see that the OLS estimator is more efficient than the SUUR estimator when

$$R_{1\cdot[1]}^2 < \left(\frac{M-1}{n-M}\right) , \tag{4.173}$$

while the SUUR estimator is relatively more efficient than is the OLS estimator if inequality (4.173) is reversed.

In order to examine the effect of a change in the number of equations in the model on the relative efficiency of the SUUR estimator, we observe that as M increases, $R_{1\cdot[1]}$ also increases so that no definite statement can be made about the change in efficiency arising from a change in M. It may increase or decrease depending upon the resultant impact on the expression (4.172) due to the changes in M and $R_{1\cdot[1]}$.

4.8.2 Unconstrained Regressors

When there are no constraints on the regressor matrices in the SURE model, Phillips (1985) worked out the joint probability density function of the elements of the SUUR estimator, but the derivation is quite involved and therefore is not presented here. However, in this general case the algebra is not that complicated if we consider the moments of Conniffe's estimator which, as noted earlier, amounts to the first iteration of Telser's (1964) estimator. This estimator for β_1 in a M equation model is (Satchell, 1983):

$$\overset{*}{\beta}_{(1)SU} = (X_1'X_1)^{-1}X_1'y_1 - (X_1'X_1)^{-1}X_1'(\tilde{s}_{[1]}'\tilde{S}_{[1]}^{-1} \otimes I_T)\bar{P}_{x_{[1]}}y_{[1]}$$

$$= (X_1'X_1)^{-1}X_1'y_1 - (X_1'X_1)^{-1}X_1'(d' \otimes I_T)\bar{P}_{x_{[1]}}y_{[1]} \,. \qquad (4.174)$$

Substituting (4.165) and (4.166) in this expression, we get

$$(\overset{*}{\beta}_{(1)SU} - \beta_1) = (X_1'X_1)^{-1}X_1'u_1 - (X_1'X_1)^{-1}X_1'(d' \otimes I_T)\bar{P}_{x_{[1]}}u_{[1]}$$

$$= (X_1'X_1)^{-1}X_1'v + (X_1'X_1)^{-1}X_1'U_1\delta - (X_1'X_1)^{-1}X_1'\hat{U}_1 d \,, \qquad (4.175)$$

where $\hat{U}_1$ is the same as U_1 except that u_i in U_1 is replaced by $\bar{P}_{x_i}u_i$ $(i = 2, 3, \ldots, M)$.

If we write

$$\bar{U}_1 = (U_1 - \hat{U}_1)$$

so that $\bar{U}_1$ is just U_1 with u_i replaced by $P_{x_i}u_i$ $(i = 2, 3, \ldots, M)$, we have

$$(\overset{*}{\beta}_{(1)SU} - \beta_1) = (X_1'X_1)^{-1}X_1'v + (X_1'X_1)^{-1}X_1'\bar{U}_1\delta - (X_1'X_1)^{-1}X_1'\hat{U}_1(d-\delta) \,. \qquad (4.176)$$

Defining δ_i as the i'th element of δ and $\sigma_{[1]}^{ij}$ as the (i, j)'th element of $\Sigma_{[1]}^{-1}$, we observe that

$$E[X_1'vv'X_1] = \sigma_{11\cdot[1]}(X_1'X_1) \qquad (4.177)$$

$$E[X_1'\bar{U}_1\delta\delta'\bar{U}_1'X_1] = X_1'\left[\sum_{i=2}^{M}\sum_{j=2}^{M}\delta_i\delta_j P_{x_i}E(u_iu_j')P_{x_j}\right]X_1$$

$$= X_1'\left[\sum_{i=2}^{M}\sum_{j=2}^{M}\delta_i\delta_j\sigma_{ij}P_{x_i}P_{x_j}\right]X_1 \qquad (4.178)$$

$$
\begin{aligned}
E[X_1'\hat{U}_1(d-\delta)(d-\delta)'\hat{U}_1'X_1] &= E[X_1'\hat{U}_1 E\{(d-\delta)(d-\delta)' \mid U_1\}\hat{U}_1'X_1] \\
&= \sigma_{11\cdot[1]}\, E[X_1'\hat{U}_1 S_{[1]}^{-1}\hat{U}_1'X_1] \\
&= \frac{\sigma_{11\cdot[1]}}{(n-M)} E[X_1'\hat{U}_1\Sigma_{[1]}^{-1}\hat{U}_1'X_1] \\
&= \frac{\sigma_{11\cdot[1]}}{(n-M)} X_1'\left[\sum_{i=2}^{M}\sum_{j=2}^{M}\sigma_{[1]}^{ij}\bar{P}_{x_i}E(u_iu_j')\bar{P}_{x_j}\right]X_1 \\
&= \frac{\sigma_{11\cdot[1]}}{(n-M)} X_1'\left[\sum_{i=2}^{M}\sum_{j=2}^{M}\sigma_{[1]}^{ij}\sigma_{ij}\bar{P}_{x_i}\bar{P}_{x_j}\right]X_1 , \qquad (4.179)
\end{aligned}
$$

provided that $n > M$.

Similarly, it can be shown that

$$E[X_1'v\delta'\bar{U}_1'X_1] = 0 \qquad (4.180)$$

$$E[X_1'v(d-\delta)'\hat{U}_1'X_1] = 0 \qquad (4.181)$$

$$E[X_1'\bar{U}_1\delta(d-\delta)'\hat{U}_1'X_1] = 0 . \qquad (4.182)$$

Using (4.176) with the results (4.177)-(4.182), the expression for the variance covariance matrix of Conniffe's estimator can be straightforwardly obtained as

$$
\begin{aligned}
V(\overset{*}{\beta}_{(1)SU}) &= E(\overset{*}{\beta}_{(1)SU} - \beta_1)(\overset{*}{\beta}_{(1)SU} - \beta_1)' \\
&= \sigma_{11\cdot[1]}(X_1'X_1)^{-1} + (X_1'X_1)^{-1}X_1'AX_1(X_1'X_1)^{-1} , \qquad (4.183)
\end{aligned}
$$

where

$$A = \sum_{i=2}^{M}\sum_{j=2}^{M}\sigma_{ij}\left[\delta_i\delta_j P_{x_i}P_{x_j} + \left(\frac{\sigma_{11\cdot[1]}}{n-M}\right)\sigma_{[1]}^{ij}\bar{P}_{x_i}\bar{P}_{x_j}\right] . \qquad (4.184)$$

The above result was first derived by Satchell (1983, pp. 4-5). It is interesting to see that for $X_i'X_j = 0$ $(i \neq j)$ the matrix A reduces to:

$$A = \left(\frac{\sigma_{11\cdot[1]}}{n-M}\right)\sum_{i=2}^{M}\sum_{j=2}^{M}\sigma_{ij}\sigma_{[1]}^{ij}(I_T - P_{x_i} - P_{x_j}) ,$$

which when substituted in (4.183) yields

$$V(\beta^*_{(1)SU}) = \sigma_{11\cdot[1]}\left[1 + \left(\frac{M-1}{n-M}\right)\right](X_1'X_1)^{-1} . \tag{4.185}$$

Notice that this expression is the same as (4.172). As a second special case, note that if all of the regressor matrices except X_1 are identical, i.e., $X_2 = X_3 = \cdots = X_M = X_0$ (say), then

$$A = \left(\frac{M-1}{n-M}\right)\sigma_{11\cdot[1]}\bar{P}_{x_0} + (\sigma_{11} - \sigma_{11\cdot[1]})P_{x_0}$$

$$= \left(\frac{M-1}{n-M}\right)\sigma_{11\cdot[1]}I_T + \left[\sigma_{11} - \left(\frac{n-1}{n-M}\right)\sigma_{11\cdot[1]}\right]P_{x_0}$$

which on substitution in (4.183) provides

$$V(\beta^*_{(1)SU})$$

$$= \sigma_{11\cdot[1]}\left(\frac{n-1}{n-M}\right)(X_1'X_1)^{-1} + \left[\sigma_{11} - \left(\frac{n-1}{n-M}\right)\sigma_{11\cdot[1]}\right](X_1'X_1)^{-1}X_1'P_{x_0}X_1(X_1'X_1)^{-1} . \tag{4.186}$$

Comparing this with the variance covariance matrix of the OLS estimator $b_{(1)0}$ of β_1, we find

$$V(b_{(1)0}) - V(\beta^*_{(1)SU}) = \left[\sigma_{11} - \left(\frac{n-1}{n-M}\right)\sigma_{11\cdot[1]}\right](X_1'X_1)^{-1}X_1'P_{x_0}X_1(X_1'X_1)^{-1} \tag{4.187}$$

from which it follows that Conniffe's estimator will be more efficient than the OLS estimator, when $X_2 = X_3 = \cdots = X_M$, at least as long as

$$\left[\sigma_{11} - \left(\frac{n-1}{n-M}\right)\sigma_{11\cdot[1]}\right] > 0$$

or

$$R^2_{1\cdot[1]} > \left(\frac{M-1}{n-1}\right) .$$

In addition to the restriction $X_2 = X_3 = \cdots = X_M = X_0$, if X_1 is a submatrix of X_0, then $(X_1' P_{X_0} X_1)$ is equal to $(X_1'X_1)$ and consequently both OLS and Conniffe's estimator are equally efficient (Satchell, 1983; pp. 6-7).

Finally, we note that Hillier and Satchell (1985) derived the exact density function of $\overset{*}{\alpha} = a'\overset{*}{\beta}_{(1)SU}$ as an estimator of the linear function $\alpha = a'\beta_1$, where a is any arbitrary known vector. Using zonal polynomials, and results of James (1964), they showed that the probability density function of $\overset{*}{\alpha}$ is a variance-mixture of Normal densities, is symmetric, and has greater kurtosis than a Normal density. Although the density expression obtained by Hillier and Satchell is extremely difficult to interpret in the general case, it simplifies greatly in certain special cases. For example, under orthogonal regressors it accords with the earlier results of Zellner (1963) and Kataoka (1974).

EXERCISES

4.1 Suppose that FGLS estimators of the coefficients in a two equation SURE model with orthogonal regressors and normally distributed disturbances are obtained from the following two choices of the elements of S:

Case A

$$s_{11} = \frac{1}{T} y_1' \bar{P}_{X_1} y_1 , \quad s_{22} = \frac{1}{T} y_2' \bar{P}_{X_2} y_2 , \quad s_{12} = \frac{1}{T} y_1'(\bar{P}_{X_1} - P_{X_2}) y_2$$

Case B

$$s_{11} = \frac{1}{(T - K_1)} y_1' \bar{P}_{X_1} y_1 , \quad s_{22} = \frac{2}{(T - K_2)} y_2' \bar{P}_{X_2} y_2 ,$$

$$s_{12} = \frac{1}{(T - K_1 - K_2)} y_1'(\bar{P}_{X_1} - P_{X_2}) y_2 .$$

Notice that the first choice provides a biased estimator of Σ while the second choice gives an unbiased estimator. Taking the performance criterion to be the variance covariance matrix of the FGLS estimators of the coefficients, which one of the two choices of S should be preferred? Can you suggest the optimal choice?

4.2 Let $\hat{\beta}_{(1)SR}$ and $\hat{\beta}_{(2)SR}$ be the SURR estimators of β_1 and β_2 respectively in a two equation SURE model having orthogonal regressors across the equations. If the disturbances are normally distributed, prove that $(\hat{\beta}_{(1)SR}\hat{\beta}'_{(2)SR})$ is an unbiased estimator of $\beta_1\beta_2'$ while $(\hat{\beta}_{(1)SR}\hat{\beta}'_{(1)SR})$ is a biased estimator of $\beta_1\beta_1'$.

4.3 Using the unbiased estimators of Σ, the SUUR and SURR estimators of β_1 in a two equation SURE model with $X_1'X_2 = 0$ are obtained.

Assuming that the distribution of the disturbances is Normal, find an expression for the ratio of the generalized variances of the two estimators and comment on their relative efficiency.

4.4 Let V_T and V_{T+1} denote the variance covariance matrix of the SUUR estimator, based on T and (T+1) observations respectively, of β_1 in a two equation SURE model with normally distributed disturbances. Further, suppose that x_{T+1} is a row vector giving the (T+1)'th observations on the explanatory variables in the first equation. If $X_1'X_2 = 0$ and $x_{T+1}'X_2 = 0$, prove that the ratio of the determinants of V_T and V_{T+1} is given by

$$\frac{|V_T|}{|V_{T+1}|} = \left[1 + \frac{1}{(T-K)(T-K-2)}\right]^{K_1}\left[1 + x_{T+1}(X_1'X_1)^{-1}x_{T+1}'\right].$$

4.5 In the context of a two equation SURE model with orthogonal regressors and normally distributed disturbances, prove that the mean and variance of the ratio

$$\frac{y_1'(I_T - P_{X_1} - P_{X_2})y_2/\sigma_{12}}{y_2'(I_T - P_{X_1} - P_{X_2})y_2/\sigma_{22}}$$

are 1 and $(1 - \rho_{12}^2)/m\rho_{12}^2$ respectively, provided that $m = (T - K_1 - K_2 - 2)$ is positive.

4.6 Consider the following two equation SURE model:

$$\begin{aligned} y_1 &= \beta_1 x_1 + u_1 \\ y_2 &= \beta_2 x_2 + u_2 \end{aligned} \qquad (x_1'x_1 = T,\ x_1'x_2 = 0)$$

in which the disturbances are bivariate Normal with means 0 and correlation coefficient ρ_{12}. In order to study the efficiency of the SUUR estimator relative to the OLS estimator for the coefficient β_1, the criterion of concentration probability is chosen. Taking $\sigma_{11} = \sigma_{22} = 1$, carry out a numerical evaluation for some specific values of ρ_{12} and T, and comment on this efficiency.

4.7 Suppose that the regressors in the second equation of a two equation SURE model form a subset of the regressors in the first equation. Assuming normality of the disturbances, show that the estimators $d_1 = (y_1'\bar{P}_{X_1}y_2)/(y_2'\bar{P}_{X_1}y_2)$ and $d_2 = (y_1'\bar{P}_{X_1}y_2)/(y_2'\bar{P}_{X_2}y_2)$ are unbiased for the ratio $(\sigma_{12}/\sigma_{22})$. Obtain and compare their variances.

4.8 In a two equation SURE model, suppose that we take

$$\hat{s}_{ij} = \frac{1}{\alpha_{ij}} y_i' \bar{P}_{x_i} \bar{P}_{x_j} y_j$$

where α_{ij} is any fixed scalar (i, j = 1, 2). Assuming that X_2 is a submatrix of X_1 and the disturbances are normally distributed, obtain the variance covariance matrix of the SURR estimator of β_1 and determine the value of the ratio $(\alpha_{22}/\alpha_{12})$ that minimizes it.

4.9 Show that the sampling distribution of the consistent SUUR estimators of coefficients in a two equation SURE model with normally distributed disturbances remains unaltered whether the estimator of Σ is unbiased or biased. Under what condition will such a result hold true in the case of the SURR estimators?

4.10 Evaluate the expectation $E(\tilde{\beta}_{(1)SU} - \beta_1)(\tilde{\beta}_{(2)SU} - \beta_2)'$ assuming that the disturbances are normally distributed and that the SURE model comprises merely two equations.

4.11 For a two equation SURE model having normally distributed disturbances, prove that the absolute value of the difference between the asymptotic and exact probability density functions of the scalar random variable $\sqrt{T}\, h'(\tilde{\beta}_{SU} - \beta)$ cannot exceed $\sqrt{(2/m\pi)}$ where h is any column vector with nonzero (fixed) elements and $m = (T - K - 3)^2$. [Hint: See Kariya and Maekawa (1982).]

4.12 Prove that the lower and upper bounds for the variance covariance matrix of the OLS estimators of coefficients in a two equation SURE model with Normal disturbances are $[X'(\Sigma^{-1} \otimes I_T)X]^{-1}$ and $[\mu^2 X'(\Sigma^{-1} \otimes I_T)X]^{-1}$ respectively, where μ is the harmonic mean of the characteristic roots of Σ.

4.13 Give an example of the two equation SURE model in which you suspect that X_1 (X_2) will not be independent of y_2 (y_1). Suggest a method for testing the hypothesis of independence between X_1 and y_2. [Hint: See Conniffe (1982c).]

4.14 Let $\tilde{S}$, based on unrestricted OLS residuals, be an unbiased estimator of Σ in the context of a multi-equation SURE model. If the distribution of the disturbances is multivariate Normal, prove that the matrix

$$E[X'(\tilde{S}^{-1} \otimes I_T)X]^{-1} - \left(1 - \frac{M+1}{T-K}\right)[X'(\Sigma^{-1} \otimes I_T)X]^{-1}$$

is nonnegative definite. [Hint: See Srivastava (1970).]

REFERENCES

Conniffe, D. (1982a). "Covariance analysis and seemingly unrelated regressions," The American Statistician 36, 169-171.

Conniffe, D. (1982b). "A note on seemingly unrelated regressions," Econometrica 50, 229-233.

Conniffe, D. (1982c). "Testing the assumptions of seemingly unrelated regressions," Review of Economics and Statistics 64, 172-174.

Don, F. J. H., and J. R. Magnus (1980). "On the unbiasedness of iterated GLS estimators," Communications in Statistics A9, 519-527.

Fuller, W. A., and G. E. Battese (1973). "Transformations for estimation of linear models with nested-error structure," Journal of the American Statistical Association 68, 626-632.

Hall, A. D. (1977). "Further finite sample results in the context of two seemingly unrelated regression equations," Working Paper No. 39, Department of Economics, Australian National University, Canberra.

Hillier, G. H., and S. E. Satchell (1985). "Finite sample properties of a two-stage single equation estimator in the SUR model," mimeo., forthcoming in Econometric Theory.

James, A. T. (1964). "Distributions of matrix variates and latent roots derived from normal samples," Annals of Mathematical Statistics 35, 475-501.

Kakwani, N. C. (1967). "The unbiasedness of Zellner's seemingly unrelated regression equations estimators," Journal of the American Statistical Association 62, 141-142.

Kakwani, N. C. (1974). "A note on the efficiency of Zellner's seemingly unrelated regressions estimator," Annals of the Institute of Statistical Mathematics 26, 361-362.

Kariya, T. (1981). "Bounds for the covariance matrices of Zellner's estimator in the SUR model and the 2SAE in a heteroscedastic model," Journal of the American Statistical Association 76, 975-979.

Kariya, T., and K. Maekawa (1982). "A method for approximations to the pdf's and cdf's of GLSE's and its application to the seemingly unrelated regression model," Annals of the Institute of Statistical Mathematics 34, 281-297.

Kataoka, Y. (1974). "The exact finite sample distribution of joint least squares estimators of seemingly unrelated regressions," Economic Studies Quarterly 25, 36-44.

Kunitomo, N. (1977). "A note on the efficiency of Zellner's estimator for the case of two seemingly unrelated regression equations," Economic Studies Quarterly 28, 73-77.

Mehta, J. S., and P. A. V. B. Swamy (1976). "Further evidence on the relative efficiencies of Zellner's seemingly unrelated regressions estimator," Journal of the American Statistical Association 71, 634-639.

Phillips, P. C. B. (1985). "The exact distribution of the SUR estimator," Econometrica 53, 745-756.

Rao, C. R. (1973). Linear Statistical Inference and its Applications (Wiley, New York).

Revankar, N. S. (1974). "Some finite sample results in the context of two seemingly unrelated regression equations," Journal of the American Statistical Association 69, 187-190.

Revankar, N. S. (1976). "Use of restricted residuals in SUR systems: some finite sample results," Journal of the American Statistical Association 71, 183-188.

Satchell, S. E. (1983). "Some properties of Telser's estimator for seemingly unrelated regression models," Discussion Paper No. 226, Department of Economics, University of Essex, Colchester.

Srivastava, A. K. (1982). "Improved estimation of linear regression models," Ph.D. thesis, Department of Statistics, Lucknow University, Lucknow.

Srivastava, V. K. (1970). "On the expectation of the inverse of a matrix," Sankhyā A 32, 236.

Srivastava, V. K., and B. Raj (1979). "The existence of the mean of the estimator in seemingly unrelated regressions," Communications in Statistics A8, 713-717.

Srivastava, V. K., and S. Upadhyaya (1978). "Large-sample approximations in seemingly unrelated regression equations," Annals of the Institute of Statistical Mathematics 30, 89-96.

Telser, L. G. (1964). "Iterative estimation of a set of linear regression equations," Journal of the American Statistical Association 59, 845-862.

Ullah, A., and M. Rafiquzzaman (1977). "A note on the skewness and kurtosis coefficients of Zellner's SUR estimator," Indian Economic Review 12, 181-184.

Zellner, A. (1963). "Estimators for seemingly unrelated regression equations: some exact finite sample results," Journal of the American Statistical Association 58, 977-992.

Zellner, A. (1972). "Corrigenda," Journal of the American Statistical Association 67, 255.

5
Iterative Estimators

5.1 INTRODUCTION

In the preceding three chapters, the efficiency properties of three estimators (OLS, SUUR and SURR) were studied. It was shown that the OLS estimator is unbiased while the SUUR and SURR estimators also are unbiased under a mild restriction on the distribution of the disturbances. Assuming this distribution to be Normal, the relative efficiencies of the OLS, SUUR and SURR estimators were analyzed and it was observed that no estimator uniformly dominates the remaining two in this sense. This suggests asking whether or not the performances of the SUUR and SURR estimators can be improved, and whether more efficient estimators can be constructed. This chapter describes one approach to answering these questions by considering the possibility of iterating with respect to the choice of an observable replacement matrix for Σ. Modifications to the existing estimation procedures are suggested and some new estimators are proposed.

5.2 THE ITERATIVE FEASIBLE GENERALIZED LEAST SQUARES ESTIMATOR

Let us consider the SURE model:

$$y = X\beta + u$$
$$E(u) = 0, \quad E(uu') = (\Sigma \otimes I_T) = \Psi . \tag{5.1}$$

Recognizing that the GLS estimator of β is obtained by minimizing the quantity

$$\begin{aligned}\bar{Q} &= u'(\Sigma^{-1} \otimes I_T)u \\ &= (y - X\beta)'(\Sigma^{-1} \otimes I_T)(y - X\beta) \\ &= y'(\Sigma^{-1} \otimes I_T)y - 2\beta'X'(\Sigma^{-1} \otimes I_T)y + \beta'X'(\Sigma^{-1} \otimes I_T)X\beta ,\end{aligned} \tag{5.2}$$

this minimization can be achieved by setting the first-order derivatives of $\bar{Q}$ with respect to the elements of β to zero and checking the usual second-order condition. Thus, the first-order condition

$$\frac{\partial \bar{Q}}{\partial \beta} = -2X'(\Sigma^{-1} \otimes I_T)y + 2X'(\Sigma^{-1} \otimes I_T)X\beta = 0$$

yields the estimating equations

$$[X'(\Sigma^{-1} \otimes I_T)X]\beta = X'(\Sigma^{-1} \otimes I_T)y . \tag{5.3}$$

Similarly, the second-order condition requires the matrix $(\partial^2\bar{Q}/\partial\beta\partial\beta')$ to be positive definite. Since

$$\frac{\partial^2 \bar{Q}}{\partial\beta\partial\beta'} = X'(\Sigma^{-1} \otimes I_T)X$$

is always a positive definite matrix, this ensures that $\bar{Q}$ attains a minimum for the solution of (5.3).

If Σ is unknown, the FGLS estimator is obtained from the solution of (5.3) after replacing Σ by an estimator S. This motivates an iterative procedure (see Zellner, 1962). For instance, in order to obtain a feasible version of the GLS estimator, suppose that we replace Σ by I_M in (5.3) so that we get the OLS estimator of β. In the absence of any knowledge about Σ, this proposition is the simplest of several alternatives, though it is not appealing because I_M is not a consistent estimator of Σ. However, beginning with this, we can then estimate Σ from the residuals based on the OLS estimate of the coefficient vector, β. This estimator of Σ can now be used to replace Σ in (5.3). This yields what we call an FGLS estimator. Likewise, the residuals based on the FGLS estimates of the coefficients can be used to develop another estimate of Σ. This, when substituted for Σ in (5.3), provides yet another estimate of the coefficient vector, β. Repeating this process leads to an iterative feasible generalized least squares (IFGLS) estimator of β.

Let this estimator at the k'th round be denoted by $b^{(k)}$. It is then easy to see that

$$b^{(k)} = [X'(S^{(k)^{-1}} \otimes I_T)X]^{-1}X'(S^{(k)^{-1}} \otimes I_T)y \tag{5.4}$$

where $S^{(k)}$ is a consistent estimator of Σ constructed from the residuals based on $b^{(k-1)}$.

Obviously, we have

$$S^{(0)} \equiv I_M , \quad b^{(0)} \equiv b_0$$

$$S^{(1)} \equiv S , \quad b^{(1)} \equiv b_F ,$$

so that $S^{(1)} = \tilde{S}$ yields the SUUR estimator at the first iteration, while $S^{(1)} = \hat{S}$ provides the SURR estimator of β at this step.

The iterative estimator outlined by Zellner (1962) is defined by (5.4) with the initial choice of $S^{(1)}$ taken as $\hat{S}$. When this iterative procedure converges (i.e., the proportional changes in the estimates of the elements of β become negligibly different at two successive iterations), the converged quantity may be labelled the ISURR estimator. If we take $S^{(1)} = \tilde{S}$ and iterate, the result may be termed the ISUUR estimator. Notice that in this case $S^{(1)}$ is based on unrestricted residuals but subsequent estimators $S^{(k)}$ (k = 2, 3, ...) will be based on residuals which are restricted in a particular way.

5.3 THE ITERATIVE ORDINARY LEAST SQUARES ESTIMATOR

We now consider an iterative estimator based on the application of OLS to each equation sequentially, as proposed by Telser (1964). Recalling that the disturbances in the different equations of the SURE model are correlated, we can express the disturbances in a particular equation as a linear function of the disturbances in the remaining equations, plus an additional disturbance term. If U_i denotes the $[T \times (M-1)]$ matrix having column vectors $u_1, u_2, \ldots, u_M$ (except u_i), we can write

$$u_i = U_i\delta_i + v_i$$

$$= \sum_{g=1}^{i-1} u_g\delta_{ig} + \sum_{h=i+1}^{M} u_h\delta_{ih} + v_i \quad (i = 1, 2, \ldots, M) \tag{5.5}$$

where v_i is a $(T \times 1)$ vector of stochastic elements and δ_i is an $[(M-1) \times 1]$ vector of coefficients that are so determined that

$$E(U_i'v_i) = 0 \,. \tag{5.6}$$

The condition (5.6) implies that

$$E(U_i'u_i) - E(U_i'U_i)\delta_i = 0$$

or

$$\sigma_{(i)} - \Sigma_{ii}\delta_i = 0$$

or

$$\delta_i = \Sigma_{ii}^{-1}\sigma_{(i)} \tag{5.7}$$

where Σ_{ii} is obtained from Σ by deleting the i'th row and i'th column, while $\sigma_{(i)}$ is derived from the i'th column of Σ by deleting the element σ_{ii}. For example, if we partition Σ as

$$\Sigma = \begin{bmatrix} \sigma_{11} & \sigma'_{(1)} \\ \sigma_{(1)} & \Sigma_{11} \end{bmatrix} \begin{matrix} (1) \\ (M-1) \end{matrix} ,$$
$$\begin{matrix} (1) & (M-1) \end{matrix}$$

then δ_1 is equal to $\Sigma_{11}^{-1}\sigma_{(1)}$.

It is now easy to see that

$$E(v_i) = E(u_i) - E(U_i)\delta_i$$

$$= 0$$

$$\frac{1}{T}E(v_i'v_i) = \frac{1}{T}[E(u_i'u_i) - 2E(u_i'U_i)\delta_i + \delta_i'E(U_i'U_i)\delta_i]$$

$$= \sigma_{ii} - 2\sigma_{(i)}'\delta_i + \delta_i'\Sigma_{11}\delta_i$$

$$= \sigma_{ii} - \sigma_{(i)}'\Sigma_{11}^{-1}\sigma_{(i)} = \frac{1}{\sigma^{ii}} ,$$

where σ^{ii} denotes the i'th diagonal element of the matrix Σ^{-1}.

Using (5.5), the i'th equation of the SURE model (5.1) can be written as

$$y_i = X_i\beta_i + U_i\delta_i + v_i$$

$$= X_i\beta_i + \sum_{g=1}^{i-1} u_g\delta_{ig} + \sum_{h=i+1}^{M} u_h\delta_{ih} + v_i . \tag{5.8}$$

Now, first of all we shall obtain the restricted residuals. This stage will be referred to as the first round. If $\hat{u}_j^{(1)}$ denotes the vector of restricted residuals, i.e.,

$$\hat{u}_j^{(1)} = [I_T - X_j(X_j'X_j)^{-1}X_j']y_j , \tag{5.9}$$

the second round starts with the replacement of $u_2, u_3, \cdots, u_M$ by $\hat{u}_2^{(2)}, \hat{u}_3^{(2)}, \ldots, \hat{u}_M^{(2)}$ in the first equation to yield the following:

$$y_1 = X_1\beta_1 + \hat{U}_1^{(1)}\delta_1 + \epsilon_1$$

$$= X_1\beta_1 + \sum_{h=2}^{M} \hat{u}_h^{(1)}\delta_{1h} + \epsilon_1$$

where ϵ_1 is now the disturbance vector.

Using OLS to estimate β_1 and δ_1, we get

$$\begin{bmatrix} b_{(1)0}^{(2)} \\ \delta_{(1)0} \end{bmatrix} = \begin{bmatrix} X_1'X_1 & X_1'\hat{U}_1^{(1)} \\ \hat{U}_1^{(1)'}X_1 & \hat{U}_1^{(1)'}\hat{U}_1^{(1)} \end{bmatrix}^{-1} \begin{bmatrix} X_1'y_1 \\ \hat{U}_1^{(1)'}y_1 \end{bmatrix} ,$$

so that

$$b_{(1)0}^{(2)} = (X_1'\bar{P}_{\hat{U}_1^{(1)}}X_1)^{-1}X_1'\bar{P}_{\hat{U}_1^{(1)}}y_1 .$$

Now writing

$$\hat{u}_1^{(2)} = y_1 - X_1 b_{(1)0}^{(2)} \tag{5.10}$$

we substitute $\hat{u}_1^{(2)}$ in place of u_1 and $\hat{u}_h^{(1)}$ in place of u_h ($h = 3, 4, \ldots, M$) in the second equation, so that we obtain

$$y_2 = X_2\beta_2 + \hat{u}_1^{(2)}\delta_{21} + \sum_{h=3}^{M} \hat{u}_h^{(1)}\delta_{2h} + \epsilon_2 .$$

We can then obtain the OLS estimator $b_{(2)0}^{(2)}$ of β_2 and

$$\hat{u}_2^{(2)} = y_2 - X_2 b_{(2)0}^{(2)} . \tag{5.11}$$

Next, replacing u_1 and u_2 in the third equation by $\hat{u}_1^{(2)}$ and $\hat{u}_2^{(2)}$ respectively and the remaining $u_4, u_5, \ldots, u_M$ by $\hat{u}_4^{(1)}, \hat{u}_5^{(1)}, \ldots, \hat{u}_M^{(1)}$, we get the third equation as follows:

$$y_3 = X_3\beta_3 + \sum_{g=1}^{2} \hat{u}_g^{(2)}\delta_{3g} + \sum_{h=4}^{M} \hat{u}_h^{(1)}\delta_{3h} + \epsilon_3 ,$$

from which the application of OLS yields $b_{(3)0}^{(2)}$ for β_3. This estimator provides the residual vector $\hat{u}_3^{(2)}$, and so on. For instance, the i'th equation is replaced by

$$y_i = X_i\beta_i + \sum_{g=1}^{i} \hat{u}_g^{(2)}\delta_{ig} + \sum_{h=i+1}^{M} \hat{u}_h^{(1)}\delta_{ih} + \epsilon_i .$$

This process is continued until we reach the last (viz., M'th) equation of the model. This completes the second round.

The third round starts with the substitution of $\hat{u}_j^{(2)}$ for u_j ($j=2, 3, \ldots, M$) in the first equation and the coefficient vector β_1 is estimated by OLS. This implies $\hat{u}_1^{(3)} = (y_1 - X_1 b_{(1)0}^{(3)})$ which is used to replace u_1 in the second equation. Now, in the same manner as indicated for the second round, the OLS estimators of β_2 in the second equation, β_3 in the third equation, ..., β_M in the M'th equation may be derived successively. This completes the third round, and so on. These rounds are repeated until they lead to a converged estimator, referred to as the iterative ordinary least squares (IOLS) estimator. This iterative procedure was proposed and discussed by Telser (1964). He studied its convergence as the number of rounds increases and showed that these iterations will almost always converge, provided T is large enough (see Telser, 1964; p. 860). Further, it may be shown that the converged estimator is identical to the ISURR estimator.

5.4 THE MAXIMUM LIKELIHOOD ESTIMATOR

If the disturbances are Normal, then the likelihood function associated with (5.1) is

$$L = (2\pi)^{-\frac{TM}{2}} \,|(\Sigma \otimes I_T)|^{-\frac{1}{2}} \exp\{-\tfrac{1}{2}(y - X\beta)'(\Sigma^{-1} \otimes I_T)(y - X\beta)\} \quad (5.12)$$

where Σ is an unknown matrix. The maximization of (5.12) with respect to the unknown parameters provides likelihood equations which may be solved to yield the parameters' maximum likelihood (ML) estimators. For this purpose, we consider the natural logarithm of the above likelihood function:

$$\mathscr{L} = -\frac{TM}{2}\log(2\pi) - \frac{T}{2}\log|\Sigma| - \frac{1}{2}(y - X\beta)'(\Sigma^{-1} \otimes I_T)(y - X\beta) \quad (5.13)$$

which can be maximized by setting the partial derivatives of $\mathscr{L}$ with respect to the elements of β and the distinct elements of Σ to zero, and solving the resulting equations for β and Σ. At the same time we need to check the Hessian matrix to see that these solutions correspond to a maximum of $\mathscr{L}$.

Differentiating $\mathscr{L}$ with respect to β and equating to 0, we get the following set of likelihood equations:

$$X'(\Sigma^{-1} \otimes I_T)X\beta = X'(\Sigma^{-1} \otimes I_T)y \,. \quad (5.14)$$

Similarly, noting from Magnus (1978; p. 282) that

$$\begin{aligned}(y - X\beta)'(\Sigma^{-1} \otimes I_T)(y - X\beta) &= \mathrm{vec}(U)'(\Sigma^{-1} \otimes I_T)\mathrm{vec}(U) \\ &= \mathrm{tr}\,(U'U\Sigma^{-1}) \,,\end{aligned}$$

where the $(T \times M)$ matrix U is given by $U = (u_1, u_2, \ldots, u_M)$, and noting that

$$\frac{\partial \log |\Sigma|}{\partial \Sigma^{-1}} = -\Sigma ,$$

and (5.15)

$$\frac{\partial \operatorname{tr}(U'U\Sigma^{-1})}{\partial \Sigma^{-1}} = U'U ,$$

then equating the derivative of $\mathscr{L}$ with respect to Σ^{-1} to zero yields

$$T\Sigma = U'U . \tag{5.16}$$

Next, assembling the $\frac{1}{2}M(M+1)$ distinct elements of Σ in the form of a column vector σ (say) and following Magnus (1978), the Hessian matrix can be derived and from its form it can be checked that the solution to the likelihood equations (5.14) and (5.16) corresponds to a maximum of $\mathscr{L}$. Further, the negative of the expectation of this Hessian matrix provides the Fisher information matrix, which may be shown to be of the form:

$$-\begin{bmatrix} E\left(\frac{\partial^2 \mathscr{L}}{\partial\beta\,\partial\beta'}\right) & E\left(\frac{\partial^2 \mathscr{L}}{\partial\beta\,\partial\sigma'}\right) \\ E\left(\frac{\partial^2 \mathscr{L}}{\partial\sigma\,\partial\beta'}\right) & E\left(\frac{\partial^2 \mathscr{L}}{\partial\sigma\,\partial\sigma'}\right) \end{bmatrix} = \begin{bmatrix} X'(\Sigma^{-1} \otimes I_T)X & 0 \\ 0 & \mathscr{I} \end{bmatrix} \tag{5.17}$$

where a typical element of $\mathscr{I}$ is given by

$$\frac{\partial^2 \mathscr{L}}{\partial\sigma_{ij}\,\partial\sigma_{k\ell}} = \begin{cases} 2T(\sigma^{ik}\sigma^{j\ell} + \sigma^{i\ell}\sigma^{jk}) & \text{for } i \neq j \text{ and } k \neq \ell \\ 2T\sigma^{ik}\sigma^{jk} & \text{for } i \neq j \text{ and } k = \ell \\ 2T\sigma^{ik}\sigma^{i\ell} & \text{for } i = j \text{ and } k \neq \ell \\ T(\sigma^{ik})^2 & \text{for } i = j \text{ and } k = \ell \end{cases} .$$

The likelihood equations (5.14) and (5.16) are nonlinear in the elements of β and Σ, so it is not possible to obtain analytical closed-form expressions for the corresponding ML estimators. Numerical procedures are therefore required to obtain ML estimates of the unknown parameters.

Suppose that $S^{(1)}$ is any initial estimate of Σ. Substituting it for Σ in (5.14) provides an estimator of β:

$$b^{(1)} = [X'(S^{(1)^{-1}} \otimes I_T)X]^{-1}X'(S^{(1)^{-1}} \otimes I_T)y ,$$

from which we can estimate u by

$$u^{(2)} = y - Xb^{(1)} .$$

From these residuals we can obtain an estimator $S^{(2)}$ for Σ through (5.16). Substituting $S^{(2)}$ for Σ in (5.14) yields another estimator of β:

$$b^{(2)} = [X'(S^{(2)^{-1}} \otimes I_T)X]^{-1}X'(S^{(2)^{-1}} \otimes I_T)y ,$$

and so on. Thus, the estimator at the k'th iteration is

$$b^{(k)} = [X'(S^{(k)^{-1}} \otimes I_T)X]^{-1}X'(S^{(k)^{-1}} \otimes I_T)y . \tag{5.18}$$

Continuing these iterations until convergence provides the ML estimator, b_M, of β. Following Oberhofer and Kmenta (1974), it can be shown that such an iterative procedure converges to yield a solution to the estimating equations provided that

$$\inf_{\beta}(\hat{\sigma}_{ii}) \geq \eta_0 > 0$$
$$\det((\hat{\rho}_{ij})) \geq \eta_1 > 0$$

for some arbitrarily chosen positive numbers η_0 and η_1, where

$$\hat{\sigma}_{ij} = \frac{1}{T}(y_i - X_i\beta_i)'(y_j - X_j\beta_j)$$

$$\hat{\rho}_{ij} = \frac{\hat{\sigma}_{ij}}{\sqrt{(\hat{\sigma}_{ii}\hat{\sigma}_{jj})}} \qquad (i, j = 1, 2, \ldots, M) .$$

Further, if $S^{(1)}$ is a consistent estimator of Σ, it follows from Rao (1973; Chap. 5) that b_M is unique. Obviously, if $S^{(1)}$ is based on restricted residuals, i.e., $S^{(1)} = \hat{S}$, the ML estimator coincides with the ISURR estimator.

Taking a Taylor's expansion of $\mathscr{L}$ around $\beta = b^{(1)}$, we have

$$\mathscr{L} = \mathscr{L}^{(1)} + (\beta - b^{(1)})'\left(\frac{\partial \mathscr{L}}{\partial \beta}\right)_{\beta=b^{(1)}} + \frac{1}{2}(\beta - b^{(1)})'\left(\frac{\partial^2 \mathscr{L}}{\partial \beta\, \partial \beta'}\right)_{\beta=b^{(1)}}(\beta - b^{(1)}) + \cdots \tag{5.19}$$

where $\mathscr{L}^{(1)}$ is the value of $\mathscr{L}$ at $\beta = b^{(1)}$. Ignoring higher order terms in this expansion and differentiating with respect to β, we have

$$\frac{\partial \mathscr{L}}{\partial \beta} = \left(\frac{\partial \mathscr{L}}{\partial \beta}\right)_{\beta=b^{(1)}} + \left(\frac{\partial^2 \mathscr{L}}{\partial \beta\, \partial \beta'}\right)_{\beta=b^{(1)}}(\beta - b^{(1)}) \,. \tag{5.20}$$

If we define

$$Q = T\left(\frac{\partial^2 \mathscr{L}}{\partial \beta\, \partial \beta'}\right), \quad q = T\left(\frac{\partial \mathscr{L}}{\partial \beta}\right)$$

and partition these as

$$Q = \begin{bmatrix} Q_{11} & \cdots & Q_{1M} \\ \vdots & & \vdots \\ Q_{M1} & \cdots & Q_{MM} \end{bmatrix}, \quad q = \begin{bmatrix} q_1 \\ \vdots \\ q_M \end{bmatrix},$$

it is easy to see that

$$Q_{ij} = \frac{1}{T} X_i' \left[\sum_{g,\ell}^{M} (s^{ig} s^{\ell j} + s^{ij} s^{\ell g}) u_g u_\ell' \right] X_j - \frac{1}{T} s^{ij} X_i' X_j$$

$$q_i = \frac{1}{T} \sum_{g}^{M} s^{ig} X_i' u_g \,.$$

Thus the first-order condition obtained by setting the expression in (5.20) equal to a null vector yields

$$(\beta - b^{(1)}) = -\left[\left(\frac{\partial^2 \mathscr{L}}{\partial \beta\, \partial \beta'}\right)_{\beta=b^{(1)}}\right]^{-1} \left(\frac{\partial \mathscr{L}}{\partial \beta}\right)_{\beta=b^{(1)}}$$

$$= -Q^{(1)^{-1}} q^{(1)}$$

where $Q^{(1)}$ and $q^{(1)}$ denote the values of Q and q respectively at $\beta = b^{(1)}$. This provides the second round estimator of β as follows:

$$b^{(2)} = b^{(1)} - Q^{(1)^{-1}} q^{(1)}$$

which once again can be used in place of $b^{(1)}$. Repeating this process until convergence yields the ML estimator of β. When the disturbances are not Normal, the ML estimator is often referred to as a quasi-maximum likelihood estimator.

If we substitute (5.16) into (5.13), we get the concentrated log likelihood function

$$\begin{aligned}\mathcal{L}^* &= -\frac{TM}{2}\log(2\pi) - \frac{T}{2}\log\left|\frac{1}{T}U'U\right| - \frac{MT}{2} \\ &= -\frac{TM}{2}\log(2\pi) - \frac{T}{2}\log\left|\frac{1}{T}(Y - \underline{X}\underline{\beta})'(Y - \underline{X}\underline{\beta})\right| - \frac{MT}{2}, \qquad (5.21)\end{aligned}$$

where

$$\underset{(T \times K^*)}{\underline{X} = (X_1, X_2, \ldots, X_M)}, \qquad \underset{(T \times M)}{Y = (y_1, y_2, \ldots, y_M)}$$

$$\underset{(K^* \times M)}{\underline{\beta} = \begin{bmatrix} \beta_1 & 0 & \cdots & 0 \\ 0 & \beta_2 & \cdots & 0 \\ \vdots & \vdots & & \vdots \\ 0 & 0 & \cdots & \beta_M \end{bmatrix}}.$$

Maximization of $\mathcal{L}^*$ is equivalent to the minimization of $\left|\frac{1}{T}U'U\right|$ with respect to the unknown elements of $\underline{\beta}$, but this quantity is the generalized residual variance so the maximum likelihood estimator is identical to the least generalized residual variance estimator. Notice that the least generalized residual variance principle does not require the assumption of normality of the disturbances in order to be defined.

5.5 COMPUTATIONAL ISSUES

Recognizing that the SURE model may be interpreted as a special form of the simultaneous equations model, it is clear that the iterative three stage least squares, iterative limited information maximum likelihood and full information maximum likelihood estimators in simultaneous equation models correspond to the ISURR, IOLS and ML estimators respectively in the SURE model. Thus, the various computational procedures used in the context of simultaneous equation models can be particularized in the present context. Other numerical methods for solving estimating equations also can be used to compute these estimates. Ruble (1968) presented an interesting account of computational layouts that ensure rapid convergence, minimize round-off errors and are easy to program. Commenting on the rapidity of convergence, Ruble (1968; p. 294) stated that although the IOLS and ISURR procedures are easier to program in comparison with the ML procedure, in most situations the latter may exhibit faster convergence and thus may save substantial computer time in comparison with the IOLS and ISURR procedures. In the context of a Monte Carlo study (see Section 3.5) carried out by Kmenta and Gilbert (1968), it was pointed out that with a coefficient convergence criterion of 0.000001, the number of rounds for the convergence of the ML procedure was about one-third of the number of rounds required for the convergence of the IOLS and ISURR procedures in all of the cases considered with two equation models. The number of rounds needed for convergence showed more variability for the IOLS estimator than for the ISURR estimator. Comparing the IOLS and ISURR procedures, considerably more rounds were needed for convergence in the case of the IOLS estimator. In the numerical results reported by Kmenta and Gilbert (1968), all of the IOLS, ISURR and ML procedures were found to converge to identical estimates.

5.6 UNBIASEDNESS

To begin our discussion of the properties of the iterative estimators, let us consider IFGLS, which encompasses the ISUUR and ISURR estimators. From (5.4), we see that $S^{(k)}$ $(k = 1, 2, \ldots)$ is an even function of the disturbances, so that $(b^{(k)} - \beta)$ is an odd function for all k including $k = 0$. Therefore the expectation of $(b^{(k)} - \beta)$ will be a null vector provided that the disturbances are symmetrically distributed and $E(b^{(k)})$ exists. Following the line of proof for the unbiasedness of the SUUR and SURR estimators in Section 4.2, it can be shown that the mean vector of $b^{(k)}$ exists if the disturbances have at least fourth-order moments and the expectation of the inverse of $S^{(k)}$ is finite. Since the unbiasedness of $b^{(k)}$ $(k = 1, 2, \ldots)$ ensures the unbiasedness of the IFGLS estimator, we find that the ISUUR and ISURR estimators are unbiased when the disturbances follow a symmetric distribution with finite moments of order four and $E(S^{(k)^{-1}})$ exists for all k $(k = 1, 2, \ldots)$. Similar sufficient conditions also may be derived for the IOLS estimator. For the ML estimator, the disturbances are assumed to be Normal so that this estimator is unbiased if $E(S^{(k)^{-1}})$ is finite and the iterative procedure converges. Don and Magnus (1980) derived the condition that ensures both the convergence and unbiasedness of the ML estimator. This condition is as follows.

Suppose V denotes the set of values of the elements of Σ such that the sum of squares of these elements is bounded above, while $|\Sigma|$ is bounded below, by a non-negative number. Let S^* be the value of Σ that maximizes the log likelihood function, $\mathscr{L}$. Now, with probability one there exists a number n_0 such that for all β and for all Σ the elements of which belong to V, $\mathscr{L} \geq n_0$ implies that the elements of S^* belong to V. It is generally difficult to test if this condition holds. Further, when it is violated, it was recommended by Don and Magnus (1980) that the iterative procedure should be modified so as to satisfy it, for instance, in the light of work reported by Sargan (1978). A detailed discussion of this work is not given here.

5.7 EFFICIENCY PROPERTIES

Now let us consider the efficiencies of the iterative estimators. We observed in Section 2.6 that the asymptotic distribution of $\sqrt{T}$ times the deviation of each of the SUUR and SURR estimators from β is $N(0, Q_\Sigma)$, with

$$Q_\Sigma = \begin{bmatrix} \sigma^{11}Q_{11} & \cdots & \sigma^{1M}Q_{1M} \\ \vdots & & \vdots \\ \sigma^{M1}Q_{M1} & \cdots & \sigma^{MM}Q_{MM} \end{bmatrix}; \quad Q_{ij} = \lim_{T\to\infty} \left(\frac{1}{T} X_i'X_j\right) .$$

Likewise it can be shown that the asymptotic distributions of the ISUUR and ISURR estimators are also $N(0, Q_\Sigma)$. The same is true for the IOLS estimator (see Telser, 1964). Following Magnus (1978), it can be further shown that the ML estimator has the same asymptotic distribution under some mild conditions. Thus, the asymptotic distributions of all the iterative estimators are identical and therefore they are equally efficient asymptotically. This being the case, a consideration of asymptotic properties does not permit us to choose one estimator over the other. Let us therefore look at the large-sample asymptotic approximations to the distributions of these estimators. For this purpose, we consider the ISURR estimator. This estimator at the second round is

$$\hat{\beta}_{SR}^{(2)} = [X'(\hat{S}^{(2)^{-1}} \otimes I_T)X]^{-1}X'(\hat{S}^{(2)^{-1}} \otimes I_T)y . \tag{5.22}$$

Notice that the second round starts with the estimation of the disturbance vector u by

$$\hat{u}^{(2)} = y - X\hat{\beta}^{(1)}_{SR}$$

$$= u - X(\hat{\beta}^{(1)}_{SR} - \beta) \,, \tag{5.23}$$

where $\hat{\beta}^{(1)}_{SR}$ is the SURR estimator, $\hat{\beta}_{SR}$, given by

$$\hat{\beta}^{(1)}_{SR} \equiv \hat{\beta}_{SR} = [X'(\hat{S}^{-1} \otimes I_T)X]^{-1}X'(\hat{S}^{-1} \otimes I_T)y \,.$$

Defining

$$\Delta = (\hat{S} \otimes I_T) - (\Sigma \otimes I_T)$$

$$\Omega = [X'(\Sigma^{-1} \otimes I_T)X]^{-1} \tag{5.24}$$

$$Q = (\Sigma^{-1} \otimes I_T) - (\Sigma^{-1} \otimes I_T)X[X'(\Sigma^{-1} \otimes I_T)X]^{-1}X'(\Sigma^{-1} \otimes I_T)$$

and following the procedure adopted in Section 3.2, we can write

$$(\hat{\beta}^{(1)}_{SR} - \beta) = \xi_{-1/2} + \xi_{-1} + \xi_{-3/2} + O_p(T^{-g}) \qquad (g \geq 2) \tag{5.25}$$

where (Srivastava, 1970; p. 487):

$$\xi_{-1/2} = \Omega X'(\Sigma^{-1} \otimes I_T)u$$

$$\xi_{-1} = -\Omega X'(\Sigma^{-1} \otimes I_T)\Delta Q u \tag{5.26}$$

$$\xi_{-3/2} = \Omega X'(\Sigma^{-1} \otimes I_T)\Delta Q \Delta Q u \,.$$

If $\hat{u}_j^{(2)}$ denotes the j'th subvector of $\hat{u}^{(2)}$, we can write this subvector as

$$\hat{u}_j^{(2)} = \Gamma_j \hat{u}^{(2)} \,,$$

where $\Gamma_j = (0, \ldots, 0, I_T, 0, \ldots, 0)$ is a $(T \times MT)$ matrix containing $(M-1)$ null matrices of order $(T \times T)$, and with I_T occurring at the j'th place.

Thus, we find

$$\hat{s}_{ij}^{(2)} = \frac{1}{T}\hat{u}_i^{(2)'}\hat{u}_j^{(2)}$$
$$= \frac{1}{T}\hat{u}^{(2)'}\Gamma_i'\Gamma_j\hat{u}^{(2)} . \qquad (5.27)$$

Using (5.23), (5.24), (5.25) and (5.27), we can express the estimation error of $\hat{\beta}_{SR}^{(2)}$ in the form (5.25) by proceeding in the same manner as in Section 3.2. Then the variance covariance matrix to order $O(T^{-2})$ can be evaluated as for the SURR estimator in Section 3.3. Interestingly, this matrix turns out to be the same as in the case of $\hat{\beta}_{SR}^{(1)}$ (see Srivastava, 1970). The same holds true if we consider further rounds. The implication of this is that iterations beyond the first round do not bring any gain in efficiency relative to the second-order asymptotic approximation for the variance covariance matrix of the ISUUR estimator.

Kmenta and Gilbert (1968) conducted a Monte Carlo study (described in Section 3.5) to compare the performances of the IOLS, ISURR and ML estimators in finite samples. They found that the three estimation procedures gave rise to numerically identical estimates. Labelling these three as iterative estimators, it was found (as expected) that the iterative estimators and the noniterative estimator (SURR) have the same variance for large T. For small T, no clear pattern emerged, so the usefulness of iterating these estimation procedures might be questioned. In fact, in some cases, iterations resulted in a loss in efficiency. Situations favourable to the iterative estimators over the SURR estimator were found to be a low correlation between regressors across the equations, and a high correlation between disturbances across the equations.

The empirical evidence that iterations may not always be worthwhile, in the sense of improving estimator efficiency, motivated Srivastava, Upadhyaya and Dwivedi (1976) to investigate this point analytically. Their results for a two equation model with subset regressors are now presented. If we take X_2 to be a submatrix of X_1, the FGLS estimator of the coefficient vector β_2 in the second equation is the same as the OLS estimator, so that iterations will have no influence on the efficiency of the estimation of β_2. So, confining attention to the estimation of the coefficient vector β_1 in the first equation, the IFGLS estimator at the k'th round is given by

$$b_{(1)F}^{(k)} = (X_1'X_1)^{-1}X_1'y_1 - d^{(k)}(X_1'X_1)^{-1}X_1'\bar{P}_{x_2}y_2 \qquad (5.28)$$

where

$$d^{(k)} = \frac{s_{12}^{(k)}}{s_{22}^{(k)}} = \frac{\left(y_1 - X_1 b_{(1)F}^{(k-1)}\right)'\left(y_2 - X_2 b_{(2)F}^{(k-1)}\right)}{\left(y_2 - X_2 b_{(2)F}^{(k-1)}\right)'\left(y_2 - X_2 b_{(2)F}^{(k-1)}\right)}$$

$$= \frac{\left(y_1 - X_1 b_{(1)F}^{(k-1)}\right)'\left(y_2 - X_2 b_{(2)0}\right)}{(y_2 - X_2 b_{(2)0})'(y_2 - X_2 b_{(2)0})}$$

$$= \frac{\left(y_1 - X_1 b_{(1)F}^{(k-1)}\right)'\bar{P}_{x_2} y_2}{y_2'\bar{P}_{x_2} y_2} \qquad (k = 2,\ 3,\ \ldots) . \tag{5.29}$$

Substituting

$$b_{(1)F}^{(k-1)} = (X_1'X_1)^{-1}X_1'y_1 - d^{(k-1)}(X_1'X_1)^{-1}X_1'\bar{P}_{x_2} y_2$$

in (5.29), we get

$$d^{(k)} = \frac{(y_1'\bar{P}_{x_1} + d^{(k-1)}y_2'\bar{P}_{x_2} P_{x_1})\bar{P}_{x_2} y_2}{y_2'\bar{P}_{x_2} y_2}$$

$$= \frac{y_1'\bar{P}_{x_1} y_2 + d^{(k-1)}y_2'(\bar{P}_{x_2} - \bar{P}_{x_1})y_2}{y_2'\bar{P}_{x_2} y_2}$$

$$= d^{(1)} + pd^{(k-1)} \tag{5.30}$$

where

$$p = \frac{y_2'(\bar{P}_{x_2} - \bar{P}_{x_1})y_2}{y_2'\bar{P}_{x_2} y_2} = \frac{u_2'(\bar{P}_{x_2} - \bar{P}_{x_1})u_2}{u_2'\bar{P}_{x_2} u_2} . \tag{5.31}$$

Thus we find

$$\begin{aligned} d^{(k)} &= d^{(1)} + p(d^{(1)} + pd^{(k-2)}) \\ &= d^{(1)} + pd^{(1)} + p^2(d^{(1)} + pd^{(k-3)}) \\ &\vdots \\ &= d^{(1)} \sum_{j=0}^{k} p^j \end{aligned} \tag{5.32}$$

so that from (5.28) we have

$$v^{(k)}_{(1)F} - \beta_1) = (X_1'X_1)^{-1}X_1'u_1 - d^{(1)}\Big(\sum_{j=0}^{k} p^j\Big)(X_1'X_1)^{-1}X_1'\bar{P}_{x_2}u_2 \,.$$

Under the assumption of normality of the disturbances, it is easy to see that $b^{(k)}_{(1)F}$ is an unbiased estimator whether $d^{(1)}$ is based on restricted residuals or on unrestricted residuals. Further, the variance covariance matrix of $b^{(k)}_{(1)F}$ is given by

$$\begin{aligned} V^{(k)}_{(1)F} &= E(b^{(k)}_{(1)F} - \beta_1)(b^{(k)}_{(1)F} - \beta_1)' \\ &= (X_1'X_1)^{-1}E\Big[X_1'u_1u_1'X_1 - d^{(1)}\Big(\sum_{j=0}^{k} p^j\Big)(X_1'u_1u_2'\bar{P}_{x_2}X_1 + X_1'\bar{P}_{x_2}u_2u_1'X_1) \\ &\quad + \Big\{d^{(1)}\Big(\sum_{j=0}^{k} p^j\Big)\Big\}^2 X_1'\bar{P}_{x_2}u_2u_2'\bar{P}_{x_2}X_1\Big](X_1'X_1)^{-1} \\ &= \sigma_{11}(X_1'X_1)^{-1} - (X_1'X_1)^{-1}X_1'E\Big[d^{(1)}\Big(\sum_{j=0}^{k} p^j\Big)(u_1u_2'\bar{P}_{x_2} + \bar{P}_{x_2}u_2u_1')\Big]X_1(X_1'X_1)^{-1} \\ &\quad + (X_1'X_1)^{-1}X_1'\bar{P}_{x_2}E\Big[\Big\{d^{(1)}\Big(\sum_{j=0}^{k} p^j\Big)\Big\}^2 u_2u_2'\Big]\bar{P}_{x_2}X_1(X_1'X_1)^{-1} \,. \end{aligned} \tag{5.33}$$

In order to evaluate the expectations in (5.33) we note that for the two equation SURE model we have, from (5.5):

$$u_1 = \delta_1 u_2 + v_1 \tag{5.34}$$

where $\delta_1 = \rho_{12}(\sigma_{11}/\sigma_{22})^{\frac{1}{2}}$ is now a scalar, and u_2 and v_1 are stochastically independent with $u_2 \sim N(0, \sigma_{22}I_T)$ and $v_1 \sim N(0, \sigma_{11}(1 - \rho_{12}^2)I_T)$. We now consider particular IFGLS estimators, first taking the case where restricted residuals are used to construct S and hence $d^{(1)}$, so that

$$
\begin{aligned}
d^{(1)} &= \frac{y_1'\bar{P}_{x_1}\bar{P}_{x_2}y_2}{y_2'\bar{P}_{x_2}y_2} \\
&= \frac{u_1'\bar{P}_{x_1}u_2}{u_2'\bar{P}_{x_2}u_2} \\
&= \delta_1\frac{u_2'\bar{P}_{x_1}u_2}{u_2'\bar{P}_{x_2}u_2} + \frac{v_1'\bar{P}_{x_1}u_2}{u_2'\bar{P}_{x_2}u_2} \\
&= \delta_1(1 - p) + \frac{v_1'\bar{P}_{x_1}u_2}{u_2'\bar{P}_{x_2}u_2} .
\end{aligned}
\tag{5.35}
$$

As X_2 is a sub-matrix of X_1, let us assume without loss of generality that $X_1 = (X_2 \quad X_0)$. Thus, we have

$$
X_1'\bar{P}_{x_2}u_2u_2'\bar{P}_{x_2}X_1 = \begin{bmatrix} 0 & 0 \\ 0 & X_0'\bar{P}_{x_2}u_2u_2'\bar{P}_{x_2}X_0 \end{bmatrix} . \tag{5.36}
$$

Further, let Γ be a nonsingular matrix such that

$$
\Gamma'X_0'\bar{P}_{x_2}X_0\Gamma = I_{(K_1-K_2)} . \tag{5.37}
$$

Now if we define

$$
\chi^2 = (u_2'\bar{P}_{x_1}u_2)/\sigma_{22} \tag{5.38}
$$

$$
w = (\Gamma'X_0'\bar{P}_{x_2}u_2)/(\sigma_{22})^{1/2} ,
$$

we have

$$w'w = (u_2'\bar{P}_{x_2}X_0\Gamma\Gamma'X_0'\bar{P}_{x_2}u_2)/\sigma_{22}$$

$$= (u_2'\bar{P}_{x_2}X_0(X_0'\bar{P}_{x_2}X_0)^{-1}X_0'\bar{P}_{x_2}u_2)/\sigma_{22}$$

$$= (u_2'(\bar{P}_{x_2} - \bar{P}_{x_1})u_2)/\sigma_{22}$$

so that from (5.31) we observe that

$$p = (w'w)/(\chi^2 + w'w) \ . \tag{5.39}$$

Using (5.34) and (5.35) to evaluate the expectations in (5.33), we find that

$$X_1'E\left[d^{(1)}\left(\sum_{j=0}^{k} p^j\right)u_1u_2'\right]\bar{P}_{x_2}X_1$$

$$= \delta_1^2X_1'E\left[(1-p)\left(\sum_{j=0}^{k} p^j\right)u_2u_2'\right]\bar{P}_{x_2}X_1$$

$$+ \delta_1X_1'E\left[(1-p)\left(\sum_{j=0}^{k} p^j\right)E(v_1)u_2'\right]\bar{P}_{x_2}X_1$$

$$+ \delta_1X_1'E\left[\frac{E(v_1')\bar{P}_{x_1}u_2}{u_2'\bar{P}_{x_2}u_2}\left(\sum_{j=0}^{k} p^j\right)u_2u_2'\right]\bar{P}_{x_2}X_1$$

$$+ X_1'E\left[\left(\sum_{j=0}^{k} p^j\right)\frac{1}{u_2'\bar{P}_{x_2}u_2}E(v_1v_1')\bar{P}_{x_1}u_2u_2'\right]\bar{P}_{x_2}X_1$$

$$= \delta_1^2X_1'E\left[(1-p^{k+1})u_2u_2'\right]\bar{P}_{x_2}X_1$$

$$= \delta_1^2X_1'E\left[(1-p^{k+1})(P_{x_2} + \bar{P}_{x_2})u_2u_2'\right]\bar{P}_{x_2}X_1$$

$$= \delta_1^2X_1'\bar{P}_{x_2}E\left[(1-p^{k+1})u_2u_2'\right]\bar{P}_{x_2}X_1 \ , \tag{5.40}$$

because $P_{x_2}u_2$ is stochastically independent of the vectors $\bar{P}_{x_2}u_2$ and $\bar{P}_{x_1}u_2$. (Notice that $\bar{P}_{x_1}u_2$ is involved in the expression for p.)

Now using (5.38) and (5.39) together with the results (B.2) of the Appendix, we get

$$X_0'\bar{P}_{x_2} E\left[(1 - p^{k+1})u_2u_2'\right]\bar{P}_{x_2}X_0$$

$$= \sigma_{22}\Gamma'^{-1}E\left[ww' - \left(\frac{w'w}{\chi^2 + w'w}\right)^{k+1} ww'\right]\Gamma^{-1}$$

$$= \sigma_{22}\left[1 - \frac{\Gamma\left(\frac{T-K_2}{2}+1\right)\Gamma\left(\frac{K_1-K_2}{2}+k+2\right)}{\Gamma\left(\frac{T-K_2}{2}+k+2\right)\Gamma\left(\frac{K_1-K_2}{2}+1\right)}\right](\Gamma\Gamma')^{-1}$$

$$= \sigma_{22}\left[1 - \frac{\Gamma\left(\frac{T-K_2}{2}+1\right)\Gamma\left(\frac{K_1-K_2}{2}+k+2\right)}{\Gamma\left(\frac{T-K_2}{2}+k+2\right)\Gamma\left(\frac{K_1-K_2}{2}+1\right)}\right]X_0'\bar{P}_{x_2}X_0 \quad . \tag{5.41}$$

Using (5.36) and (5.41) in (5.40), we obtain

$$X_1'E\left[d^{(1)}\left(\sum_{j=0}^{k} p^j\right)u_1u_2'\right]\bar{P}_{x_2}X_1$$

$$= \rho_{12}^2\sigma_{11}\left[1 - \frac{\Gamma\left(\frac{T-K_2}{2}+1\right)\Gamma\left(\frac{K_1-K_2}{2}+k+2\right)}{\Gamma\left(\frac{T-K_2}{2}+k+2\right)\Gamma\left(\frac{K_1-K_2}{2}+1\right)}\right]X_1'\bar{P}_{x_2}X_1 \quad . \tag{5.42}$$

Similarly, it can be shown that

$$X_1'\bar{P}_{x_2} E\left[\left\{d^{(1)}\left(\sum_{j=0}^{k} p^j\right)\right\}^2 u_2u_2'\right]\bar{P}_{x_2}X_1$$

$$= \sigma_{11}\left[\rho_{12}^2 + \rho_{12}^2 \frac{\Gamma\left(\frac{T-K_2}{2}+1\right)}{\Gamma\left(\frac{K_1-K_2}{2}+1\right)}\left[\frac{\Gamma\left(\frac{K_1-K_2}{2}+2k+3\right)}{\Gamma\left(\frac{T-K_2}{2}+2k+3\right)} - \frac{2\Gamma\left(\frac{K_1-K_2}{2}+k+2\right)}{\Gamma\left(\frac{T-K_2}{2}+k+2\right)}\right]\right.$$

$$\left. + \frac{1}{2}(1-\rho_{12}^2)\frac{\Gamma\left(\frac{T-K_2}{2}\right)}{\Gamma\left(\frac{K_1-K_2}{2}+1\right)} \sum_{j=0}^{k}\left[\frac{\Gamma\left(\frac{K_1-K_2}{2}+j+1\right)}{\Gamma\left(\frac{T-K_2}{2}+j+1\right)} - \frac{\Gamma\left(\frac{K_1-K_2}{2}+j+k+2\right)}{\Gamma\left(\frac{T-K_2}{2}+j+k+2\right)}\right]\right]$$

$$\cdot\ X_1'\bar{P}_{x_2}X_1 \tag{5.43}$$

where the following result is used:

$$E\left[\left\{d^{(1)}\left(\sum_{j=0}^{k} p^j\right)\right\}^2 \mid u_2\right]$$

$$= E\left[\delta_1^2\left((1-p)\sum_{j=0}^{k} p^j\right)^2 + 2\delta_1(1-p)\left(\sum_{j=0}^{k} p^j\right)^2 \frac{E(v_1')\bar{P}_{x_1}u_2}{u_2'\bar{P}_{x_2}u_2}\right.$$

$$\left. + \left(\sum_{j=0}^{k} p^j\right)^2 \frac{u_2'\bar{P}_{x_1}E(v_1v_1')\bar{P}_{x_1}u_2}{(u_2'\bar{P}_{x_2}u_2)^2} \mid u_2\right]$$

$$= E\left[\delta_1^2(1-p^{k+1})^2 + \sigma_{11}(1-\rho_{12}^2)\left(\sum_{j=0}^{k} p^j\right)^2 \frac{u_2'\bar{P}_{x_1}u_2}{(u_2'\bar{P}_{x_2}u_2)^2} \mid u_2\right]$$

$$= E\left[\delta_1^2(1-p^{k+1})^2 + \sigma_{11}(1-\rho_{12}^2)\left(\sum_{j=0}^{k} p^j\right)^2 (1-p)\frac{1}{u_2'\bar{P}_{x_2}u_2} \mid u_2\right]$$

$$= E\left[\delta_1^2(1+p^{2k+2}-2p^{k+1}) + \sigma_{11}(1-\rho_{12}^2)\frac{1}{u_2'\bar{P}_{x_2}u_2}\sum_{j=0}^{k}(p^j - p^{j+k+1}) \mid u_2\right].$$

Substituting (5.42) and (5.43) in (5.33) and recalling that this iterative round started with restricted residuals, the variance covariance matrix of $\hat{\beta}^{(k)}_{(1)SR}$ is

$$V^{(k)}_{(1)SR} = \sigma_{11}(X_1'X_1)^{-1} + \sigma_{11}\alpha_k(X_1'X_1)^{-1}X_1'\bar{P}_{x_2}X_1(X_1'X_1)^{-1}$$
$$(k = 2, 3, \ldots) \qquad (5.44)$$

where

$$\alpha_k = \rho_{12}^2\left[\frac{\Gamma\left(\frac{T-K_2}{2}+1\right)\Gamma\left(\frac{K_1-K_2}{2}+2k+3\right)}{\Gamma\left(\frac{T-K_2}{2}+2k+3\right)\Gamma\left(\frac{K_1-K_2}{2}+1\right)} - 1\right]$$
$$+ \frac{(1-\rho_{12}^2)}{2}\cdot\frac{\Gamma\left(\frac{T-K_2}{2}\right)}{\Gamma\left(\frac{K_1-K_2}{2}+1\right)}\sum_{j=0}^{k}\left[\frac{\Gamma\left(\frac{K_1-K_2}{2}+j+1\right)}{\Gamma\left(\frac{T-K_2}{2}+j+1\right)} - \frac{\Gamma\left(\frac{K_1-K_2}{2}+j+k+2\right)}{\Gamma\left(\frac{T-K_2}{2}+j+k+2\right)}\right].$$
$$(5.45)$$

Now, suppose that unrestricted residuals are used to construct S, and hence $d^{(1)}$, and consider the evaluation of the expectations in (5.33). In this case,

$$d^{(1)} = \frac{y_1'\bar{P}_{x_1}y_2}{y_2'\bar{P}_{x_1}y_2}$$
$$= \frac{u_1'\bar{P}_{x_1}u_2}{u_2'\bar{P}_{x_1}u_2}.$$

It is easy to verify that $d^{(k)}$ $(k = 2, 3, \ldots)$, when we start with unrestricted residuals, is the same as $d^{(k)}$ when we start with restricted residuals. So, if we start with the SUUR estimator, the iterative estimator $\tilde{\beta}^{(k)}_{(1)SU}$ for $k = 2, 3, \ldots$ is unbiased and its variance covariance matrix has the form (5.44).

Writing

$$\mu_k = (\alpha_{k-1} - \alpha_k)$$

we have, from (5.44),

$$V_{(1)SR}^{(k-1)} - V_{(1)SR}^{(k)} = V_{(1)SU}^{(k-1)} - V_{(1)SU}^{(k)} = \sigma_{11}\mu_k(X_1'X_1)^{-1}X_1'\bar{P}_{X_2}X_1(X_1'X_1)^{-1}, \quad (5.46)$$

which gives the change in the variance covariance matrix arising at the k'th round (k = 2, 3, ...). Thus the k'th round will be worthwhile in the sense of a gain in efficiency if μ_k is positive. However, it is difficult to discern any clear condition for the positivity of μ_k as such.

With this in mind, Srivastava, Upadhyaya and Dwivedi (1976) calculated $\mu_2, \mu_3, \ldots, \mu_8$ for a few selected values of $(K_1 - K_2)$, $(T - K_1)$ and ρ_{12}, viz., $(K_1 - K_2) = 1, 2, 3, 5$; $(T - K_1) = 3, 5(5)20, 30$ and $\rho_{12} = 0(0.1)0.3, 0.5, 0.7(0.1)1$. A general finding of their numerical results was that iterations do not always bring about an improvement in the efficiency of these estimators. For instance, the second iteration reduced the efficiency in a little less than one-third of the total cases. Similarly, at the fourth iteration there was a loss in efficiency for more than half of the cases. This figure became roughly three quarters at the eighth iteration. Moreover, if a particular iteration resulted in a loss of efficiency, the subsequent iterations continued to result in further losses. Thus, as soon as an iteration results in an efficiency loss, one should stop and no further iterations should be carried out. In many cases, higher order iterations were found to reduce efficiency, though substantial gains may have been achieved in the first few iterations. This suggests that there exists an optimal maximum number of iterations. It was found that iterations bring an efficiency improvement for a high correlation between the disturbances across the equations, a small excess of observations over the number of regressors in the first equation, and a large number of regressors omitted in the second equation. These results relate to a two equation model with subset regressors and may not hold in more general situations. However, they do provide important clues for further investigations of the role of iterations in improving the efficiency of estimators for a general SURE model.

EXERCISES

5.1 For a two equation SURE model, develop a proof of the convergence of the IOLS estimator. [Hint: See Telser (1964).]

5.2 Assuming normality of the disturbances in the SURE model

$$y_1 = \beta_1 x_1 + u_1$$

$$y_2 = \beta_2 x_2 + u_2$$

with $x_1'x_1 = x_2'x_2 = T$, $E(u_1u_1') = E(u_2u_2') = I_T$ and $E(u_1u_2') = \sigma_{12}I_T$, obtain the ML estimators of β_1 and β_2 when σ_{12} is (i) known; (ii) unknown.

5.3 Derive the matrix of second-order partial derivatives of the log likelihood function $\mathscr{L}$, defined by (5.13), with respect to the elements of the vector β and the distinct elements of the matrix Σ. [Hint: See Magnus (1978).]

5.4 For the SURE model with normally distributed disturbances, examine the conditions under which the Maximum Likelihood estimator of β is unbiased. [Hint: See Don and Magnus (1980) and Sargan (1978).]

5.5 Obtain the estimator of β in the general SURE model by the application of the least generalized residual variance principle. [Hint: See Goldberger (1964; Chap. 4, Sec. 11).]

5.6 Examine the efficiency, as measured by the variance covariance matrix to order $O(T^{-2})$, of the successive rounds of the ISUUR estimator. [Hint: See Srivastava (1970).]

5.7 Consider the following SURE model:

$$y_1 = \beta_1 x + u_1$$

$$y_2 = \beta_2 e_T + u_2$$

in which $E(u_1u_1') = E(u_2u_2') = I_T$, $E(u_1u_2') = \sigma_{12}I_T$ and $x'e_T = 0$ where σ_{12} is an unknown scalar and e_T is a $(T \times 1)$ vector with all elements unity. Assuming that the disturbances are normally distributed, examine the efficiency of the ISURR estimator of β_1 at successive iterations.

5.8 Consider the factors which affect the rate of convergence, in the iterative sense, of the Maximum Likelihood estimator of β in the SURE model with normally distributed disturbances. [Hint: See Oberhofer and Kmenta (1974).]

REFERENCES

Don, F. J. H., and J. R. Magnus (1980). "On the unbiasedness of iterated GLS estimators," Communications in Statistics A9, 519-527.

Goldberger, A. S. (1964). Econometric Theory (Wiley, New York).

Kmenta, J., and R. F. Gilbert (1968). "Small sample properties of alternative estimators of seemingly unrelated regressions," Journal of the American Statistical Association 63, 1180-1200.

Magnus, J. R. (1978). "Maximum likelihood estimation of the GLS model with unknown parameters in the disturbance covariance matrix," Journal of Econometrics 7, 281-312.

Oberhofer, W., and J. Kmenta (1974). "A general procedure for obtaining maximum likelihood estimates in generalized regression models," Econometrica 42, 579-590.

Rao, C. R. (1973). Linear Statistical Inference and its Applications (Wiley, New York).

Ruble, W. (1968). "Improving the computation of simultaneous stochastic linear equations estimates," Agricultural Economics Report No. 116, Michigan State University, East Lansing.

Sargan, J. D. (1978). "On the existence of the moments of 3SLS estimators," Econometrica 46, 1329-1350.

Srivastava, V. K. (1970). "The efficiency of estimating seemingly unrelated regression equations," Annals of the Institute of Statistical Mathematics 22, 483-493.

Srivastava, V. K., S. Upadhyaya, and T. D. Dwivedi (1976). "Efficiency of an iterative procedure in a two equation seemingly unrelated regression model," mimeo, Concordia University, Montreal.

Telser, L. G. (1964). "Iterative estimation of a set of linear regression equations," Journal of the American Statistical Association 59, 845-862.

Zellner, A. (1962). "An efficient method of estimating seemingly unrelated regression equations and tests for aggregation bias," Journal of the American Statistical Association 57, 348-368.

6

Shrinkage Estimators

6.1 INTRODUCTION

In the last chapter we discussed the use of iterative procedures, with the objective of improving the efficiency of estimators which are unbiased under some mild restrictions on the distribution of the disturbances. In this chapter we relax our interest in unbiased estimation, and broaden the discussion to consider estimators which may be biased but which may have smaller sampling variability than their unbiased counterparts. Thus, in terms of either the mean squared error matrix criterion, or in terms of risk under quadratic loss, there may be a gain in adopting a biased estimator, at least in certain regions of the parameter space.

One important class of biased estimators which may be considered in this way is that of "shrinkage" estimators. A shrinkage estimator is essentially one which incorporates a factor which has the effect of "pulling" or "shrinking" the estimation vector toward some chosen origin. In this chapter we consider three types of shrinkage techniques, and so three classes of shrinkage estimators. The first type stems from the well known inadmissibility of the least squares estimator for the classical linear regression model (with three or more parameters), when a squared error loss function is adopted (see James and Stein, 1961). The second type of shrinkage is that associated with "ridge regression," which was developed originally to overcome the problem of multicollinearity in the linear regression model, but which later formed a basis for risk reduction in the estimation of the parameters of this model (see Hoerl and Kennard, 1970, 1981). The third and final type of shrinkage that we consider involves combining two or more estimators on the basis of their sampling performance. In this way, it is hoped to improve upon the sampling properties of either of the individual component estimators when taken in isolation (see Srivastava and Srivastava, 1984). The reasonableness of these shrinkage techniques is discussed in the last section of the chapter.

6.2 STEIN-RULE ESTIMATORS

Moving away from the class of linear and unbiased estimators for the coefficient vector in a classical linear regression model, James and Stein (1961) presented an estimator which shrinks the least squares estimator toward the origin, and at the same time performs better than the latter in terms of risk under quadratic loss. This led to the development of several families of estimators which are superior (in this sense) to the conventional unbiased estimators (see Judge and Bock, 1978, and Vinod and Ullah, 1981). Such families are popularly referred to as Stein-rule estimators.

6.2.1 Two Families

Let us consider the SURE model:

$$\begin{aligned} y &= X\beta + u \\ E(u) &= 0, \quad E(uu') = \Psi = (\Sigma \otimes I_T) . \end{aligned} \tag{6.1}$$

Assuming for the moment that Σ is known, and discarding the criterion of unbiasedness, let us consider $(1 - \theta)b_G$ as an estimator of β, where θ is any arbitrary scalar characterizing the estimator. If θ lies between 0 and 1, then this defines a shrinkage estimator. When $\theta = 0$, we have the best linear unbiased estimator b_G. At the other extreme, when $\theta = 1$, we have a null vector as the estimator of β. Thus, for $0 < \theta < 1$, the estimator $(1 - \theta)b_G$ tries to shrink the GLS estimator toward a null vector.

Taking θ to be a scalar constant lying between 0 and 1, the weighted mean squared error (WMSE) of the shrinkage estimator is

$$\begin{aligned} \text{WMSE} &= E[(1 - \theta)b_G - \beta]'\Theta[(1 - \theta)b_G - \beta] \\ &= E[(1 - \theta)(b_G - \beta) - \theta\beta]'\Theta[(1 - \theta)(b_G - \beta) - \theta\beta] \\ &= (1 - \theta)^2 \operatorname{tr}(X'\Psi^{-1}X)^{-1}\Theta + \theta^2\beta'\Theta\beta \end{aligned} \tag{6.2}$$

where the $(K^* \times K^*)$ loss matrix, Θ, is positive definite and symmetric and does not depend on θ, and $K^* = \Sigma_{i=1}^{M} K_i$.

Differentiating the expression (6.2) with respect to θ, we get

$$\frac{\partial WMSE}{\partial \theta} = 2[-(1 - \theta)\,\mathrm{tr}\,(X'\Psi^{-1}X)^{-1}\Theta + \theta\beta'\Theta\beta]$$
$$\frac{\partial^2 WMSE}{\partial \theta^2} = 2[\mathrm{tr}\,(X'\Psi^{-1}X)^{-1}\Theta + \beta'\Theta\beta]\,. \tag{6.3}$$

Setting the first of these expressions equal to 0 and solving for θ, we find the value of θ that minimizes the WMSE:

$$\theta_0 = \frac{\mathrm{tr}\,(X'\Psi^{-1}X)^{-1}\Theta}{\mathrm{tr}\,(X'\Psi^{-1}X)^{-1}\Theta + \beta'\Theta\beta}\,.$$

Thus, the minimum weighted mean squared error estimator of β is given by

$$\beta^* = (1 - \theta_0)b_G$$
$$= \left[1 - \frac{\mathrm{tr}\,(X'\Psi^{-1}X)^{-1}\Theta}{\mathrm{tr}\,(X'\Psi^{-1}X)^{-1}\Theta + \beta'\Theta\beta}\right]b_G\,. \tag{6.4}$$

Notice that β^* is not an estimator of β in the true sense because it is a function of the unknown β itself. (Recall that Σ is assumed to be known.) Therefore, as suggested by Zellner and Vandaele (1975), we might replace the expression in the denominator of θ_0 by its unbiased estimator in order to get an operational version of β^*. This yields the feasible estimator

$$\left[1 - \frac{\mathrm{tr}\,(X'\Psi^{-1}X)^{-1}\Theta}{b_G'\Theta b_G}\right]b_G\,, \tag{6.5}$$

because

$$E[b_G'\Theta b_G] = \mathrm{tr}\,(X'\Psi^{-1}X)^{-1}\Theta + \beta'\Theta\beta\,.$$

Another operational version of β^* can be derived by replacing β in θ_0 by b_G:

$$\left[1 - \frac{\mathrm{tr}\,(X'\Psi^{-1}X)^{-1}\Theta}{\mathrm{tr}\,(X'\Psi^{-1}X)^{-1}\Theta + b_G'\Theta b_G}\right]b_G\,. \tag{6.6}$$

It is easy to verify that both of the estimators (6.5) and (6.6) are asymptotically equivalent to β^* in the sense that all three have identical asymptotic properties. However, this is not the case for samples of finite size. For instance, according to the weighted mean squared error criterion, the operational estimators (6.5) and (6.6) are not uniformly superior to b_G in finite samples, while on this basis β^* is uniformly better than b_G irrespective of the sample size. (This last result is, of course, somewhat pathological.)

From (6.5), we can define a simple family of estimators of β:

$$\left[1 - g\left(\frac{\operatorname{tr}(X'\Psi^{-1}X)^{-1}\Theta}{b_G'\Theta b_G}\right)\right]b_G \tag{6.7}$$

where g is any arbitrary scalar characterizing the estimator.

Now, dropping the assumption that Σ is known, we may replace $\Psi = (\Sigma \otimes I_T)$ by its consistent estimator $(S \otimes I_T)$ so that we have the following family of Stein-rule feasible generalized least squares (SFGLS) estimators of β:

$$b_{SF} = \left[1 - g\,\frac{\operatorname{tr}[X'(S^{-1} \otimes I_T)X]^{-1}\Theta}{b_F'\Theta b_F}\right]b_F \tag{6.8}$$

where g is any non-negative scalar (see Vinod and Ullah, 1981; pp. 246-248).

Setting $S = \tilde{S}$ and $S = \hat{S}$, we obtain

$$\tilde{\beta}_{SSU} = \left[1 - g\,\frac{\operatorname{tr}[X'(\tilde{S}^{-1} \otimes I_T)X]^{-1}\Theta}{\tilde{\beta}_{SU}'\Theta\tilde{\beta}_{SU}}\right]\tilde{\beta}_{SU} \tag{6.9}$$

$$\hat{\beta}_{SSR} = \left[1 - g\,\frac{\operatorname{tr}[X'(\hat{S}^{-1} \otimes I_T)X]^{-1}\Theta}{\hat{\beta}_{SR}'\Theta\hat{\beta}_{SR}}\right]\hat{\beta}_{SR} \tag{6.10}$$

which will be referred to as the SSUUR and SSURR estimators of β respectively.

Similarly, generalizing (6.6), we have another family of SFGLS estimators for β:

$$b_{SF}^* = \left[1 - g\,\frac{\operatorname{tr}[X'(S^{-1} \otimes I_T)X]^{-1}\Theta}{\operatorname{tr}[X'(S^{-1} \otimes I_T)X]^{-1}\Theta + b_F'\Theta b_F}\right]b_F \tag{6.11}$$

where g is the characterizing scalar.

In particular, we have

$$\tilde{\beta}^*_{SSU} = \left[1 - g \frac{\operatorname{tr}[X'(\tilde{S}^{-1} \otimes I_T)X]^{-1}\Theta}{\operatorname{tr}[X'(\tilde{S}^{-1} \otimes I_T)X]^{-1}\Theta + \tilde{\beta}'_{SU}\Theta\tilde{\beta}_{SU}}\right]\tilde{\beta}_{SU} \tag{6.12}$$

$$\hat{\beta}^*_{SSR} = \left[1 - g \frac{\operatorname{tr}[X'(\hat{S}^{-1} \otimes I_T)X]^{-1}\Theta}{\operatorname{tr}[X'(\hat{S}^{-1} \otimes I_T)X]^{-1}\Theta + \hat{\beta}'_{SR}\Theta\hat{\beta}_{SR}}\right]\hat{\beta}_{SR} \,. \tag{6.13}$$

It is difficult to derive the exact expressions for the bias vectors and mean squared error matrices of these proposed families of estimators. We therefore consider large-sample asymptotic approximations.

6.2.2 Large-Sample Asymptotic Approximations

First, without making any further assumptions about the distribution of u, we shall consider the approximate distribution of b_{SF}. Let us define

$$\begin{aligned} \Delta &= (S \otimes I_T) - (\Sigma \otimes I_T) \\ &= (S \otimes I_T) - \Psi \end{aligned} \tag{6.14}$$

where Δ is an $(MT \times MT)$ matrix with elements of order $O_p(T^{-\frac{1}{2}})$.

Writing $\Omega = [X'(\Sigma^{-1} \otimes I_T)X]^{-1}$ and following the procedure outlined in Sections 3.2 and 3.3 (see also Srivastava, 1973), it is easy to see that

$$\begin{aligned} X'(S^{-1} \otimes I_T)X &= X'(\Psi + \Delta)^{-1}X \\ &= X'\Psi^{-1}X - X'\Psi^{-1}\Delta\Psi^{-1}X + \cdots \end{aligned}$$

from which we find

$$\begin{aligned} \operatorname{tr}[X'(S^{-1} \otimes I_T)X]^{-1}\Theta &= \operatorname{tr}[X'\Psi^{-1}X - X'\Psi^{-1}\Delta\Psi^{-1}X + \cdots]^{-1}\Theta \\ &= \operatorname{tr}(X'\Psi^{-1}X)^{-1}\Theta \\ &\quad + \operatorname{tr}(X'\Psi^{-1}X)^{-1}X'\Psi^{-1}\Delta\Psi^{-1}X(X'\Psi^{-1}X)^{-1}\Theta + \cdots \\ &= \operatorname{tr}\Omega\Theta + \operatorname{tr}\Omega X'\Psi^{-1}\Delta\Psi^{-1}X\Omega\Theta + \cdots \,. \end{aligned} \tag{6.15}$$

Recall, from (3.7), that the estimation error of the FGLS estimator is expressible, to order $O_p(T^{-3/2})$, as

$$(b_F - \beta) = \xi_{-1/2} + \xi_{-1} + \xi_{-3/2} \tag{6.16}$$

where

$$\xi_{-1/2} = \Omega X'\Psi^{-1}u$$

$$\xi_{-1} = -\Omega X'\Psi^{-1}\Delta Qu$$

$$\xi_{-3/2} = \Omega X'\Psi^{-1}\Delta Q\Delta Qu$$

$$Q = [\Psi^{-1} - \Psi^{-1}X\Omega X'\Psi^{-1}] .$$

Using (6.16), we observe that

$$\begin{aligned}(b_F'\Theta b_F)^{-1} &= [\beta'\Theta\beta + 2\beta'\Theta(b_F - \beta) + (b_F - \beta)'\Theta(b_F - \beta)]^{-1}\\
&= \left(\frac{1}{\beta'\Theta\beta}\right)\left[1 + 2\frac{\beta'\Theta(b_F - \beta)}{\beta'\Theta\beta} + \frac{(b_F - \beta)'\Theta(b_F - \beta)}{\beta'\Theta\beta}\right]^{-1}\\
&= \left(\frac{1}{\beta'\Theta\beta}\right)\left[1 + \frac{2\beta'\Theta(\xi_{-1/2} + \xi_{-1} + \xi_{-3/2})}{\beta'\Theta\beta}\right.\\
&\quad \left. + \frac{(\xi_{-1/2} + \xi_{-1} + \xi_{-3/2})'\Theta(\xi_{-1/2} + \xi_{-1} + \xi_{-3/2})}{\beta'\Theta\beta}\right]^{-1}\\
&= \left[\frac{1}{\beta'\Theta\beta} - 2\frac{\beta'\Theta\xi_{-1/2}}{(\beta'\Theta\beta)^2}\right]\end{aligned}$$

up to order $O_p(T^{-1/2})$.

Now, employing (6.15), (6.16) and (6.17), we get from (6.8) the following expression for the estimation error, to order $O_p(T^{-3/2})$, of b_{SF}:

$$(b_{SF} - \beta) = \xi_{-1/2} + e_{-1} + e_{-3/2} \tag{6.18}$$

where

$$e_{-1} = \xi_{-1} - g\left(\frac{\text{tr}\,\Omega\Theta}{\beta'\Theta\beta}\right)\beta$$

$$e_{-3/2} = \xi_{-3/2} - g\left(\frac{\mathrm{tr}\,\Omega\Theta}{\beta'\Theta\beta}\,D\right)\xi_{-1/2} + g\left(\frac{\mathrm{tr}\,\Omega X'\Psi^{-1}\Delta\Psi^{-1}X\Omega\Theta}{\beta'\Theta\beta}\right)\beta$$

with $D = \left[I - \frac{2}{\beta'\Theta\beta}\beta\beta'\Theta\right]$, and I of order $(K^* \times K^*)$ (Srivastava, 1973; p. 347).

If we consider a first-order asymptotic approximation (i.e., approximate $(b_{SF} - \beta)$ by $\xi_{-1/2}$ alone) we have:

$$E(\xi_{-1/2}) = 0$$
$$E(\xi_{-1/2}\xi'_{-1/2}) = \Omega \tag{6.19}$$

implying that $\sqrt{T}(b_{SF} - \beta)$ has the same asymptotic distribution as $\sqrt{T}(b_F - \beta)$ which, in turn, is identical to the asymptotic distribution of $\sqrt{T}(b_G - \beta)$.

Next, following Srivastava (1973; p. 248), let us consider second-order asymptotic approximations to the distributions of particular SFGLS estimators, assuming that the disturbances are normally distributed. For the SUUR estimator defined by (6.9), it is easy to see that the bias vector to order $O(T^{-1})$ is

$$\begin{aligned} E(\tilde{\beta}_{SSU} - \beta) &= E(\xi_{-1/2}) + E(e_{-1}) \\ &= E(\xi_{-1/2}) + E(\xi_{-1}) - g\left(\frac{\mathrm{tr}\,\Omega\Theta}{\beta'\Theta\beta}\right) \\ &= -g\left(\frac{\mathrm{tr}\,\Omega\Theta}{\beta'\Theta\beta}\right) . \end{aligned} \tag{6.20}$$

Similarly, its mean squared error matrix to order $O(T^{-2})$ is given by

$$\begin{aligned} &\mathrm{MSE}(\tilde{\beta}_{SSU}) \\ &= E(\tilde{\beta}_{SSU} - \beta)(\tilde{\beta}_{SSU} - \beta)' \\ &= V(\tilde{\beta}_{SU}) - g\left(\frac{\mathrm{tr}\,\Omega\Theta}{\beta'\Theta\beta}\right)E[\beta\xi'_{-1/2} + \xi_{-1/2}\beta'] \\ &\quad - g\frac{\mathrm{tr}\,\Omega\Theta}{\beta'\Theta\beta}E[\beta\xi'_{-1} + \xi_{-1}\beta' + D\xi_{-1/2}\xi'_{-1/2} + \xi_{-1/2}\xi'_{-1/2}D'] \\ &\quad + g\frac{1}{\beta'\Theta\beta}E[(\mathrm{tr}\,\Omega X'\Psi^{-1}\Delta\Psi^{-1}X\Omega\Theta)(\beta\xi'_{-1/2} + \xi_{-1/2}\beta')] + g^2\left(\frac{\mathrm{tr}\,\Omega\Theta}{\beta'\Theta\beta}\right)^2\beta\beta' , \end{aligned} \tag{6.21}$$

where $V(\tilde{\beta}_{SU})$ denotes the variance covariance matrix, to order $O(T^{-2})$, of the SUUR estimator $\tilde{\beta}_{SU}$. The expression for this matrix is, from Section 3.3,

$$V(\tilde{\beta}_{SU}) = \left(1 + \frac{M}{T}\right)\Omega - \frac{1}{T}\Omega X'[(\Sigma^{-1} \otimes W) - P]X\Omega \tag{6.22}$$

where the matrices W and P are defined as follows.

If we partition

$$Q = \begin{bmatrix} Q_{11} & \cdots & Q_{1M} \\ \vdots & & \vdots \\ Q_{M1} & \cdots & Q_{MM} \end{bmatrix}, \quad (X\Omega X'\Psi^{-1}) = \begin{bmatrix} R_{11} & \cdots & R_{1M} \\ \vdots & & \vdots \\ R_{M1} & \cdots & R_{MM} \end{bmatrix}$$

then

$$W = \sum_{i=1}^{M} R_{ii}$$

$$P = \begin{bmatrix} Q_{11} & \cdots & Q_{M1} \\ \vdots & & \vdots \\ Q_{1M} & \cdots & Q_{MM} \end{bmatrix}.$$

Next, by virtue of the assumed normality of the disturbances, it is easy to see that

$$E[(\mathrm{tr}\, \Omega X'\Psi^{-1}\Delta\Psi^{-1}X\Omega\Theta)\beta\xi'_{-1/2}] = 0\,. \tag{6.23}$$

Using (6.22) and (6.23), we obtain from (6.21) the following expression for the mean squared error matrix of the SSUUR estimator:

$$\mathrm{MSE}(\tilde{\beta}_{SSU}) = V(\tilde{\beta}_{SU}) - g\left(\frac{\mathrm{tr}\,\Omega\Theta}{\beta'\Theta\beta}\right)\left[D\Omega + \Omega D' - g\left(\frac{\mathrm{tr}\,\Omega\Theta}{\beta'\Theta\beta}\right)\beta\beta'\right], \tag{6.24}$$

to order $O(T^{-2})$.

Postmultiplying both sides of (6.24) by Θ and then taking the trace, we get the expression for the weighted mean squared error, to order $O(T^{-2})$, of this estimator:

$$\text{WMSE}(\tilde{\beta}_{SSU}) = \text{tr}\,[\text{MSE}(\tilde{\beta}_{SSU})\Theta]$$

$$= \text{tr}\,[V(\tilde{\beta}_{SU})\Theta] - g\left(\frac{\text{tr}\,\Omega\Theta}{\beta'\Theta\beta}\right)\left[(2-g)\,\text{tr}\,\Omega\Theta - 4\left(\frac{\beta'\Theta\Omega\Theta\beta}{\beta'\Theta\beta}\right)\right]. \quad (6.25)$$

It is interesting to note that we get exactly the same expressions for the bias vector, the mean squared error matrix and the weighted mean squared error, to the orders of our approximations, if we consider the SSURR estimator given by (6.10) (see Srivastava, 1973). It should also be noted that the same expressions may be found for the estimators $\tilde{\beta}^*_{SSU}$ and $\hat{\beta}^*_{SSR}$. Thus, all of the four families (6.9), (6.10), (6.12) and (6.13) share the same sampling properties according to a second-order asymptotic approximation. We conjecture that the evaluation of higher order asymptotic approximations may provide a basis for discriminating among the sampling performances of these four families of shrinkage estimators.

6.2.3 Efficiency Comparisons

We now compare these biased estimators, with weighted mean squared error to order $O(T^{-2})$ as the performance criterion, with the regular SUUR and SURR estimators. As just noted, on this basis all of the four families (6.9), (6.10), (6.12) and (6.13) are identical, and so we give them the common label $\underline{b}_{SF}$. Similarly, on this basis the SUUR and SURR estimators are identical, so we refer to them as $\underline{b}_F$. Then, from (6.25) we have

$$\text{WMSE}(\underline{b}_F) - \text{WMSE}(\underline{b}_{SF}) = E(\underline{b}_F - \beta)'\Theta(\underline{b}_F - \beta) - E(\underline{b}_{SF} - \beta)'\Theta(\underline{b}_{SF} - \beta)$$

$$= g\left(\frac{\text{tr}\,\Omega\Theta}{\beta'\Theta\beta}\right)\left[(2-g)\,\text{tr}\,\Omega\Theta - 4\left(\frac{\beta'\Theta\Omega\Theta\beta}{\beta'\Theta\beta}\right)\right], \quad (6.26)$$

which is positive when

$$0 < g < 2\left[1 - 2\left(\frac{\beta'\Theta\Omega\Theta\beta}{\beta'\Theta\beta\,\text{tr}\,\Omega\Theta}\right)\right]; \quad \left(\frac{\beta'\Theta\Omega\Theta\beta}{\beta'\Theta\beta\,\text{tr}\,\Omega\Theta}\right) < \frac{1}{2}. \quad (6.27)$$

These inequalities will definitely hold if

$$0 < g < 2(1 - 2\phi); \quad \phi < \frac{1}{2} \quad (6.28)$$

where ϕ denotes the ratio of the largest characteristic root to the sum of the characteristic roots of the matrix $(\Omega\Theta)$. This follows by noting that from Rao (1973; p. 74, result 22.1), the quantity $\left(\frac{\beta'\Theta\Omega\Theta\beta}{\beta'\Theta\beta}\right)$ cannot exceed the largest characteristic root of $(\Omega\Theta)$, and the sum of all of the characteristic roots is equal to $\text{tr}\,(\Omega\Theta)$. Thus, we see that the Stein-rule estimators $(\underline{b}_{SF})$

will have smaller weighted mean squared error, to the order $O(T^{-2})$, than the feasible estimators ($\underline{b}_F$) when the largest characteristic root of the matrix ($\Omega\Theta$) is less than half of the sum of all the characteristic roots, and g is chosen to satisfy (6.28) (see also Srivastava, 1973; pp. 345-346).

If we choose $\Theta = \Omega^{-1} = (X'\Psi X)$, then condition (6.27) reduces to the following:

$$0 < g < 2\left(1 - \frac{2}{K^*}\right) ; \quad K^* > 2 . \tag{6.29}$$

To compare the Stein-rule families with the OLS estimator, b_0, we observe that the weighted mean squared error of the latter is

$$\begin{aligned} \text{WMSE}(b_0) &= E(b_0 - \beta)'\Theta(b_0 - \beta) \\ &= \text{tr } \Theta V(b_0) \\ &= \text{tr } \Theta(X'X)^{-1}X'(\Sigma \otimes I_T)X(X'X)^{-1} . \end{aligned} \tag{6.30}$$

From (6.25) and (6.30), it is rather difficult to deduce any clear general result regarding the dominance of the Stein-rule estimators over the OLS estimator. So, taking a special case, let us postulate a two equation model with orthogonal regressors (i.e., $X_1'X_2 = 0$) and choose $\Theta = \Omega^{-1}$. For this case, we observe that

$$\text{WMSE}(\underline{b}_{SF}) = (K_1 + K_2)\left[1 + \frac{1}{T} - \frac{g}{\beta'\Omega^{-1}\beta}\{(2-g)(K_1+K_2) - 4\}\right]$$

$$\text{WMSE}(b_0) = \left(\frac{K_1 + K_2}{1 - \rho_{12}^2}\right)$$

where $\rho_{12} = \sigma_{12}/(\sigma_{11}\sigma_{22})^{\frac{1}{2}}$.

Thus, WMSE($\underline{b}_{SF}$) is smaller than WMSE(b_0) when

$$\rho_{12}^2 > \left(\frac{1}{T\alpha + 1}\right) \tag{6.31}$$

where

$$\alpha = \frac{1}{1 + \dfrac{(K_1 + K_2)}{(\beta'\Omega^{-1}\beta)} gT\left(g - 2 + \dfrac{4}{K_1 + K_2}\right)} . \tag{6.32}$$

Notice that

$$\beta'\Omega^{-1}\beta = \frac{1}{(1-\rho_{12}^2)}\left[\frac{1}{\sigma_{11}}\beta_1'X_1'X_1\beta_1 + \frac{1}{\sigma_{22}}\beta_2'X_2'X_2\beta_2\right] \quad . \tag{6.33}$$

Setting $g = 0$ in α, we find that $\alpha = 1$, so that the condition (6.31) is

$$\rho_{12}^2 > \left(\frac{1}{T+1}\right) \quad , \tag{6.34}$$

which is just the condition for the dominance of $\underline{b}_F$ (i.e., the SUUR and SURR estimators) over b_0.

If we choose g to satisfy

$$0 < g < 2\left(1 - \frac{2}{K_1 + K_2}\right) ; \quad (K_1 + K_2) > 2 \tag{6.35}$$

the quantity α exceeds 1 and consequently it follows from (6.32) and (6.34) that the range of ρ_{12}^2 values over which the SUUR and SURR estimators dominate the OLS estimator is shorter than the corresponding range of ρ_{12}^2 values for the dominance of the Stein-rule estimators over the OLS estimator, with respect to the weighted mean squared error criterion (to the order of our approximation). Thus, we see that the performance of the SUUR and SURR estimators with respect to the OLS estimator can be improved through the use of Stein-rule versions of the former estimators by choosing the characterizing scalar g appropriately.

6.2.4 Further Remarks

The Stein-rule estimators presented above suggest several other shrinkage estimators. For example, we may use an exponential function to shrink the FGLS estimators, such as

$$[1 - g_1 \exp\{-g_2 b_F' X'(S^{-1} \otimes I_T)Xb_F\}]b_F \tag{6.36}$$

where $0 < g_1 < 1$ and $g_2 > 0$ are two adjustable constants to be chosen suitably.

Looking at the Stein-rule families obtained from the FGLS estimators, it is possible that the associated shrinking of an FGLS estimator can change its sign. To overcome this undesirable feature, the families (6.8) and (6.11) can be modified as follows:

$$b_{SF}^{+} = \left[1 - g\,\frac{\mathrm{tr}\,[X'(S^{-1} \otimes I_T)X]^{-1}\Theta}{b_F'\Theta b_F}\right]^{+} b_F \tag{6.37}$$

$$b_{SF}^{*+} = \left[1 - g\,\frac{\mathrm{tr}\,[X'(S^{-1} \otimes I_T)X]^{-1}\Theta}{\mathrm{tr}\,[X'(S^{-1} \otimes I_T)X]^{-1}\Theta + b_F'\Theta b_F}\right]^{+} b_F$$

where $[a]^+$ takes the value a if $a > 0$ and 0 if $a \leq 0$. These are examples of Stein-rule positive part estimators, and they ensure that sign changes of the type mentioned are avoided. The sampling properties of the estimators (6.37) for the SURE model are yet to be investigated. Assuming Σ to be known, Judge and Bock (1978) indicated several other Stein-rule estimators but they are obviously not useful in any given application due to the absence of knowledge about Σ in practice. Modifications are therefore needed to overcome this limitation. This is a potential area for further work in connection with the SURE model.

6.3 RIDGE-TYPE ESTIMATORS

The ridge regression estimator was developed primarily to circumvent the problem of multicollinearity in the classical linear regression model. Later it became used as a tool for designing improved estimators even in the absence of multicollinearity. See Dwivedi and Srivastava (1978), Hoerl and Kennard (1981), Smith and Campbell (1980), Vinod (1978), and Vinod and Ullah (1981) for a survey of progress in this area and an appraisal of the development of ridge estimators.

Extending the analysis of Hoerl and Kennard (1970), the general ridge estimator of β in the SURE model can be defined as:

$$\begin{aligned} b_{OR} &= [X'X + G]^{-1}X'y \\ &= [X'X + G]^{-1}X'Xb_0 \\ &= [I_{\Sigma K_i} + (X'X)^{-1}G]^{-1}b_0 \end{aligned} \tag{6.38}$$

where G is a $(K^* \times K^*)$ matrix with nonnegative elements (stochastic and/or nonstochastic) characterizing the estimator. Notice that this ridge estimator fails to take into account the cross-equation correlations between the disturbances. Accordingly, to generalise the ridge estimator further with this in mind we transform the model (6.1) in the following way:

$$(\Sigma^{-\frac{1}{2}} \otimes I_T)y = (\Sigma^{-\frac{1}{2}} \otimes I_T)X\beta + (\Sigma^{-\frac{1}{2}} \otimes I_T)u \;. \tag{6.39}$$

Assuming Σ to be known for the moment and applying the ridge technique, we get the ridge version of the GLS estimator:

$$
\begin{aligned}
b_{GR} &= [X'(\Sigma^{-1} \otimes I_T)X + G]^{-1}X'(\Sigma^{-1} \otimes I_T)y \\
&= [X'\Psi^{-1}X + G]^{-1}X'\Psi^{-1}y \\
&= [X'\Psi^{-1}X + G]^{-1}X'\Psi^{-1}Xb_G \\
&= [I_{\Sigma K_i} + (X'\Psi^{-1}X)^{-1}G]^{-1}b_G \ . \qquad (6.40)
\end{aligned}
$$

When G is nonstochastic, it is easy to see that the bias vector and the mean squared error matrix of b_{GR} are:

$$E(b_{GR} - \beta) = -[X'\Psi^{-1}X + G]^{-1}G\beta \qquad (6.41)$$

$$
\begin{aligned}
MSE(b_{GR}) &= E(b_{GR} - \beta)(b_{GR} - \beta)' \\
&= [X'\Psi^{-1}X+G]^{-1}(G\beta\beta'G' + X'\Psi^{-1}X)[X'\Psi^{-1}X+G]^{-1} \ . \qquad (6.42)
\end{aligned}
$$

If we choose G to be diagonal, the estimator b_{GR} can be regarded as an extension of the general ridge regression estimator. Further, if we take $G = gI_{\Sigma K_i}$ with g a nonstochastic scalar, the estimator b_{GR} is like the ordinary ridge regression estimator. Comparing the ridge and regular GLS estimators of β for this choice of G, we have

$$
\begin{aligned}
N &= E(b_G - \beta)(b_G - \beta)' - E(b_{GR} - \beta)(b_{GR} - \beta)' \\
&= (X'\Psi^{-1}X)^{-1} - [X'\Psi^{-1}X+gI_{\Sigma K_i}]^{-1}(g^2\beta\beta' + X'\Psi^{-1}X)[X'\Psi^{-1}X+gI_{\Sigma K_i}]^{-1} \\
&= g^2[X'\Psi^{-1}X+gI_{\Sigma K_i}]^{-1}\left[\frac{2}{g}I_{\Sigma K_i} + (X'\Psi^{-1}X)^{-1} - \beta\beta'\right][X'\Psi^{-1}X + gI_{\Sigma K_i}]^{-1} , \qquad (6.43)
\end{aligned}
$$

which is at least positive semidefinite if the matrix

$$\left[\frac{2}{g}I_{\Sigma K_i} + (X'\Psi^{-1}X)^{-1} - \beta\beta'\right]$$

is at least so. That is, if for all non-null column vectors η,

$$\eta'\left[\frac{2}{g}I_{\Sigma K_i} + (X'\Psi^{-1}X)^{-1} - \beta\beta'\right]\eta \geq 0$$

or

$$\frac{(\eta'\beta)^2}{\eta'\left[\frac{2}{g}I_{\Sigma K_i} + (X'\Psi^{-1}X)^{-1}\right]\eta} \leq 1 . \tag{6.44}$$

Now, this inequality will definitely be satisfied for all non-null η if the maximum value of the expression on the left side of the inequality (6.44), over all such η, does not exceed unity. From Rao (1973; p. 60), this yields the following condition:

$$\beta'\left[\frac{2}{g}I_{\Sigma K_i} + (X'\Psi^{-1}X)^{-1}\right]^{-1}\beta \leq 1 . \tag{6.45}$$

To explore this condition further, suppose that Γ is an orthogonal matrix such that

$$\Gamma'(X'\Psi^{-1}X)\Gamma = \Lambda$$

where Λ is a diagonal matrix with diagonal elements $\lambda_1 \geq \lambda_2 \geq \cdots \geq \lambda_{\Sigma K_i} > 0$, these being the characteristic roots of $(X'\Psi^{-1}X)$. In this case the condition (6.45) can be written as

$$\beta'\Gamma\left[\frac{2}{g}I_{\Sigma K_i} + \Lambda^{-1}\right]^{-1}\Gamma'\beta \leq 1$$

or, from Rao (1973; p. 33):

$$\left(\frac{g}{2}\right)\beta'\Gamma[I_{\Sigma K_i} - g(gI_{\Sigma K_i} + 2\Lambda)^{-1}]\Gamma'\beta \leq 1 , \tag{6.46}$$

which will definitely hold if

$$\left(\frac{g}{2}\right)\beta'\Gamma\Gamma'\beta = \left(\frac{g}{2}\right)(\beta'\beta) < 1 ; \; g > 0$$

or

$$0 < g < \left(\frac{2}{\beta'\beta}\right) . \tag{6.47}$$

This provides a sufficient condition, and demonstrates the existence of a nonzero value of g, for the nonnegative definiteness of the matrix N. This, of course, is a condition for the superiority of the generalized least squares ridge (GLSR) estimator over the GLS estimator with respect to the mean squared error matrix criterion.

Notice that if we define

$$N_G = E(b_0 - \beta)(b_0 - \beta)' - E(b_G - \beta)(b_G - \beta)'$$
$$N_{GR} = E(b_0 - \beta)(b_0 - \beta)' - E(b_{GR} - \beta)(b_{GR} - \beta)' , \tag{6.48}$$

then it follows from (6.43) that

$$N = N_{GR} - N_G . \tag{6.49}$$

It may be observed that the matrix N_G provides a measure of the efficiency (in terms of the mean squared error matrix) of the GLS estimator relative to the OLS estimator. Efficiency is gained or lost depending upon whether the matrix N_G is positive definite or negative definite. Similarly, the matrix N_{GR} determines the efficiency of the GLSR estimator relative to the OLS estimator. In this way, the positive definiteness of the matrix N ensures a larger gain in efficiency of the GLSR estimator over the OLS estimator as compared with the gain in the efficiency of the GLS estimator over the OLS estimator. It thus follows from (6.47) that we can choose g suitably so that the matrix N is positive definite, reflecting the superiority of the GLSR estimator over both the OLS and GLS estimators with respect to the mean squared error matrix criterion.

The above discussion suffers from two serious limitations. First, the upper bound of g, as specified by (6.47), requires knowledge of the value of the elements of β. Merely demonstrating the existence of a nonempty interval for g does not serve any useful purpose in any given application. An applied researcher will require an exact specification of the value of g to be used (see Vinod and Ullah 1981; Chapter 7, for an interesting discussion of the choice of this scalar). The second point is that the formulation of the RGLS estimator rests upon the assumption that Σ is known, and so in practice this estimator usually will not be feasible. One simple solution is to replace Σ by its consistent estimator S. The thus obtained estimator can be termed a feasible generalized least squares ridge (FGLSR) estimator:

$$\begin{aligned} b_{FR} &= [X'(S^{-1} \otimes I_T)X + gI_{\Sigma K_i}]^{-1}X'(S^{-1} \otimes I_T)y \\ &= [X'(S^{-1} \otimes I_T)X + gI_{\Sigma K_i}]^{-1}X'(S^{-1} \otimes I_T)Xb_F \\ &= [I_{\Sigma K_i} + g\{X'(S^{-1} \otimes I_T)X\}^{-1}]^{-1}b_F . \end{aligned} \tag{6.50}$$

Setting $S = \tilde{S}$ and $S = \hat{S}$ provides the ridge-type SUUR and SURR estimators:

$$\tilde{\beta}_{SUR} = [X'(\tilde{S}^{-1} \otimes I_T)X + gI_{\Sigma K_i}]^{-1}X'(\tilde{S}^{-1} \otimes I_T)y$$

$$\hat{\beta}_{SRR} = [X'(\hat{S}^{-1} \otimes I_T)X + gI_{\Sigma K_i}]^{-1}X'(\hat{S}^{-1} \otimes I_T)y \tag{6.51}$$

where g is the characterizing scalar, and may be either stochastic or nonstochastic. Replacing $gI_{\Sigma K_i}$ by more general choices of G in (6.51) yields more general formulations of the ridge-type SUUR and SURR estimators. Exact expressions for the bias vectors and mean squared error matrices of the estimators (6.51) are rather difficult to derive. However, their large-sample asymptotic approximations can be worked out with little effort and these can be used to analyze the relative efficiencies of these estimators. Appropriate procedures for the optimal choice of g also can be designed. Such an investigation may suggest new estimators which are efficient in those cases where the OLS estimator dominates the SUUR and SURR estimators, and in this way we may be able to reduce the inefficiency of the SUUR and SURR estimators relative to the OLS estimator, and also construct possibly more efficient estimators. This is an area that deserves careful exploration.

All that has been said above with regard to the choice $G = gI_{\Sigma K_i}$ in the estimator (6.40) applies to other choices of G. The theory of general ridge estimation needs to be extended to cover the SURE model (see Vinod and Ullah, 1981; pp. 244-245, for an attempt). Further, along the lines of the work reported in Casella (1980), Srivastava and Chaturvedi (1982) and Strawderman (1978), it would be interesting to develop families of adaptive ridge-type estimators for the SURE model and to study their sampling properties. This topic is a promising area for future research.

6.4 WEIGHTED-COMBINATION ESTIMATORS

The poor performance of the SUUR and SURR estimators relative to the OLS estimator when there are very weak correlations between the disturbances across the equations led Srivastava and Srivastava (1984) to develop estimators that are weighted combinations of the SUUR/SURR and OLS estimators. Their work is now discussed in detail.

Suppose that W_F and W_0 are two matrices of order $(K^* \times K^*)$. Then a weighted combination of the OLS and FGLS estimators is defined by

$$b_{0F} = W_0 b_0 + W_F b_F \,. \tag{6.52}$$

If we choose the matrices W_0 and W_F in such a manner that b_0 receives lesser weight for higher correlations between the disturbances across the equations and vice-versa, it is obvious that the performance of the FGLS estimator can perhaps be improved by using the estimator b_{0F}. Accordingly, let us consider a simple case by choosing

$$\begin{aligned} W_0 &= (1 - g) I_{\Sigma K_i} \\ W_F &= I_{\Sigma K_i} - W_0 = g I_{\Sigma K_i} \ , \end{aligned} \tag{6.53}$$

so that the estimator (6.52) defines a class of shrinkage estimators for β:

$$b_{0F} = b_0 - g(b_0 - b_F) \ , \tag{6.54}$$

where in general g may be a stochastic or nonstochastic positive scalar.

Now, assuming g to be a fixed positive scalar and postulating a two equation model, we may analyze the sampling properties of b_{0F}. It is easy to see that the weighted-combination estimator of β_1 (the coefficient vector in the first equation of the model) is given by

$$b_{(1)0F} = (X_1'X_1)^{-1}X_1'y_1 - gR_{12}\left[r^2 X_1'P_{x_2}\bar{P}_{x_1}y_1 + \left(\frac{s_{12}}{s_{22}}\right)X_1'\bar{P}_{x_2}y_2\right] \tag{6.55}$$

where

$$r^2 = \frac{s_{12}^2}{s_{11}s_{22}}$$

$$R_{12} = [X_1'(I_T - r^2 P_{x_2})X_1]^{-1} \ .$$

Under the assumption of normality of the disturbances, it may be shown that the estimator $b_{(1)0F}$ is unbiased for the SUUR and SURR members of the FGLS estimator family. Further, in the case of a combination of the OLS and SUUR estimators, the exact expression for the variance covariance matrix of the weighted-combination estimator can be derived precisely in the same manner as was indicated for the SUUR estimator in Section 4.6. This expression turns out to be rather complicated in form, and does not permit us to draw any clear conclusions. We therefore place another restriction on the model, namely that X_1 and X_2 are orthogonal matrices. In this case the estimator (6.55) for the SUUR choice of the FGLS estimator becomes

$$\tilde{\beta}_{(1)0SU} = (X_1'X_1)^{-1}X_1'y_1 - g\frac{\tilde{s}_{12}}{\tilde{s}_{22}}(X_1'X_1)^{-1}X_1'y_2 \; . \tag{6.56}$$

It is easy to show that

$$\begin{aligned} V(\tilde{\beta}_{(1)0SU}) &= E(\tilde{\beta}_{(1)0SU} - \beta_1)(\tilde{\beta}_{(1)0SU} - \beta_1)' \\ &= \sigma_{11}\left[1 - 2g\rho_{12}^2 + g^2\left(\rho_{12}^2 + \frac{1 - \rho_{12}^2}{T - K_1 - K_2 - 2}\right)\right](X_1'X_1)^{-1} \end{aligned} \tag{6.57}$$

provided that $(T - K_1 - K_2) > 2$.

Notice that setting $g = 0$ in (6.57) gives the expression for the variance covariance matrix of the OLS estimator while putting $g = 1$ provides the corresponding expression for the SUUR estimator. Thus, we find that

$$V(b_{(1)0}) - V(\tilde{\beta}_{(1)0SR}) = \sigma_{11}g\left[2\rho_{12}^2 - g\left(\rho_{12}^2 + \frac{1 - \rho_{12}^2}{T - K_1 - K_2 - 2}\right)\right](X_1'X_1)^{-1} , \tag{6.58}$$

from which it follows that the weighted-combination estimator is more efficient than the OLS estimator if

$$0 < g < \frac{2(T - K_1 - K_2 - 2)\rho_{12}^2}{1 + (T - K_1 - K_2 - 3)\rho_{12}^2} \; ; \quad (T - K_1 - K_2) > 2 \tag{6.59}$$

or, equivalently, if

$$\rho_{12}^2 > \frac{g}{g + (T - K_1 - K_2 - 2)(2 - g)} \; ; \quad (T - K_1 - K_2) > 2 \; . \tag{6.60}$$

Similarly, we have

$$\begin{aligned} &V(\tilde{\beta}_{(1)SU}) - V(\tilde{\beta}_{(1)0SU}) \\ &= \sigma_{11}(1 - g)\left[-\rho_{12}^2 + \frac{1 - \rho_{12}^2}{T - K_1 - K_2 - 2} + g\left(\rho_{12}^2 + \frac{1 - \rho_{12}^2}{T - K_1 - K_2 - 2}\right)\right](X_1'X_1)^{-1} \end{aligned} \tag{6.61}$$

from which we find that the weighted-combination estimator is more efficient than the SUUR estimator when

$$\frac{(T - K_1 - K_2 - 1)\rho_{12}^2 - 1}{(T - K_1 - K_2 - 3)\rho_{12}^2 + 1} < g < 1 ; \quad \rho_{12}^2 > \frac{1}{(T - K_1 - K_2 - 1)} \tag{6.62}$$

or, equivalently, when

$$\rho_{12}^2 < \frac{1 + g}{(T - K_1 - K_2 - 1) - g(T - K_1 - K_2 - 3)} ; \quad 0 < g < 1 \tag{6.63}$$

provided that $(T - K_1 - K_2) > 2$.

If we minimize the expression (6.57) with respect to g, we obtain

$$g = \frac{\rho_{12}^2 (T - K_1 - K_2 - 2)}{1 + (T - K_1 - K_2 - 3)\rho_{12}^2} = g_{opt} \quad \text{(say)} , \tag{6.64}$$

which enables us to define the minimum variance unbiased weighted-combination estimator. Note that this estimator is not feasible, as ρ_{12}^2 is unobservable. The variance covariance matrix of such an estimator is found by substituting $g = g_{opt}$ in (6.57), which yields

$$\sigma_{11}(1 - \rho_{12}^2) \left[\frac{1 + (T - K_1 - K_2 - 2)\rho_{12}^2}{1 + (T - K_1 - K_2 - 3)\rho_{12}^2}\right] (X_1' X_1)^{-1} . \tag{6.65}$$

Recalling that the variance covariance matrix of the GLS estimator of β_1 is

$$\sigma_{11}(1 - \rho_{12}^2)(X_1' X_1)^{-1} , \tag{6.66}$$

and comparing it with the expression (6.65), we find that the latter is very close to the expression (6.66).

When the OLS estimator is combined with the SURR estimator, the weighted-combination estimator is

$$\hat{\beta}_{(1)OSR} = (X_1'X_1)^{-1} X_1'y_1 - g\left(\frac{\hat{s}_{12}}{\hat{s}_{22}}\right) (X_1'X_1)^{-1} X_1'y_2 . \tag{6.67}$$

For nonstochastic g, the properties of this estimator can be studied in the same way as for the estimator (6.56). Note that if we assume that X_2 is a submatrix of X_1, instead of assuming the orthogonality of X_1 and X_2, a similar analysis can be carried out without any difficulty and a corresponding set of conclusions can be drawn.

Finally, let us consider the case of unrestricted regressors and combine the OLS estimator with Conniffe's estimator (discussed in Section 4.7.4), given by

$$\overset{*}{\beta}_{(1)SU} = (X_1'X_1)^{-1}X_1'y_1 - \frac{\tilde{s}_{12}}{\tilde{s}_{22}}(X_1'X_1)^{-1}X_1'\bar{P}_{x_2}y_2 . \tag{6.68}$$

Thus the weighted-combination estimator of β_1 is

$$\begin{aligned}\overset{*}{\beta}_{(1)0SU} &= (1-g)b_{(1)0} + g\overset{*}{\beta}_{(1)SU} \\ &= (X_1'X_1)^{-1}X_1'y_1 - g\frac{\tilde{s}_{12}}{\tilde{s}_{22}}(X_1'X_1)^{-1}X_1'\bar{P}_{x_2}y_2 .\end{aligned} \tag{6.69}$$

Taking g to be nonstochastic and assuming normality of the disturbances, it is easy to see that the estimator (6.69) is unbiased. Further, its variance covariance matrix is given by

$$\begin{aligned}V(\overset{*}{\beta}_{(1)0SU}) &= E(\overset{*}{\beta}_{(1)0SU} - \beta)(\overset{*}{\beta}_{(1)0SU} - \beta)' \\ &= \sigma_{11}(X_1'X_1)^{-1} - \sigma_{11}g\left[2\rho_{12}^2 - g\left(\rho_{12}^2 + \frac{1-\rho_{12}^2}{T-K_1-K_2-2}\right)\right] \\ &\quad \cdot (X_1'X_1)^{-1}X_1'\bar{P}_{x_2}X_1(X_1'X_1)^{-1} ,\end{aligned} \tag{6.70}$$

provided that $(T - K_1 - K_2) > 2$.

Comparing this matrix with $\sigma_{11}(X_1'X_1)^{-1}$, the variance covariance matrix of the OLS estimator, we observe that the weighted-combination estimator is not dominated by the OLS estimator as long as (6.59) or (6.60) holds. Similarly, the weighted-combination estimator is not inferior to Conniffe's estimator, (6.68), as long as (6.62) or (6.63) is satisfied. Interestingly enough the estimator (6.69), with $g = g_{opt}$ given by (6.64), turns out to be the minimum variance unbiased estimator of β_1. Thus, it is clear that we can always pick a more efficient estimator from the proposed family of weighted-combination estimators by selecting an appropriate value of g. However, the range of g as well as the optimal value g_{opt} requires knowledge of the value of ρ_{12}^2. A simple alternative is then to replace ρ_{12}^2 by its sample analogue, r^2. If we estimate g_{opt} in this way and use it to obtain the weighted-combination estimator, then this estimator will not have the same properties as those enjoyed by the estimator characterized by g_{opt} itself. It would be interesting to analyze the statistical consequences of such a substitution.

The key feature of the above discussion is that g is assumed to be fixed. Since g should be small for low values of ρ_{12}, we may choose g as a function of r^2, such that the first derivative of g with respect to r^2 is positive and g converges to 1 in probability. These conditions are satisfied for a variety of functions, such as

$$g = 1 - \alpha_1 \exp\left(-\alpha_2 \log\left(\frac{1 + r^2}{1 - r^2}\right)\right) , \tag{6.71}$$

with $0 < \alpha_1 < 1$ and $\alpha_2 > 0$ as adjustable constants. It would be interesting to determine the properties of estimators implied by such stochastic choices of g.

Another interesting direction for further work is to develop pooled estimators based on a preliminary test of the significance of the correlation coefficient ρ_{12}. For this purpose, one could use Kariya's (1981) locally best invariant and unbiased test of the null hypothesis, $H_0 : \rho_{12} = 0$ against the alternative $H_1 : \rho_{12} \neq 0$. His test statistic is given by

$$\left[\frac{(T - K_1)(T - K_2)(y_1'\bar{P}_{x_1}\bar{P}_{x_2}y_2)^2}{(y_1'\bar{P}_{x_1}y_1)(y_2'\bar{P}_{x_2}y_2)} - \frac{(T - K_1)y_1'\bar{P}_{x_1}\bar{P}_{x_2}\bar{P}_{x_1}y_1}{y_1'\bar{P}_{x_1}y_1} - \frac{(T - K_2)y_2'\bar{P}_{x_2}\bar{P}_{x_1}\bar{P}_{x_2}y_2}{y_2'\bar{P}_{x_2}y_2}\right] . \tag{6.72}$$

The exact distribution of this test statistic is difficult to derive. However, when X_2 is a submatrix of X_1, the problem simplifies and the critical region is defined by

$$[(T - K_2)V - (T - K_1)] \geq k \tag{6.73}$$

where

$$V = \frac{(T - K_1)(y_1'\bar{P}_{x_1}y_2)^2 - (y_1'\bar{P}_{x_1}y_1)(y_2'\bar{P}_{x_1}y_2)}{(y_1'\bar{P}_{x_1}y_1)(y_2'\bar{P}_{x_2}y_2)} \tag{6.74}$$

and k is the critical value determined by the size of the test.

Assuming that $(T - K_1) > 1$, the distribution function of V under H_0 is given by

$$
F(V \leq v) = \begin{cases} 0 & \text{if } v < -1 \\ \displaystyle\int_{-v}^{\left(\frac{T-K_1-\frac{1}{2}-v}{T-K_1-\frac{1}{2}}\right)} B\left(\frac{v+z}{1-z}\right)\phi(z)\,dz & \text{if } -1 \leq v < 0 \\ \displaystyle\int_{0}^{\left(\frac{T-K_1-\frac{1}{2}-v}{T-K_1-\frac{1}{2}}\right)} B\left(\frac{v+z}{1-z}\right)\phi(z)\,dz & \text{if } 0 \leq v < \left(T-K_1-\frac{1}{2}\right) \\ 1 & \text{if } v \geq \left(T-K_1-\frac{1}{2}\right) \end{cases} \tag{6.75}
$$

where $B\left(\frac{v+z}{1-z}\right)$ denotes the incomplete Beta function with parameters 1/2 and $(K_1 - K_2)/2$, and $\phi(z)$ is the probability density function of a Beta variable z with parameters $(T - K_1 - 1)/2$ and $(K_1 - K_2 + 1)/2$. The value of k in (6.73) can be determined by using (6.75). Using Jacobi polynomials, Kariya (1981) suggested a method of approximating the distribution of the test statistic (6.72) under H_0 without placing any restrictions on X_1 and X_2, but his method is operationally quite difficult. He also presented a locally best invariant test of H_0 against the alternative hypothesis $H_1 : \rho_{12} > 0$ or $H_1^* : \rho_{12} < 0$.

One could also define a preliminary test (PT) estimator for β:

$$
b_{PT} = \begin{cases} b_0 & \text{if } H_0 \text{ is accepted} \\ b_F & \text{if } H_0 \text{ is rejected} \end{cases} \tag{6.76}
$$

which can be expressed alternatively as

$$
\begin{aligned} b_{PT} &= I_{[o,k)} b_0 + I_{[k,\infty)} b_F \\ &= b_F - I_{[o,k)}(b_F - b_0) \end{aligned} \tag{6.77}
$$

where $I_{[o,k)}$ and $I_{[k,\infty)}$ denote indicator functions taking the value unity if the value of the test statistic falls in the interval subscripted, and the value o otherwise (see Judge and Bock, 1978, for an account of some such estimators). It should be mentioned that the properties of estimators of the form (6.76) have yet to be explored. An alternative, and perhaps more appropriate, PT estimator would be one based on a test of $H_0 : |\rho_{12}| > \rho^*$ against $H_1 : |\rho_{12}| \leq \rho^*$, where ρ^* is that value of $|\rho_{12}|$ below which the OLS estimator of β has smaller risk than has the FGLS estimator. The properties of this PT estimator also are unexplored.

We have restricted our attention to weighted-combination estimators in which the weighting matrices are scalar multiples of an identity matrix. Taking them to be diagonal or nondiagonal with some specified structure raises some interesting possibilities for developing more efficient estimators. The supposition that the sum of the two matrices W_0 and W_F should be an identity matrix can be relaxed, and biased weighted-combination estimators could be considered in a search for decision rules with reduced risks.

6.5 LINDLEY-LIKE MEAN CORRECTIONS

A feature common to most of the estimators presented in this chapter is that they attempt to shrink the FGLS estimator toward a null vector. This may appear to be somewhat unreasonable, unsatisfactory and unappealing in some situations. In many cases it may be more reasonable to shrink the estimators toward a prior vector, $\beta_{.}$, or toward the general mean vector, as advocated by Lindley (1962) and studied by Srivastava and Ullah (1980) in the context of the classical linear regression model. This strategy is referred to as a Lindley-like mean correction.

Incorporation of a Lindley-like mean correction for shrinking the FGLS estimator yields, from (6.8), the estimator

$$\bar{\beta}_{.}e + \left[1 - g\frac{\operatorname{tr}[X'(S^{-1} \otimes I_T)X]^{-1}\Theta}{(b_F - \bar{\beta}_{.}e)'\Theta(b_F - \bar{\beta}_{.}e)}\right](b_F - \bar{\beta}_{.}e) \tag{6.78}$$

where e is a column vector with all elements equal to 1,

$$\bar{\beta}_{.} = \frac{e'\Theta\beta_{.}}{e'\Theta e} \tag{6.79}$$

and $\beta_{.}$ is any known prior vector.

If we shrink the FGLS estimator toward the general mean $\bar{b}_F$, we obtain the following estimator of β:

$$\bar{b}_F e + \left[1 - g\frac{\operatorname{tr}[X'(S^{-1} \otimes I_T)X]^{-1}\Theta}{(b_F - \bar{b}_F e)'\Theta(b_F - \bar{b}_F e)}\right](b_F - \bar{b}_F e) \tag{6.80}$$

where

$$\bar{b}_F = \frac{e'\Theta b_F}{e'\Theta e}\,. \tag{6.81}$$

If we set $S = \tilde{S}$ in (6.80), we obtain the Stein-rule SUUR estimator with a Lindley-like mean correction:

$$\tilde{\beta}_{SSUL} = \bar{\bar{\beta}}_{SU} e + \left[1 - g \frac{\operatorname{tr}[X'(\tilde{S}^{-1} \otimes I_T)X]^{-1}\Theta}{(\tilde{\beta}_{SU} - \bar{\bar{\beta}}_{SU}e)'\Theta(\tilde{\beta}_{SU} - \bar{\bar{\beta}}_{SU}e)}\right](\tilde{\beta}_{SU} - \bar{\bar{\beta}}_{SU}e)$$

$$= \tilde{\beta}_{SU} - g \frac{\operatorname{tr}[X'(\tilde{S}^{-1} \otimes I_T)X]^{-1}\Theta}{\tilde{\beta}_{SU}'B'\Theta B\tilde{\beta}_{SU}} B\tilde{\beta}_{SU} \tag{6.82}$$

where

$$\bar{\bar{\beta}}_{SU} = \left(\frac{e'\Theta\tilde{\beta}_{SU}}{e'\Theta e}\right); \quad B = \left(I_{\Sigma K_i} - \frac{1}{e'\Theta e} ee'\Theta\right). \tag{6.83}$$

Similarly, we can write down the Stein-rule SURR estimator with a Lindley-like mean correction by substituting $\hat{S}$ for S in (6.80).

Using a large-sample asymptotic approximation, it is easy to obtain the bias vectors to order $O(T^{-1})$ and the mean squared error matrices to order $O(T^{-2})$ for these estimators. From these the expression for weighted mean squared error to order $O(T^{-2})$ can be derived. Comparing this quantity with (6.25), the expression for the weighted mean squared error (to the same order) of the Stein-rule SUUR estimator without a Lindley-like mean correction, we find:

$$\begin{aligned}
&\operatorname{WMSE}(\tilde{\beta}_{SSU}) - \operatorname{WMSE}(\tilde{\beta}_{SSUL}) \\
&= g\left(\frac{\operatorname{tr}\Omega\Theta}{\beta'\Theta^*\beta}\right)\left[\left(1 - \frac{1}{\alpha}\right)((\alpha + 1)g - 2\alpha)\operatorname{tr}\Omega\Theta - 2\left(\frac{e'\Theta\Omega\Theta e}{e'\Theta e}\right)\right. \\
&\quad \left. + \frac{4}{\beta'\Theta^*\beta}(\alpha^2\beta'\Theta\Omega\Theta\beta - \beta'\Theta^*\Omega\Theta^*\beta)\right]
\end{aligned} \tag{6.84}$$

where

$$\begin{aligned}
\Theta^* &= \left(\Theta - \frac{1}{e'\Theta e}\Theta ee'\Theta\right) \\
\alpha &= \frac{\beta'\Theta^*\beta}{\beta'\Theta\beta} = \left[1 - \frac{(e'\Theta\beta)^2}{(e'\Theta e)(\beta'\Theta\beta)}\right]; \quad 0 \le \alpha \le 1.
\end{aligned} \tag{6.85}$$

The expression (6.84) describes the gain or loss in efficiency attributable to the application of a Lindley-like mean correction. Let us examine the case when we choose $\Theta = \Omega^{-1}$. Substituting this in (6.84), we get

$$\frac{g(\sum_i K_i)}{\beta'\Theta^*\beta}\left[2\{(1-\alpha)(\sum_i K_i) - 3 + 2\alpha\} - g\left(\frac{1-\alpha^2}{\alpha}\right)(\sum_i K_i)\right] \tag{6.86}$$

which is positive when

$$0 < g < 2\left(\frac{\alpha}{\alpha+1}\right)\left[1 - \frac{3-2\alpha}{(1-\alpha)(\sum_i K_i)}\right]; \quad \alpha < \left[1 - \frac{1}{(\sum_i K_i) - 2}\right]. \tag{6.87}$$

Thus, (6.87) specifies the range of g over which the incorporation of a Lindley-like mean correction lowers the weighted squared error of the Stein-rule SUUR estimator, to the order of our approximation. A similar analysis can be performed for the estimators defined by (6.78). Further, proceeding in the same manner, we could apply the Lindley-like mean correction to other shrinkage estimators for the SURE model and then examine its impact on their relative sampling performances.

EXERCISES

6.1 Prove that the GLS estimator b_G of β in a SURE model with a known disturbance variance covariance matrix Ψ is uniformly dominated by the "estimator"

$$\beta^* = \left[1 - \frac{\operatorname{tr}\Theta(X'\Psi^{-1}X)^{-1}}{\operatorname{tr}\Theta(X'\Psi^{-1}X)^{-1} + \beta'\Theta\beta}\right]b_G$$

under the risk criterion with quadratic loss and Θ as the weighting matrix. If, instead of risk, the mean squared error matrix is the performance criterion, does β^* continue to dominate b_G uniformly?

6.2 If we take $(1-\theta)b_0$ as an estimator of β in a SURE model and if we choose the scalar θ so as to minimize the total mean squared error, show that the "estimator" obtained in this way is

$$\left[1 - \frac{\operatorname{tr}(X'X)^{-2}X'\Psi X}{\operatorname{tr}(X'X)^{-2}X'\Psi X + \beta'\beta}\right]b_0 .$$

Compare the efficiency of this "estimator" with the "estimator" β^* in Exercise 6.1 (setting $\Theta = I$) according to the total mean squared error criterion.

6.3 Consider a two equation SURE model in which the regressors in the second equation form a subset of the regressors in the first equation and the disturbances are normally distributed. Compare the SUUR and SSUUR estimators of the coefficients using the criterion of weighted mean squared error to order $O(T^{-2})$, assuming that $\Theta = (X'\Psi X)$.

6.4 An estimator of the conditional mean forecast vector, given a set of future values of the regressors, is derived from the GLS estimator in a SURE model. Next, a constant multiple of this estimator is taken as another estimator of the conditional mean vector and this constant is so chosen that the predictive mean squared error is minimized. Derive the estimator of the conditional mean vector obtained in this way. Suggest feasible versions of the two estimators and work out their asymptotic properties.

6.5 Show that the GLS ridge estimator b_{GR} defined by (6.40) can be regarded as the OLS estimator of β in the following model:

$$(\Sigma^{-\frac{1}{2}} \otimes I_T)y = (\Sigma^{-\frac{1}{2}} \otimes I_T)X\beta + (\Sigma^{-\frac{1}{2}} \otimes I_T)u$$

$$0 = G^{-\frac{1}{2}}\beta + v$$

where v is a column vector of disturbances.

6.6 Consider the following two families of estimators of β in the SURE model:

$$b = [X'X + G_1]^{-1}X'y$$

$$b = [(X'X) * G_2]^{-1}X'y$$

where * denotes the Hadamard product operator and G_1 and G_2 are nonstochastic matrices characterizing the estimators. Obtain the expressions of the bias vector and mean squared error matrix for these two estimators. Supposing that $(X'X)$, G_1 and G_2 are diagonal matrices, it is proposed to choose G_1 and G_2 such that the total mean squared errors of the corresponding estimators are minimized. Prove that the optimal estimators obtained in this way are identical.

6.7 Consider the following estimator of β_{1j}, the j'th coefficient in the first equation of a SURE model:

$$\beta_{1j}^{\dagger} = \frac{\lambda_{1j} b_{(1)j}^2}{\lambda_{1j} b_{(1)j}^2 + (T - K_1)(y_1 - X_1 b_{(1)})'(y_1 - X_1 b_{(1)})}$$

where $b_{(1)j}$ denotes the j'th element of the OLS estimator $b_{(1)} = (X_1'X_1)^{-1}X_1'y_1$ of β_1 and $(X_1'X_1)$ is assumed to be a diagonal matrix with j'th element as λ_{1j}. Assuming normality of the disturbances, it is observed that $\beta_{1j}^{\dagger}$ has smaller mean squared error than $b_{(1)j}$ when $(\lambda_{1j}\beta_{1j}^2/2\sigma_{11}) < 1$. Suggest a test of this condition and formulate an associated preliminary test estimator for β_{1j}. Work out the exact expressions for the bias and mean squared error of this estimator. [Hint: See Srivastava and Giles (1984).]

6.8 Consider a two equation SURE model in which X_2 is a submatrix of X_1 and the disturbances are normally distributed. If β is estimated by $[g\tilde{\beta}_{SU} + (1 - g)b_0]$, where g is a fixed scalar, obtain explicit expressions for the estimators of β_1 and β_2. Work out their variance covariance matrices and find the conditions for the superiority (with respect to the variance covariance matrix) of the proposed estimators of β_1 and β_2 over the corresponding OLS and SUUR estimators.

6.9 In a two equation SURE model with orthogonal regressors and normally distributed disturbances, consider the estimation of β_1 by an estimator $[gb_{(1)0} + k\tilde{\beta}_{(1)SU}]$ in which g and k are any two fixed scalars. Determine g and k such that the mean squared error matrix of the estimator is minimized.

6.10 Derive the likelihood ratio and Lagrange multiplier tests for the diagonality of Σ, the variance covariance matrix of the disturbances in the two equation SURE model, when $\Sigma = \begin{pmatrix} 1 & \rho \\ \rho & 1 \end{pmatrix}$. Define a pre-test estimator of β based on the outcome of this test and discuss its finite-sample properties under quadratic loss.

REFERENCES

Casella, G. (1980). "Minimax ridge regression estimation," Annals of Statistics 8, 1036-1056.

Dwivedi, T., and V. K. Srivastava (1978). "A survey of ridge estimators in linear regression model," in International Dedication Seminar on Recent Advances in Mathematics and its Applications (Invited Lectures), Banaras Hindu University, 171-181.

Hoerl, A. E., and R. W. Kennard (1970). "Ridge regression: biased estimation for nonorthogonal problems," Technometrics 12, 55-67.

Hoerl, A. E., and R. W. Kennard (1981). "Ridge regression 1980: advances, algorithms and applications," American Journal of Mathematical and Management Sciences 1, 5-83.

James, W., and C. Stein (1961). "Estimation with quadratic loss," in Proceedings of the Fourth Berkeley Symposium on Mathematical Statistics and Probability, Vol. 1 (University of California Press, Berkeley), 361-379.

Judge, G. G., and M. E. Bock (1978). The Statistical Implications of Pre-Test and Stein-Rule Estimators in Econometrics (North-Holland, Amsterdam).

Kariya, T. (1981). "Tests for the independence between two seemingly unrelated regression equations," Annals of Statistics 9, 381-390.

Lindley, D. V. (1962). "Discussion on Professor Stein's paper," Journal of the Royal Statistical Society B, 24, 285-287.

Rao, C. R. (1973). Linear Statistical Inference and its Applications (Wiley, New York).

Smith, G., and F. Campbell (1980). "A critique of some ridge regression methods," Journal of the American Statistical Association 75, 74-81.

Srivastava, V. K. (1973). "The efficiency of an improved method of estimating seemingly unrelated regression equations," Journal of Econometrics 1, 341-350.

Srivastava, V. K., and A. Chaturvedi (1982). "Bias vector and mean squared error matrix of minimax adaptive generalized ridge regression estimators," Sankhyā B, 44, 76-91.

Srivastava, V. K., and D. E. A. Giles (1984). "Exact finite-sample properties of a pre-test estimator in ridge regression," Australian Journal of Statistics 26, 323-336.

Srivastava, V. K., and A. K. Srivastava (1984). "Improved estimation in a two equation seemingly unrelated regression model," Statistica 44, 417-422.

Srivastava, V. K., and A. Ullah (1980). "On Lindley-like mean correction in the improved estimation of linear regression models," Economics Letters 6, 29-35.

Strawderman, W. E. (1978). "Minimax adaptive generalized ridge regression estimators," Journal of the American Statistical Association 73, 623-627.

Vinod, H. D. (1978). "A survey of ridge regression and related techniques for improvements over ordinary least squares," Review of Economics and Statistics 60, 121-131.

Vinod, H. D., and A. Ullah (1981). Recent Advances in Regression Methods (Marcel Dekker, New York).

Zellner, A., and W. Vandaele (1975). "Bayes-Stein estimators for k-means, regression and simultaneous equation models," in S. E. Fienberg and A. Zellner (eds.), Studies in Bayesian Econometrics and Statistics in Honor of Leonard J. Savage (North-Holland, Amsterdam), 627-641.

7
Autoregressive Disturbances

7.1 INTRODUCTION

In many applications of the SURE model, especially those involving the use of time-series data, it may be inappropriate to retain the assumption that the model's disturbances are serially independent. Instead, a formal allowance for some specific autocorrelation scheme may be required. It is important to consider this generalization of the SURE model's specification in some detail, because to incorrectly assume serial independence of the errors will have a number of implications for the properties of our estimators and inferences generally.

In this chapter we consider SURE models in which the disturbances are assumed to be generated by a first-order autoregressive scheme. Two such schemes are considered. The first is scalar, so that the only intertemporal effects in the disturbance vector involve correlations between current and lagged realizations of the same element. The second is a vector autoregressive scheme, so that cross-disturbance intertemporal correlations are allowed for. In each case we consider the implications for the estimation of the SURE model, from both large and small-sample viewpoints.

7.2 FIRST-ORDER SCALAR AUTOREGRESSIVE DISTURBANCES

Consider the SURE model

$$y = X\beta + u, \qquad E(u) = 0 \tag{7.1}$$

in which we continue to assume that the elements of u are contemporaneously correlated. In addition, we now assume that these elements follow a first-order autoregressive scheme.

If u_{ti} denotes the t'th element of the disturbance vector u_i in the i'th equation of the model, then a first-order stationary autoregressive disturbance process is defined by

$$u_{ti} = \rho_i u_{t-1i} + \epsilon_{ti} \tag{7.2}$$

$$(t = 1, 2, \ldots, T;\ i = 1, 2, \ldots, M),$$

where $|\rho_i| < 1$ is a constant and ϵ_{ti} is a stochastic variable such that

$$E(\epsilon_{ti}) = 0 \qquad \text{for all t and i}$$

$$E(\epsilon_{ti}\epsilon_{sj}) = \begin{cases} \sigma_{ij} & \text{for } t = s \text{ and all i and j} \\ 0 & \text{for } t \neq s \text{ and all i and j} \end{cases}. \tag{7.3}$$

Thus, in each equation the current value of the disturbance term depends stochastically upon only the corresponding disturbance in the preceding time-period. Further, both the u_{ti}'s and ϵ_{ti}'s are contemporaneously correlated, but now the u_{ti}'s are no longer temporally independent, although the ϵ_{ti}'s are.

Defining ϵ as an $(MT \times 1)$ vector formed from the ϵ_{ti}'s in the same manner that u is defined, it is easy to see that

$$E(\epsilon) = 0$$

$$E(\epsilon\epsilon') = (\Sigma \otimes I_T),$$

so that

$$E(uu') = \Omega = \begin{bmatrix} \Omega_{11} & \cdots & \Omega_{1M} \\ \vdots & & \vdots \\ \Omega_{M1} & \cdots & \Omega_{MM} \end{bmatrix} \tag{7.4}$$

where

$$\underset{(T\times T)}{\Omega_{ij}} = \frac{\sigma_{ij}}{(1-\rho_i\rho_j)}\begin{bmatrix} 1 & \rho_j & \cdots & \rho_j^{T-1} \\ \rho_i & 1 & \cdots & \rho_j^{T-2} \\ \vdots & \vdots & & \vdots \\ \rho_i^{T-1} & \rho_i^{T-2} & \cdots & 1 \end{bmatrix}. \tag{7.5}$$

It is assumed that in the initial time period the disturbances have the same properties as in subsequent periods. For instance, we may specify that the u_{oi}'s are stochastic variables with

$$\begin{aligned} E(u_{oi}) &= 0 && \text{for all i} \\ E(u_{oi}u_{oj}) &= \sigma_{ij} && \text{for all i and j} \end{aligned} \tag{7.6}$$

or we may assume that

$$u_{1i} = \epsilon_{1i}/(1-\rho_i^2)^{1/2} . \tag{7.7}$$

As Ω is a positive definite matrix, its inverse may be written in the form

$$\Omega^{-1} = P'(\Sigma^{-1} \otimes I_T)P , \tag{7.8}$$

where P is a suitable nonsingular matrix.

One interesting choice of P is indicated by Judge, Griffiths, Hill and Lee (1980; pp. 262-64), and is obtained as follows. Let us define

$$\Sigma_0 = \begin{bmatrix} \frac{\sigma_{11}}{1-\rho_1^2} & \frac{\sigma_{12}}{1-\rho_1\rho_2} & \cdots & \frac{\sigma_{1M}}{1-\rho_1\rho_M} \\ \frac{\sigma_{21}}{1-\rho_2\rho_1} & \frac{\sigma_{22}}{1-\rho_2^2} & \cdots & \frac{\sigma_{2M}}{1-\rho_2\rho_M} \\ \vdots & \vdots & & \vdots \\ \frac{\sigma_{M1}}{1-\rho_M\rho_1} & \frac{\sigma_{M2}}{1-\rho_M\rho_2} & \cdots & \frac{\sigma_{MM}}{1-\rho_M^2} \end{bmatrix}$$

$$A = \begin{bmatrix} \alpha_{11} & 0 & \cdots & 0 \\ \alpha_{21} & \alpha_{22} & \cdots & 0 \\ \vdots & \vdots & & \vdots \\ \alpha_{M1} & \alpha_{M2} & \cdots & \alpha_{MM} \end{bmatrix} \tag{7.9}$$

$$P_{ii} = \begin{bmatrix} \alpha_{ii} & 0 & 0 & \cdots & 0 & 0 \\ -\rho_i & 1 & 0 & \cdots & 0 & 0 \\ 0 & -\rho_i & 1 & \cdots & 0 & 0 \\ \vdots & \vdots & \vdots & & \vdots & \vdots \\ 0 & 0 & 0 & \cdots & -\rho_i & 1 \end{bmatrix}$$

$$\underset{(i \neq j)}{P_{ij}} = \begin{bmatrix} \alpha_{ij} & 0 & 0 & \cdots & 0 & 0 \\ 0 & 0 & 0 & \cdots & 0 & 0 \\ 0 & 0 & 0 & \cdots & 0 & 0 \\ \vdots & \vdots & \vdots & & \vdots & \vdots \\ 0 & 0 & 0 & \cdots & 0 & 0 \end{bmatrix}$$

where Σ_0 is a matrix of contemporaneous variances and covariances, A and P_{ii} are triangular matrices and P_{ij} is a matrix with just the (1, 1)'th element being non-zero. The matrix A is selected such that $\Sigma = A\Sigma_0 A'$. One simple way of achieving this specification is to express Σ as $H_1 H_1'$ and Σ_0 as $H_2 H_2'$ with H_1 and H_2 as lower triangular matrices, and then to take A as $H_1 H_2^{-1}$.

Now, we can choose P to be a block triangular matrix given by

$$P_{(1)} = \begin{bmatrix} P_{11} & 0 & \cdots & 0 \\ P_{21} & P_{22} & \cdots & 0 \\ \vdots & \vdots & & \vdots \\ P_{M1} & P_{M2} & \cdots & P_{MM} \end{bmatrix} . \tag{7.10}$$

If we consider the elements of $\overset{*}{y} = P_{(1)} y$, we find that

$$\begin{aligned} y^*_{1i} &= \sum_{k=1}^{i} \alpha_{ik} y_{1k} \\ y^*_{ti} &= y_{ti} - \rho_i y_{t-1i} \qquad (t = 2, 3, \ldots, T) \end{aligned} \tag{7.11}$$

from which we see that for any particular equation, all of the transformed observations except the first are linear functions of current and lagged observations in that equation, while the first transformed observation in each case depends on the first observations of the other equations too. This essentially ensures that all transformed variables have an identical covariance structure.

Another particular choice of P presented by Parks (1967) is to take it to be block-diagonal:

$$P^* = \begin{bmatrix} P^*_{11} & 0 & \cdots & 0 \\ 0 & P^*_{22} & \cdots & 0 \\ \vdots & \vdots & & \vdots \\ 0 & 0 & & P^*_{MM} \end{bmatrix} \tag{7.12}$$

where P^*_{ii} is equal to P_{ii} with $\alpha_i = (1 - \rho_i^2)^{-1/2}$. In general such a specification violates the stationarity assumption associated with the autoregressive process, but this difficulty is overcome if we assume that

$$E(\epsilon_{ti}\epsilon_{tj}) = \begin{cases} \sigma_{ij}\left[\dfrac{(1-\rho_i^2)^{1/2}(1-\rho_j^2)^{1/2}}{(1-\rho_i\rho_j)}\right] & \text{if } t = 1 \\ \sigma_{ij} & \text{if } t \neq 1. \end{cases}$$

If the first observation on each equation is ignored, the choice of P may be simplified further by taking it to be the $[M(T-1) \times MT]$ matrix $P_{(0)}$, given by

$$P_{(0)} = \begin{bmatrix} P_{(0)11} & 0 & \cdots & 0 \\ 0 & P_{(0)22} & \cdots & 0 \\ \vdots & \vdots & & \vdots \\ 0 & 0 & \cdots & P_{(0)MM} \end{bmatrix}, \tag{7.13}$$

where $P_{(0)ii}$ is a $[(T-1) \times T]$ matrix obtained by the deletion of the first row vector in P_{ii}.

Now, for a given choice of P, in order to estimate the coefficient vector β in (7.1) we transform the original SURE model to obtain

$$Py = PX\beta + Pu$$

or,

$$\overset{*}{y} = \overset{*}{X}\beta + \overset{*}{u} , \tag{7.14}$$

so that from (7.8), we have

$$E(\overset{*}{u}) = PE(u) = 0$$

$$\begin{aligned} E(\overset{*}{u}\overset{*}{u}') &= PE(uu')P' \\ &= P\Omega P' \\ &= (\Sigma \otimes I_T) . \end{aligned} \tag{7.15}$$

If we take $P = P^*$, then we obtain a set of linear regression models, each transformed according to the procedure of Prais and Winsten (1954). On the other hand, choosing P as $P_{(0)}$ yields a set of linear regression models each transformed according to the recommendation of Cochrane and Orcutt (1949). Clearly, the transformed model (7.14) is precisely in the form of the conventional SURE model, and therefore all of the estimators of β discussed previously can be applied here. For example, let us consider the OLS and FGLS estimators

$$\begin{aligned} b_0 &= (\overset{*}{X}'\overset{*}{X})^{-1}\overset{*}{X}'\overset{*}{y} \\ &= (X'P'PX)^{-1}X'P'Py , \\ b_F &= [\overset{*}{X}'(S^{-1} \otimes I_T)\overset{*}{X}]^{-1}\overset{*}{X}'(S^{-1} \otimes I_T)\overset{*}{y} \\ &= [X'P'(S^{-1} \otimes I_T)PX]^{-1}X'P'(S^{-1} \otimes I_T)Py , \end{aligned} \tag{7.16}$$

where S denotes an estimator of Σ.

It should be noted that both the OLS and FGLS estimators typically require knowledge of the autoregressive parameters, $\rho_1, \rho_2, \ldots, \rho_M$. In practice these parameters are generally unknown, of course, in which case we may replace them in (7.16) by suitable consistent estimators. Some such estimators of ρ_i that are often used in practice are:

$$\hat{\rho}_{i(PW)} = \frac{\sum_{t=2}^{T-1} \hat{u}_{ti}\hat{u}_{t-1i}}{\sum_{t=2}^{T-1} \hat{u}_{ti}^2}, \qquad \hat{\rho}_{i(CO)} = \frac{\sum_{t=2}^{T} \hat{u}_{ti}\hat{u}_{t-1i}}{\sum_{t=1}^{T-1} \hat{u}_{ti}^2},$$

$$\hat{\rho}_{i(T)} = \frac{(T - K_i)\sum_{t=2}^{T} \hat{u}_{ti}\hat{u}_{t-1i}}{(T-1)\sum_{t=1}^{T} \hat{u}_{ti}^2}, \qquad \hat{\rho}_{i(DW)} = \left(1 - \frac{1}{2}d_i\right),$$

$$\hat{\rho}_{i(TN)} = \frac{1}{(T^2 - K_i^2)}\left[T^2\left(1 - \frac{1}{2}d_i\right)^2 + K_i^2\right], \qquad \hat{\rho}_{i(KG)} = \frac{T\sum_{t=2}^{T} \hat{u}_{ti}\hat{u}_{t-1i}}{(T-1)\sum_{t=1}^{T} \hat{u}_{ti}^2}, \tag{7.17}$$

$$\hat{\rho}_i = \frac{\sum_{t=2}^{T} \hat{u}_{ti}\hat{u}_{t-1i}}{\sum_{t=2}^{T} \hat{u}_{ti}^2}$$

where $\hat{u}_{ti}$ denotes an OLS residual and d_i is the Durbin-Watson statistic for the i'th equation:

$$d_i = \frac{\sum_{t=2}^{T} (\hat{u}_{ti} - \hat{u}_{t-1i})^2}{\sum_{t=1}^{T} \hat{u}_{ti}^2} \qquad (i = 1, \ldots, M).$$

The estimators $\hat{\rho}_{i(PW)}$ and $\hat{\rho}_{i(CO)}$ are obtained by minimizing the sum of squared errors from the i'th equation transformed by P_{ii}^* and $P_{(0)ii}$ respectively, conditional on the estimate of β_i. The symbols PW and CO in the subscripts refer to the Prais-Winsten and Cochrane-Orcutt procedures respectively. The estimator $\hat{\rho}_{i(T)}$ was mentioned by Theil (1971; p. 254) while $\hat{\rho}_{i(DW)}$ refers to the estimator resulting from the use of the Durbin-Watson statistic for testing the significance of the autocorrelation coefficients. The estimator $\hat{\rho}_{i(TN)}$ was proposed by Theil and Nagar (1961), and Kmenta and Gilbert (1970) used the estimator $\hat{\rho}_{i(KG)}$ in their Monte Carlo study. The estimator $\hat{\rho}_i$ is another estimator that has been used in a number of other studies. Iterative versions of these estimators of the autoregressive parameters also can be considered. For instance, one might start with the OLS residuals and use an estimate of ρ_i to obtain the FGLS estimator of β_i. This leads to another set of residuals that can be used to obtain another estimate of ρ_i and so on, until convergence.

In addition to the above estimators we can use, equation by equation in this more general context of the SURE model, the single equation maximum likelihood estimator considered by Beach and MacKinnon (1978). This approach involves obtaining a root of the following cubic equation, the first-order condition for a maximum of the likelihood function, conditional on the estimated β_i:

$$\rho_i^3 + a_i\rho_i^2 + b_i\rho_i + c_i = 0,$$

where

$$a_i = \frac{-(T-2)\sum_{t=2}^{T} \hat{u}_{ti}\hat{u}_{t-1i}}{(T-1)\sum_{t=3}^{T} \hat{u}_{t-1i}^2}, \qquad b_i = \frac{-T\sum_{t=3}^{T} \hat{u}_{t-1i}^2 - \sum_{t=3}^{T} \hat{u}_{ti}^2 - 2\hat{u}_{1i}^2}{(T-1)\sum_{t=3}^{T} \hat{u}_{t-1i}^2},$$

$$c_i = \frac{T\sum_{t=2}^{T} \hat{u}_{ti}\hat{u}_{t-1i}}{(T-1)\sum_{t=3}^{T} \hat{u}_{ti}^2}.$$

The desired solution is given by

$$\tilde{\rho}_i = -2\left(\frac{1}{3}a_i^2 - b_i\right)^{1/2} \cos\left(\frac{\pi + \phi_i}{3}\right) - \frac{1}{3}a_i \tag{7.18}$$

with ϕ_i, lying between 0 and π radians, defined by

$$\phi_i = \cos^{-1}\left[\frac{\sqrt{3}\left(3c_i - a_i b_i + \frac{2}{9}a_i^3\right)}{2\left(b_i - \frac{1}{3}a_i^2\right)\left(\frac{1}{3}a_i^2 - b_i\right)^{1/2}}\right].$$

Now, let us consider the general strategy of estimating β by some two-stage procedure. In general, we might apply a chosen estimation method to the original model (7.1), and using this estimate of β we can obtain the residuals and use them to estimate the autocorrelation coefficients. Using the latter estimated parameters, we can apply the same or some other estimation method to the transformed model (7.14). The first stage of this overall estimation strategy ignores the autoregressive nature of the model's disturbance structure, while the second stage explicitly accounts for it. For instance, suppose that the OLS estimates of the β_i's are obtained from the original model (7.1). These estimates ignore both the contemporaneous and intertemporal correlations between the disturbances. Using the OLS estimates of the β_i's, estimates of the ρ_i's can be obtained. These estimates are then substituted for the ρ_i's in the OLS estimator (7.16). This constitutes the second stage, and so we obtain the OLS-OLS estimator of β. This estimator allows for the autocorrelated nature of the disturbances but fails to account for the contemporaneous correlations between the disturbances across the different equations of the model. Now, this estimator of β in turn implies a new set of residuals which can be used to form an estimate, S, of Σ. Using this estimate in the FGLS estimator (7.16) leads to what may be termed the OLS-OLS-FGLS estimator of β. Further iterations of this process to convergence provides a class of iterative estimators of $\boldsymbol{\beta}$. Using the maximum likelihood method, we can define a similar iterative estimator (see Magnus, 1978). Several variants of this technique can be developed by changing the estimation procedures in the different stages. Thus, by combining different choices of estimators, such as SUUR and SURR; different choices of the transformation matrix, such as $P_{(1)}$, P^* and P_0; and different choices of estimators of the ρ_i's at different stages, we can generate a variety of iterative and noniterative estimators for the parameters of the SURE model with scalar autoregressive disturbances.

Parks (1967) considered the first round (noniterative) estimator of β based on P^* and $\hat{\rho}_{i(PW)}$ along with the divisor $\sqrt{(T - K_i)(T - K_j)}$ instead of T in s_{ij}. He established that this estimator is consistent, and showed that the asymptotic distribution of $\sqrt{T}$ times the estimation error is multivariate Normal with mean vector 0 and variance covariance matrix $\lim_{T\to\infty} T(X'\Omega^{-1}X)^{-1}$. Similar results can be derived for the other estimators. Based on such asymptotic properties, we can broadly classify the estimators of β in this context into four categories. The first category comprises those estimators that take no account of either the autoregressive or contemporaneously correlated nature of the disturbances. One example of this class is the OLS estimator from the model (7.1). The second category consists of estimators such as SUUR and SURR which ignore the autoregressive nature of the disturbances, but recognize the contemporaneous correlations across the disturbances of different equations. Estimators which recognize the autoregressive disturbances, but fail to take account of the contemporaneously correlated nature of the disturbances across the equations of the model form the third category of estimators. An example of this class is the OLS-OLS estimator. Finally, estimators which recognize both the autoregressive and contemporaneously correlated nature of the disturbances constitute the fourth category. The estimators in this last category will be asymptotically more efficient than those in the other categories, while the members of the first category of estimators will be the least efficient, asymptotically. Of course, all of the estimators discussed above are consistent.

7.3 FIRST-ORDER VECTOR AUTOREGRESSIVE DISTURBANCES

The disturbance specification (7.2) limits the extent to which intertemporal correlations arise in the model. In particular, in each equation, the disturbance in a given period is related only to the disturbance associated with that same equation in the previous period. There are no cross-equation correlations between the disturbances over time. (There are, of course, contemporaneous cross-equation correlations between the disturbances.) We now generalize this aspect of the model's specification by writing

$$\begin{bmatrix} u_{t1} \\ u_{t2} \\ \vdots \\ u_{tM} \end{bmatrix} = \begin{bmatrix} \rho_{11} & \rho_{12} & \cdots & \rho_{1M} \\ \rho_{21} & \rho_{22} & \cdots & \rho_{2M} \\ \vdots & \vdots & & \vdots \\ \rho_{M1} & \rho_{M2} & \cdots & \rho_{MM} \end{bmatrix} \begin{bmatrix} u_{t-11} \\ u_{t-12} \\ \vdots \\ u_{t-1M} \end{bmatrix} + \begin{bmatrix} \epsilon_{t1} \\ \epsilon_{t2} \\ \vdots \\ \epsilon_{tM} \end{bmatrix}$$

or

$$u_{(t)} = Ru_{(t-1)} + \epsilon_{(t)} . \tag{7.19}$$

This defines a first-order vector autoregressive scheme for the model's disturbances. Here, R is an $(M \times M)$ matrix of unknown autoregressive parameters and the ϵ_{ti}'s are errors assumed to have the same properties as in (7.3).

Notice that this autoregressive process is stationary if the absolute values of all the characteristic roots of R are less than unity. Further, we have

$$E(u_{(t)}u'_{(t)}) = \Theta \ , \tag{7.20}$$

which can be expressed as

$$\Theta = R\Theta R' + \Sigma \ . \tag{7.21}$$

If $E(uu') = \Omega$, we can find a nonsingular matrix Q such that

$$\Omega^{-1} = Q'(\Sigma^{-1} \otimes I_T)Q \ . \tag{7.22}$$

Given such a Q, model (7.1) may be transformed to obtain a specification in which the disturbances are temporally independent. As in the case of a scalar autoregressive process, Judge, Griffiths, Hill and Lee (1980; pp. 269-70) indicated a possible choice of Q as follows. Remembering that Θ is a symmetric matrix, we can use (7.21) to provide a set of equations, the solution to which gives the elements of Θ in terms of those of R and Σ. Using these expressions for the elements of Θ we can determine a lower triangular matrix A as defined in (7.9) such that $\Sigma = A\Theta A'$. For instance, we may choose A as $H_1 H_2^{-1}$ where H_1 and H_2 are lower triangular matrices satisfying $\Sigma = H_1 H_1'$ and $\Theta = H_2 H_2'$. Let us define the lower triangular matrices Q_{ij}'s of order $(T \times T)$ as

$$Q_{ii} = \begin{bmatrix} \alpha_{ii} & 0 & 0 & \cdots & 0 & 0 \\ -\rho_{ii} & 1 & 0 & \cdots & 0 & 0 \\ 0 & -\rho_{ii} & 1 & \cdots & 0 & 0 \\ \vdots & \vdots & \vdots & & \vdots & \vdots \\ 0 & 0 & 0 & \cdots & -\rho_{ii} & 1 \end{bmatrix} , \tag{7.23}$$

$$\underset{(i \neq j)}{Q_{ij}} = \begin{bmatrix} \alpha_{ij} & 0 & 0 & \cdots & 0 & 0 \\ -\rho_{ij} & 0 & 0 & \cdots & 0 & 0 \\ 0 & -\rho_{ij} & 0 & \cdots & 0 & 0 \\ \vdots & \vdots & \vdots & & \vdots & \vdots \\ 0 & 0 & 0 & \cdots & -\rho_{ij} & 0 \end{bmatrix} . \tag{7.24}$$

Now, one interesting choice of Q is

$$Q_{(1)} = \begin{bmatrix} Q_{11} & Q_{12} & \cdots & Q_{1M} \\ Q_{21} & Q_{22} & \cdots & Q_{2M} \\ \vdots & \vdots & & \vdots \\ Q_{M1} & Q_{M2} & \cdots & Q_{MM} \end{bmatrix} . \qquad (7.25)$$

If $Q_{(0)ij}$ denotes the $[(T-1) \times T]$ matrix obtained from Q_{ij} by the deletion of the first row, another choice of Q considered by Guilky and Schmidt (1973) is

$$Q_{(0)} = \begin{bmatrix} Q_{(0)11} & Q_{(0)12} & \cdots & Q_{(0)1M} \\ Q_{(0)21} & Q_{(0)22} & \cdots & Q_{(0)2M} \\ \vdots & \vdots & & \vdots \\ Q_{(0)M1} & Q_{(0)M2} & \cdots & Q_{(0)MM} \end{bmatrix} . \qquad (7.26)$$

It is clear that the matrices $Q_{(1)}$ and $Q_{(0)}$ collapse to $P_{(1)}$ in (7.10) and $P_{(0)}$ in (7.13) respectively when R is diagonal.

Now, transforming the model (7.1) by the matrix Q, we get

$$Qy = QX\beta + Qu$$

or

$$\overset{\oplus}{y} = \overset{\oplus}{X}\beta + \overset{\oplus}{u} \qquad (7.27)$$

where, from (7.22),

$$E(\overset{\oplus}{u}) = 0 , \quad E(\overset{\oplus}{u}\overset{\oplus}{u}{}') = (\Sigma \otimes I_T) . \qquad (7.28)$$

This transformed model is now in the form of the conventional SURE model, and the various estimators described in earlier chapters can be applied directly. For example, the OLS and FGLS estimators of β are given by

$$
\begin{aligned}
b_0 &= (\overset{\oplus}{X}{}'\overset{\oplus}{X})^{-1}\overset{\oplus}{X}{}'\overset{\oplus}{y} \\
&= (X'Q'QX)^{-1}X'Q'Qy \\
b_F &= [\overset{\oplus}{X}{}'(S^{-1} \otimes I_T)\overset{\oplus}{X}]^{-1}\overset{\oplus}{X}{}'(S^{-1} \otimes I_T)\overset{\oplus}{y} \\
&= [X'Q'(S^{-1} \otimes I_T)QX]^{-1}X'Q'(S^{-1} \otimes I_T)Qy \, .
\end{aligned}
\qquad (7.29)
$$

If we choose $Q = Q_{(0)}$, the transformed variables in $\overset{\oplus}{y}$ depend upon the unknown autoregressive parameters (ρ_{ij}), while if we take $Q = Q_{(1)}$, they depend not only on these autoregressive parameters, but also on the unknown elements of Σ. Further, it is clear that neither the OLS nor FGLS estimators are directly computable because each of them depends on the matrix Q, whose elements are unknown. The obvious solution in such situations is to replace these unknowns by suitable estimates. Guilky and Schmidt (1973) considered this issue and suggested one such estimation procedure when the choice of Q is $Q_{(0)}$. Their estimator consists of first applying the OLS estimator to the original model (7.1) and obtaining the OLS residuals. These residuals are then substituted in place of the unknown disturbances, the u_{ti}'s, in (7.19) and again the OLS method is used to estimate the elements of R. These estimates are then substituted for the ρ_{ij}'s in $Q_{(0)}$. Let us denote this estimated matrix by $\hat{Q}_{(0)}$. Setting $Q = \hat{Q}_{(0)}$ in the formula for b_0 in (7.29) provides the adaptive counterpart to that OLS estimator:

$$
b_{A0(0)} = (X'\hat{Q}'_{(0)}\hat{Q}_{(0)}X)^{-1}X'\hat{Q}'_{(0)}\hat{Q}_{(0)}y \, . \qquad (7.30)
$$

To obtain the adaptive version of the FGLS estimator b_F in (7.28), we use $b_{A0(0)}$ to develop a consistent estimator, say $\tilde{S}$, of Σ. Setting $S = \tilde{S}$ in b_F, we get the SURR type estimator of β:

$$
b_{ASR(0)} = [X'\hat{Q}'_{(0)}(\tilde{S}^{-1} \otimes I_T)\hat{Q}_{(0)}X]^{-1}X'\hat{Q}'_{(0)}(\tilde{S}^{-1} \otimes I_T)\hat{Q}_{(0)}y \, . \qquad (7.31)
$$

If the transformation matrix Q is chosen to be $Q_{(1)}$, the adaptive versions of the OLS and FGLS estimators (7.29) are given by

$$
\begin{aligned}
b_{A0(1)} &= (X'\hat{Q}'_{(1)}\hat{Q}_{(1)}X)^{-1}X'\hat{Q}'_{(1)}\hat{Q}_{(1)}y \\
b_{ASR(1)} &= [X'\hat{Q}'_{(1)}(\tilde{S}^{-1} \otimes I_T)\hat{Q}_{(1)}X]^{-1}X'\hat{Q}'_{(1)}(\tilde{S}^{-1} \otimes I_T)\hat{Q}_{(1)}y \, .
\end{aligned}
\qquad (7.32)
$$

As indicated earlier in the context of scalar first-order autoregressive disturbances, several adaptive estimators of both the iterative and noniterative types can be developed in the present context too, and each will be consistent for β.

Finally, let us make use of the analysis of Beach and MacKinnon (1979), in order to obtain the maximum likelihood estimator of β in this context. We can write

$$u_{(1)} = D^{-1}\epsilon_{(1)} , \tag{7.33}$$

where D is the nonsingular matrix chosen to ensure the stationarity of the autoregressive process (7.19). Combining (7.19) and (7.33),

$$\begin{bmatrix} \epsilon_{(1)} \\ \epsilon_{(2)} \\ \epsilon_{(3)} \\ \vdots \\ \epsilon_{(T)} \end{bmatrix} = \begin{bmatrix} D & 0 & 0 & \cdots & 0 & 0 \\ -R & I_M & 0 & \cdots & 0 & 0 \\ 0 & -R & I_M & \cdots & 0 & 0 \\ \vdots & \vdots & \vdots & & \vdots & \vdots \\ 0 & 0 & 0 & \cdots & -R & I_M \end{bmatrix} \begin{bmatrix} u_{(1)} \\ u_{(2)} \\ u_{(3)} \\ \vdots \\ u_{(T)} \end{bmatrix} . \tag{7.34}$$

As $E(u_{(t)}u'_{(t)}) = \Theta$ for all t, it follows from (7.21) and (7.33) that

$$\Theta = D^{-1}\Sigma D'^{-1} = R\Theta R' + \Sigma , \tag{7.35}$$

from which we obtain

$$|D| = [|\Theta - R\Theta R'| / |\Theta|]^{\frac{1}{2}} . \tag{7.36}$$

Assuming normality of the disturbances, the log likelihood function of $(\epsilon_{(1)}, \epsilon_{(2)}, \ldots, \epsilon_{(T)})$ is

$$\text{Constant} - \frac{T}{2} \log |\Sigma| - \frac{1}{2} \sum_{t=1}^{T} \epsilon'_{(t)} \Sigma^{-1} \epsilon_{(t)} \tag{7.37}$$

which, using (7.34) and (7.36), provides the log likelihood function of $(u_{(1)}, u_{(2)}, \ldots, u_{(T)})$ as

$$L_{(1)} = \text{Constant} - \frac{T}{2}\log|\Theta - R\Theta R'| + \frac{1}{2}\log\left(\frac{|\Theta - R\Theta R'|}{|\Theta|}\right)$$

$$- \frac{1}{2}u'_{(1)}\Theta^{-1}u_{(1)} - \frac{1}{2}\sum_{t=2}^{T}(u_{(t)} - Ru_{(t-1)})'(\Theta - R\Theta R')^{-1}(u_{(t)} - Ru_{(t-1)}) \,. \tag{7.38}$$

In practice, one often discards the first observation in the sample. Notice that if the initial observation is dropped, the log likelihood function is simply

$$L_{(0)} = \text{Constant} - \left(\frac{T-1}{2}\right)\log|\Theta - R\Theta R'|$$

$$- \frac{1}{2}\sum_{t=2}^{T}(u_{(t)} - Ru_{(t-1)})'(\Theta - R\Theta R')^{-1}(u_{(t)} - Ru_{(t-1)}) \,. \tag{7.39}$$

Now, to consider the stationarity of (7.19), let Γ be an orthogonal matrix that diagonalizes R, so that $\Gamma R\Gamma' = \Lambda$ is a diagonal matrix with nonzero diagonal elements $\lambda_1, \lambda_2, \cdots, \lambda_M$ which are the characteristic roots of R. Consider the determinant

$$\begin{aligned}|D| &= [|\Theta - \Gamma'\Lambda\Gamma\Theta\Gamma'\Lambda\Gamma| / |\Theta|]^{\frac{1}{2}} \\ &= [|\Gamma\Theta\Gamma' - \Lambda\Gamma\Theta\Gamma'\Lambda| / |\Gamma\Theta\Gamma'|]^{\frac{1}{2}} \,.\end{aligned} \tag{7.40}$$

From Rao (1973; p. 74, results 20.2),

$$|\Gamma\Theta\Gamma' - \Lambda\Gamma\Theta\Gamma'\Lambda| \le \prod_{i=1}^{M}(1 - \lambda_i^2)\overset{*}{\theta}_i$$

$$|\Gamma\Theta\Gamma'|^{-1} \ge \prod_{i=1}^{M}(\overset{*}{\theta}_i)^{-1}$$

where $\overset{*}{\theta}_i$ denotes the i'th diagonal element of the matrix $\overset{*}{\Theta} = \Gamma\Theta\Gamma'$. Thus, we have

$$|D| \le \prod_{i=1}^{M}(1 - \lambda_i^2) \,, \tag{7.41}$$

from which it follows that if the determinant of D is bounded away from 0, all of the characteristic roots of R lie within the unit circle, which is the condition for the autoregressive process (7.19) to be stationary. With this result in mind, let us compare $L_{(1)}$ and $L_{(0)}$. We see that the third term on the right hand side of $L_{(1)}$ given by (7.38) is just the logarithm of $|D|$, and this determinant is the Jacobian of the transformation from the $\epsilon_{(t)}$'s to the $u_{(t)}$'s. This constrains the numerator and denominator of this term to be of the same sign, and this can be so only if all of the roots of R lie in the unit circle (see Beach and MacKinnon, 1979; pp. 461-464). Further, $L_{(0)}$ differs from $L_{(1)}$ with respect to the term $u'_{(1)}\Sigma^{-1}u_{(1)}$, indicating that the initial disturbances do have an influence on the form of the likelihood function, and hence on the form of the maximum likelihood estimator of β.

The maximum likelihood estimator of β can be derived by maximizing $L_{(1)}$ with respect to the elements of Θ, R and β. The associated first-order conditions are, unfortunately, nonlinear in a rather complicated manner, and therefore do not lead to any neat analytical solution. Numerical methods have to be used to solve them. The same situation arises if we choose to maximize $L_{(0)}$, rather than $L_{(1)}$, but the maximization of $L_{(1)}$ may involve a somewhat heavier computational cost than does the maximization of $L_{(0)}$. Despite the fact that the estimators of β arising from the maximization of $L_{(1)}$ and $L_{(0)}$ have identical asymptotic properties, Beach and MacKinnon (1979) recommended the use of $L_{(1)}$ in preference to $L_{(0)}$, because the former takes account of the stationarity condition associated with the autoregressive error structure, fully uses all of the observations in the sample, and simplifies the problem of constructing tests of hypotheses concerning the elements of R.

7.4 EFFICIENCY COMPARISONS

Apart from the asymptotic properties of the above estimators of β in the presence of autoregressive disturbances, no analytical results relating to their statistical properties are available in the literature. However, some light on their finite-sample properties has been shed by Monte Carlo studies, to which we turn next.

Let us first consider the case in which R is diagonal, so that the disturbances in each equation follow a first-order scalar autoregressive scheme. Kmenta and Gilbert (1970) considered a two equation model in which each equation contained only one regressor besides the intercept term, and they examined three such models, characterized by different degrees of correlation between the disturbances across the equations. Their Monte Carlo study was based on 100 replications with samples of 10, 20 and 100 observations, and values of the disturbances were drawn randomly from a Normal distribution. The values of ρ_1 and ρ_2 were 0.8 and 0.6 respectively, characterizing the autoregressive nature of the disturbances in the two equations. To estimate the regression coefficients, the OLS estimator from the first category and the SURR estimator from the second of the four categories distinguished at the end of Section 7.2 were chosen. From the third of these categories, two estimators were chosen. One was based on the application of OLS at all three steps, while the second was derived by minimizing the sum of the squared errors with respect to the autocorrelation coefficient, the regression coefficient and the intercept term. This we shall term the nonlinear ordinary least squares (NOLS) estimator. Lastly, four estimators were selected from the fourth category of estimators. Following Parks (1967), three of these were OLS-OLS-SURR, SURR-SURR and NOLS-SURR. The fourth estimator was obtained by minimizing the weighted sum of squared errors (as in the case of the general linear regression model) with respect to all unknown parameters, and will be called the nonlinear generalised least squares (NGLS) estimator. In all cases the transformation matrix P* and the autocorrelation coefficient estimator $\hat{\rho}_{i(KG)}$ were used.

Their Monte Carlo study supported the asymptotic result that members of the fourth category of estimators that we have distinguished are more efficient than the others. Virtually no loss in efficiency occurred in applying these estimators when the disturbances were neither autocorrelated nor contemporaneously correlated, nor both (see also Kmenta and Gilbert, 1968). When the disturbances were autoregressive the efficiencies of the estimators that account for the non-zero cross-equation correlations, relative to those estimators which ignore it, declined as the degree of contemporaneous correlation increased and as T decreased. The SURR-SURR estimator was found to be superior to the OLS-OLS-SURR estimator, but the magnitude of the gain in relative efficiency was usually small, and approached zero for large samples. A similar pattern was observed with the NOLS-SURR and NGLS estimators, with the latter performing better than the former in terms of efficiency.

Using the above simulation design, Maeshiro (1980) carried out another Monte Carlo study which incorporated some additional features. In particular, it included a set of observations on trended variables. Samples of size 50 were also considered, and the autoregressive parameters ρ_1 and ρ_2 were assigned values 0.1 (0.1) 0.9. Two schemes were specified for the generation of the initial observations. The first scheme assumed the u_{0i}'s to be independent of the ϵ_{0i}'s, and $E(u_{0i}u_{0j}) = \sigma_{ij}/(1 - \rho_i\rho_j)$, while the second scheme assumed that

$$u_{1i} = \left(\frac{\sigma_{ii}}{1-\rho_i^2}\right)^{1/2} \epsilon_{1i} .$$

In all, five estimators were considered, with ρ_1 and ρ_2 assumed to be known. The first two estimators were obtained by the application of OLS to the original model and the transformed model respectively. The next two estimators were similarly obtained by the application of the SURR estimator instead of

OLS. The fifth estimator was derived by applying GLS to the transformed model.

Maeshiro's study revealed marked disparities between the findings based on nontrended and trended data. For instance, when the variables were not trended, the SURR estimator from the transformed model was more or less as efficient as the GLS estimator obtained from the transformed model, but was clearly superior to the OLS estimators from the original as well as the transformed models. The reverse was observed when the variables were trended. In the context of the original model with trended observations, the OLS estimator for the coefficient in the first equation outperformed the SURR estimator when a high value of ρ_1 was accompanied by a low to moderate value of ρ_2. Further, looking at the pattern of the efficiency gain that can be achieved with estimators that recognize the presence of autocorrelation and contemporaneous correlation across the equations, it was found that the SURR estimators obtained from the original model or from the transformed model often were inferior to the OLS estimator from the transformed model, especially when the variables were trended. Maeshiro offered an interesting explanation for this finding. In the trended data case, the transformation of the model results in high multicollinearity which, in turn, reduces the efficiency of the estimators. This loss tends to reduce the gain in efficiency that would otherwise be obtained by taking into account the autoregressive nature of the disturbances. Consequently, an estimator taking account of the disturbance autocorrelation may often turn out to be less efficient than one which ignores it. Similarly, the effect of inter-equation multicollinearity may have a much greater effect on estimator efficiency when the regressors are highly trended in the sample, than otherwise would be the case.

When the observations are trended, the gain in efficiency of the SURR estimator applied to the transformed model has two components, one due to the fact that the disturbances are autoregressive, and the other due to the fact that they are correlated across the equations. The total gain in efficiency due to these two factors frequently may be smaller than the loss in efficiency due to the reduction in the variation of autoregressively transformed variables. Further, ignoring both the autoregressive as well as the contemporaneously correlated nature of the disturbances appears to be a better estimation strategy than one which ignores only one of these two features of the disturbances, especially in small samples. The most appropriate strategy is to apply some estimator to the transformed model that simultaneously takes care of the autoregression and the contemporaneous correlation across the equations.

Guilkey and Schmidt (1973) considered the case of first-order vector autoregression in the disturbances of a two equation SURE model in which each equation contained an intercept term and a regressor. In all, eight specifications of the model were taken, these differing with regard to both contemporaneous and intertemporal correlations between the disturbances.

The values of the two regressors in the model were the same as those chosen by Kmenta and Gilbert (1970), and normally distributed disturbances were generated. Five estimators were examined. The first two were OLS and the SURR estimator, while the third and fourth were the OLS-OLS-SURR estimator with R taken to be diagonal and nondiagonal respectively. The fifth estimator was obtained by first applying OLS to the original model, and then applying the SURR estimator, where the elements of Σ and R were estimated from the OLS residuals. The entire investigation was based on 100 replications for $T = 20$, 50 and 100. This study also revealed that the asymptotic properties of the various estimators generally remain a reliable guide for judging the performance of the estimators in samples as small as 20. There appeared to be only a small loss in efficiency by unnecessarily using estimators in the fourth category (noted at the end of Section 7.2) when in fact there was no autocorrelation. On the other hand, when autocorrelation was present, it was found that a substantial gain in efficiency could be achieved by using these estimators in preference to the others mentioned.

Doran and Griffiths (1983) considered a three equation SURE model in which each equation contained an intercept term and two regressors. The observations on these regressors were generated in three ways. The first method produced smooth and identical regressors while the second method ensured smooth but not identical regressors. The third method provided neither smooth nor identical regressors. By smoothness of a regressor, we mean that the difference between the current and preceding observations on it is of smaller order than the observation itself. Such a condition holds, for instance, when the regressor has a slow trend. One thousand replications were used for the two cases $T = 20$ and $T = 40$, and two choices of Σ and seven of R were made. The OLS and SURR estimators applied to the original model were considered, as were the estimators $b_{ASR(0)}$ and $b_{ASR(1)}$ defined in (7.31) and (7.32), together with their counterparts when R is diagonal. Similar estimators with $\tilde{S}$ replaced by Σ were also considered in order to analyze the loss in efficiency attributable to the estimation of Σ by $\tilde{S}$.

When R is known to be a null matrix, implying the absence of any disturbance autocorrelation, this study revealed that the performance of the OLS estimator worsened as the contemporaneous correlations increased, provided that the regressors were not smooth. However, its performance improved for smooth regressors and became excellent when they were identical in each equation, as expected, as the OLS estimator coincides with the GLS estimator in this case. The SURR estimator emerged as the best choice in all cases. Among the estimators from the fourth category noted at the end of Section 7.2, their comparative performance suggested that assuming R to be diagonal when there is no disturbance autocorrelation at all is preferable to assuming it to be nondiagonal. When R was known to be diagonal, implying that the disturbances follow a scalar autoregressive scheme, the OLS estimator turned out to be the worst choice among those considered. An improvement in its performance was seen when the magnitude of the contemporaneous correlations declined or the regressors were smooth and identical across equations. Accounting for these contemporaneous correlations through the SURR estimator brought a considerable improvement, especially when the regressors were smooth in the sample. Not surprisingly, correctly specifying R to be diagonal resulted in a performance which was superior to those of estimators which wrongly regard R to be nondiagonal.

When R was known to be nondiagonal, implying that the disturbances follow a vector autoregressive scheme, the OLS estimator was found to be quite inefficient when the regressors were not smooth, but its performance improved when they were smooth. Those estimators which assume R to be diagonal, when R was actually nondiagonal, surprisingly turned out to be always superior to the remaining estimators considered, including those based on the correct specification of R. The loss in efficiency arising from the estimation of Σ was found to be small when the contemporaneous correlations between the disturbances were low and the regressors were smooth and identical. This loss increased as R departed from a null matrix. Comparing these estimators under a diagonal R with the corresponding estimators under a nondiagonal R, the former estimators exhibited comparatively smaller efficiency losses than the latter when Σ was estimated. An interesting observation emerging from this is that in general it may be better to take R to be diagonal irrespective of the correctness of the specification, at least as long as R is not in fact null. Doran and Griffiths pointed out that the nondiagonal specification of R reduces the number of degrees of freedom and the use of such imprecise estimates of the elements of the matrix R in the estimation of β may adversely affect the efficiency of the latter estimator.

As a final point we note that the estimators of β suggested in the context of the SURE model with autocorrelated disturbances can be grouped broadly into two classes—one involving estimators that retain the initial observations, and the other involving ones which discard them. Discarding the initial observations means using transformation matrices such as $P_{(0)}$ or $Q_{(0)}$, while retaining the initial observations means using transformation matrices such as $P_{(1)}$, P^* or $Q_{(1)}$. Practical considerations favour ignoring the initial observations, because this leads to some computational simplifications. On the other hand, theoretical considerations demand their inclusion in order to fully utilize the sample information. Both of these considerations have implications for the numerical estimates as well as for the finite-sample properties of the estimators. The impact of the deletion of initial observations on the estimators can be studied by obtaining a mathematical relationship connecting the estimator that discards the initial observations with the estimator which retains them. This exercise is not difficult for single equation models (see, for example, Poirier, 1978) but is a much more complicated task in the case of SURE models. Such a relationship would, however, enable us to investigate any changes to the estimators resulting from the deletion of the initial observations. Kmenta and Gilbert (1970) touched on this issue and remarked that the inclusion of the initial observations is not important from the point of view of efficiency. Doran and Griffiths (1983) examined this point in more detail. They observed that the loss in efficiency due to the deletion of the initial observations declined as the degree of contemporaneous correlation in the disturbances increased, provided that the matrix R is diagonal. For the nondiagonal case, no clear pattern emerged. If R is diagonal and if an estimator from the fourth (and most appropriate) of the categories noted at the end of Section 7.2 is based on this diagonal R specification, there was very little efficiency loss due to the deletion of the initial observations when Σ and R are known. For the nondiagonal R case, no systematic pattern could be detected. When Σ and R were estimated from the data in order to get feasible estimators, substantial losses in efficiency were found. These losses were much larger than the loss in efficiency due to the deletion of the initial observations.

7.5 TESTING THE AUTOREGRESSIVE STRUCTURE

The Monte Carlo evidence on the efficiency properties of alternative estimators of the SURE model with autoregressive disturbances clearly illustrates that this efficiency depends upon the form of the matrix R specifying the autocorrelation process. Therefore it is important to be able to test any hypothesis about the structure of R, and to choose the appropriate estimation procedure accordingly. Guilkey (1974) considered this problem and presented some asymptotic tests. To discuss his test, let us write

$$\underset{[(T-1)\times M]}{V} = \begin{bmatrix} u_{11} & u_{12} & \cdots & u_{1M} \\ u_{21} & u_{22} & \cdots & u_{2M} \\ \vdots & \vdots & & \vdots \\ u_{T-11} & u_{T-12} & \cdots & u_{T-1M} \end{bmatrix}$$

$$\underset{(M\times 1)}{\delta_i} = \begin{bmatrix} \rho_{i1} \\ \rho_{i2} \\ \vdots \\ \rho_{iM} \end{bmatrix}, \qquad \underset{(M^2\times 1)}{\delta} = \begin{bmatrix} \delta_1 \\ \delta_2 \\ \vdots \\ \delta_M \end{bmatrix}$$

so that from (7.19) we can write

$$u_i^o = V\delta_i + \epsilon_i^o \qquad (i = 1, 2, \ldots, M)$$

or

$$u^o = (I_M \otimes V)\delta + \epsilon^o \tag{7.42}$$

where u_i^o and ϵ_i^o are the column vectors obtained by deleting the first element in u_i and ϵ_i respectively, and

$$u^o = \begin{bmatrix} u_i^o \\ u_2^o \\ \vdots \\ u_M^o \end{bmatrix}, \qquad \epsilon^o = \begin{bmatrix} \epsilon_1^o \\ \epsilon_2^o \\ \vdots \\ \epsilon_M^o \end{bmatrix}.$$

Notice that the vector δ is formed by stacking rows in the matrix R as columns.

Now, to construct an asymptotic test of the structure of R we first obtain the residuals associated with the OLS estimator of the original model (7.1), and form the OLS estimators $\hat{u}^o$ and $\hat{V}$ by substituting these residuals in place of the corresponding disturbances in u^o and V respectively. Replacing u^o and V in (7.42) by $\hat{u}^o$ and $\hat{V}$ respectively and applying OLS, we get an estimator of δ as follows:

$$\hat{\delta} = [I_M \otimes (\hat{V}'\hat{V})^{-1}\hat{V}']\hat{u}^o . \tag{7.43}$$

It can be verified easily that $\hat{\delta}$ is a consistent estimator of δ. Further, the asymptotic distribution of $\sqrt{T}(\hat{\delta} - \delta)$ is multivariate Normal with mean vector 0 and variance covariance matrix

$$\Delta = \left[\Sigma \otimes \text{plim}\left(\frac{1}{T}\hat{V}'\hat{V}\right)^{-1}\right] , \tag{7.44}$$

from which it follows that the asymptotic distribution of the stochastic variable

$$\chi_0^2 = \hat{\delta}'\hat{\Delta}^{-1}\hat{\delta} \tag{7.45}$$

is χ^2 with M^2 degrees of freedom, provided that $R = 0$, where

$$\hat{\Delta} = [S \otimes (\hat{V}'\hat{V})^{-1}]$$

with S being any consistent estimator of Σ. Thus the quantity (7.45) provides a test statistic for the null hypothesis $H_0 : \delta = 0$, or equivalently $H_0 : R = 0$. That is, this statistic may be used to test for serial independence of the disturbances of the SURE model (see Guilkey, 1974; p. 97).

To test the null hypothesis that R is a diagonal matrix, we delete all of the diagonal elements of R present in the vector δ. Let us denote the resulting vector by δ_D, and suppose that $\hat{\delta}_D$ is the estimator of δ_D. Now, to test the null hypothesis $H_0 : \delta_D = 0$, or equivalently the null hypothesis that R is diagonal, we may use the statistic

$$\chi_D^2 = \hat{\delta}_D' \hat{\Delta}_D^{-1} \hat{\delta}_D , \tag{7.46}$$

where $\hat{\Delta}_D$ is obtained from $\hat{\Delta}$ by deleting the rows and columns bearing the following serial numbers: 1, M+2, 2M+3, 3M+4, ..., M(M - 1) + M. It can be shown that the asymptotic distribution of χ_D^2, under the null hypothesis of diagonal R, is χ^2 with M(M - 1) degrees of freedom. This provides an asymptotic test of scalar autocorrelation, against the alternative of a vector autoregressive process for the model's disturbances (see Guilkey, 1974; p. 97).

Next, observe that a consistent estimator $S = ((s_{ij}))$ of Σ may be obtained by taking

$$s_{ij} = \frac{1}{(T-1)}(\hat{u}_i^o - \hat{V}\hat{\delta}_i)'(\hat{u}_j^o - \hat{V}\hat{\delta}_j) . \tag{7.47}$$

If R is known to be a null matrix, then the elements of Σ may be estimated consistently by

$$s_{ij}^o = (\hat{u}_i^o{}'\hat{u}_i^o)/(T-1) \tag{7.48}$$

thus providing S^o, an estimator of Σ.

So, to test the hypothesis $H_0 : \delta = 0$, or $H_0 : R = 0$, we may use the statistic

$$F_o = |S| / |S^o| . \tag{7.49}$$

It may be observed that $F_o^{(T-1)/2}$ is rather like the likelihood ratio statistic, and that the asymptotic distribution of the quantity

$$\begin{aligned} \chi_0^{*2} &= -2 \log F_0^{(T-1)/2} \\ &= (T-1)[\log |S^o| - \log |S|] \end{aligned} \tag{7.50}$$

is χ^2 with M^2 degrees of freedom, if $R = 0$. Thus χ_O^{*2} may serve as a test statistic for the hypothesis $H_0: \delta = 0$, or $H_0: R = 0$.

Similarly, suppose that S^D denotes a consistent estimator, like S^o, of Σ constructed from the OLS residuals in (7.42) under the assumption that the nondiagonal elements of R are zero (i.e., δ is replaced by a vector in which all of the elements are zero except for those bearing the serial numbers 1, M+2, 2M+3, 3M+4, ..., M(M−1)+M). Now, the diagonality of the matrix R can be tested by using the statistic (Guilkey, 1974; p. 97)

$$F_D = |S|/|S_D| , \tag{7.51}$$

because the asymptotic distribution of the quantity

$$\chi_D^{*2} = (T-1)[\log |S^D| - \log |S|] , \tag{7.52}$$

under the hypothesized diagonality of R, is χ^2 with M(M − 1) degrees of freedom.

In order to compare the performances of the asymptotic tests for the nullity and diagonality of R, Guilkey (1974) carried out a Monte Carlo study under the framework specified in Guilkey and Schmidt (1973). Taking two choices of Σ and five of R, eight model specifications were considered for the two equation SURE model, and two test sizes (1% and 5%) were selected. This Monte Carlo study, however, did not reveal any clear-cut preference ordering of the tests; they performed almost equally well for large T. Guilkey remarked that the tests can be assumed to be very reliable for sample sizes exceeding 50. The literature is devoid of any results regarding the performance and properties of these tests of hypotheses in finite samples.

7.6 RESEARCH SUGGESTIONS

A considerable amount of research work is needed into the estimation of SURE models with autocorrelated disturbances. For example, the exact finite-sample properties of various estimators should be studied by means of analytical methods. Some initial efforts may be made to obtain large-sample asymptotic approximations to the properties of the various estimators, along the lines of Magee, Ullah and Srivastava (1987), and the issue of discarding or retaining the initial observations deserves more detailed attention. The application of shrinkage techniques may provide some interesting estimators for the SURE model with autocorrelated disturbances, hopefully with improved sampling properties. Exact tests for first-order autoregressive structures may be developed, and their power functions would need to be studied in detail in finite-sample situations. Based on the outcome of such tests, preliminary test estimators may be designed, for example, in the manner discussed by King and Giles (1984), and their small-sample properties would need to be examined in detail.

Higher-order autoregressive processes for the model's disturbances are sometimes more appropriate than a first-order autoregressive process, but the use of such processes in the context of the SURE model is as yet unexplored. In this regard, it may be interesting to carry out some investigations similar to that in the work reported by Byron (1977), Beach and MacKinnon (1978a), King and Giles (1978) and Szroeter (1978). Another interesting research direction would be to extend the analysis to the case where the disturbances are generated by some trigonometric functions, a moving average process, or an autoregressive moving average process (see Harvey, 1981; Sec. 6.8). Suitable tests may be developed in order to discriminate among various specifications for the disturbance process of the SURE model. The work conducted by Anderson and de Gooijer (1980), Ansley and Newbold (1980), King (1983), Mentz (1977) and Nicholls and Hall (1979) may serve as a useful guide in this regard.

EXERCISES

7.1 Suppose that the disturbances in a SURE model are generated by a first-order vector autoregressive scheme and the same regressors appear in each equation. Obtain the conditions under which the OLS and GLS estimators of β are identical. [Hint: See Harvey (1981; 218-219).]

7.2 Consider a SURE model in which the disturbances follow a second-order autoregressive scheme:

$$u_{ti} = \rho_i u_{t-1i} + \lambda_i u_{t-2i} + \epsilon_i$$

$$(t = 1, 2, \ldots, T; \quad i = 1, 2, \ldots, M)$$

where

$$E(\epsilon_{ti}) = 0 \qquad \text{for all } t \text{ and } i$$

$$E(\epsilon_{ti}\epsilon_{t+sj}) = \begin{cases} \sigma_{ij} & \text{for } s = 0 \text{ and all } t, i \text{ and } j \\ 0 & \text{for } s \neq 0 \text{ and all } t, i \text{ and } j . \end{cases}$$

Derive the variance covariance matrix for the u's and discuss the estimation of β in this model. [Hint: See Judge, Griffiths, Hill and Lee (1980; Sec. 5.2.2).]

7.3 Consider a two equation SURE model:

$$y_1 = x\beta_1 + u_1$$

$$y_2 = e_T\beta_2 + u_2$$

in which e_T is a $(T \times 1)$ vector with all elements unity, the sum of the elements of the vector x is zero, and the disturbances follow a second-order autoregression as specified in Exercise 7.2. Write down the variance covariance matrix of the GLS estimator of the coefficients, and compare it with the variance covariance matrix of the GLS estimator obtained under the incorrect specification that the disturbances follow a first-order autoregression:

$$u_{ti} = \rho_i u_{t-1i} + \epsilon_i$$

$$(t = 1, 2, \ldots, T; \quad i = 1, 2, \ldots, M).$$

7.4 Suppose that quarterly data are available for estimating the parameters in a SURE model and that the disturbances are assumed to be generated by the process

$$u_{ti} = \rho_i u_{t-4i} + \epsilon_i$$

$$(T = 1, 2, \ldots, T; \quad i = 1, 2, \ldots, M)$$

where the ϵ_{ti}'s are temporally independent but contemporaneously correlated. Derive the variance covariance matrix of the disturbances and suggest an appropriate estimator for β.

7.5 If $\Theta = R\Theta R' + \Sigma$, as specified by (7.21) in the context of a SURE model with first-order vector autoregressive disturbances, prove that

$$\Theta = R^m \Theta R'^m + \sum_{k=0}^{m-1} R^k \Sigma R'^k$$

where m is any positive integer. [Hint: Add together equation (7.21) and the equation obtained by pre-multiplying (7.21) by R and post-multiplying by R'.]

7.6 Show that a solution of equation (7.21) for Θ is given by

$$\text{vec}(\Theta) = [I_{M^2} - R \otimes R)]^{-1} \text{vec}(\Sigma) ,$$

and obtain explicit expressions for the elements of Θ in terms of the elements of R and Σ in the case of a two equation SURE model. Determine the matrix $H_1H_2^{-1}$ where H_1 and H_2 are lower triangular matrices satisfying $\Sigma = H_1H_1'$ and $\Theta = H_2H_2'$.

7.7 Obtain the likelihood ratio and Lagrange multiplier statistics for testing the hypothesis of serial independence against the alternative of first-order scalar autoregression in the disturbances of a SURE model.

7.8 Specify a SURE model in which the disturbances follow an autoregressive moving average process of a specified order and discuss how one would obtain the maximum likelihood estimates of the parameters. [Hint: See Harvey (1981; Chap. 6, Sec. 8) and Pagan and Byron (1978).]

REFERENCES

Anderson, O. D., and J. G. de Gooijer (1980). "Distinguishing between IMA (1,) and ARMA (1,1) models: a large scale simulation study of two particular Box-Jenkins time processes," in O. D. Anderson (ed.), Time Series (North Holland, Amsterdam), 15-40.

Ansley, C. F., and P. Newbold (1980). "Finite sample properties of estimators for autoregressive moving average models," Journal of Econometrics 13, 159-183.

Beach, C. M., and J. G. MacKinnon (1978). "A maximum likelihood procedure for regression with autocorrelated errors," Econometrica 46, 51-58.

Beach, C. M., and J. G. MacKinnon (1978a). "Full maximum likelihood estimation of second-order autoregressive error models," Journal of Econometrics 7, 187-198.

Beach, C. M., and J. G. MacKinnon (1979). "Maximum likelihood estimation of singular equation systems with autoregressive disturbances," International Economic Review 20, 459-464.

Byron, R. P. (1977). "Efficient estimation and inference in large econometric systems," Econometrica 45, 1499-1515.

Cochrane, D., and G. H. Orcutt (1949). "Application of least squares regression to relationships containing autocorrelated error terms," Journal of the American Statistical Association 44, 32-61.

Doran, H. E., and W. E. Griffiths (1983). "On the relative efficiency of estimators which include the initial observations in the estimation of seemingly unrelated regressions with first order autoregressive disturbances," Journal of Econometrics 23, 165-191.

Guilkey, D. K. (1974). "Alternative tests for a first-order vector autoregressive error specification," Journal of Econometrics 2, 95-104.

Guilkey, D. K., and P. Schmidt (1973). "Estimation of seemingly unrelated regressions with vector autoregressive errors," Journal of the American Statistical Association 68, 642-647.

Harvey, A. C. (1981). The Econometric Analysis of Time Series (Philip Allan, Oxford).

Judge, G. G., W. E. Griffiths, R. C. Hill and T. S. Lee (1980). The Theory and Practice of Econometrics (Wiley, New York).

King, M. L. (1983). "Testing for autoregressive against moving average errors in the linear regression model," Journal of Econometrics 21, 35-51.

King, M. L., and D. E. A. Giles (1978). "A comparison of some tests for fourth-order autocorrelation," Australian Economic Papers 17, 323-333.

King, M. L., and D. E. A. Giles (1984). "Autocorrelation pre-testing in the linear model: estimation, testing and prediction," Journal of Econometrics 25, 35-48.

Kmenta, J., and R. F. Gilbert (1968). "Small sample properties of alternative estimators of seemingly unrelated regressions," Journal of the American Statistical Association 63, 1180-1200.

Kmenta, J., and R. F. Gilbert (1970). "Estimation of seemingly unrelated regressions with autoregressive disturbances," Journal of the American Statistical Association 65, 186-197.

Maeshiro, A. (1980). "New Evidence on the small properties of estimators of SUR models with autocorrelated disturbances: things done half-way may not be done right," Journal of Econometrics 12, 177-187.

Magee, L., A. Ullah and V. K. Srivastava (1987). "Efficiency of estimators in the regression model with first order autoregressive errors," forthcoming in M. L. King and D. E. A. Giles (eds.), Specification Analysis in the Linear Model (In Honour of Donald Cochrane) (Routledge and Kegan-Paul, London), 81-98.

Magnus, J. R. (1978). "Maximum likelihood estimation of the GLS model with unknown parameters in the disturbance covariance matrix," Journal of Econometrics 7, 281-312.

Mentz, R. P. (1977). "Estimation in the first-order moving average model through finite autogressive approximation: some asymptotic results," Journal of Econometrics 6, 225-236.

Nicholls, D. F., and A. D. Hall (1979). "The exact maximum likelihood function of multivariate autoregressive moving average models," Biometrika 66, 259-264.

Pagan, A. R., and R. P. Byron (1978). "A synthetic approach to the estimation of models with autocorrelated disturbance terms," in A. R. Bergstrom, A. J. L. Catt, M. H. Peston and B. D. J. Silverstone (eds.), Stability and Inflation (Essays in Honour of A. W. H. Phillips) (Wiley, New York), Chapter 14.

Parks, R. W. (1967). "Efficient estimation of a system of regression equations when disturbances are both serially and contemporaneously correlated," Journal of the American Statistical Association 62, 500-509.

Poirier, D. J. (1978). "The effect of the first observation in regression models with first-order autoregressive disturbances," Applied Statistics 27, 67-68.

Prais, G. J., and C. B. Winsten (1954). "Trend estimators and serial correlation," Cowles Commission Discussion Paper No. 383, University of Chicago.

Rao, C. R. (1973). Linear Statistical Inference and its Applications (Wiley, New York).

Szroeter, J. (1978). "Generalized variance-ratio tests for serial correlation in multivariate regression models," Journal of Econometrics 8, 47-59.

Theil, H. (1971). Principles of Econometrics (Wiley, New York).

Theil, H., and A. L. Nagar (1961). "Testing the independence of regression disturbances," Journal of the American Statistical Association 56, 793-806.

8

Heteroscedastic Disturbances

8.1 INTRODUCTION

In the preceding chapter, we considered a SURE model in which the disturbances follow an autoregressive process. Another interesting generalization of the basic SURE model arises when the disturbances exhibit heteroscedasticity. Such a specification may be relevant, for instance, when observations in the full sample are known to have been generated from separate regimes with possibly unequal variances. Low (1982; Chapter 1) cited some other instances where heteroscedasticity may enter the disturbances of the SURE model (see also Duncan, 1983). Here we specify a simple type of heteroscedasticity for the disturbances in the SURE model and discuss the estimation of the parameters under this more general specification. To the best of our knowledge, Low's work is the only existing analysis providing a systematic discussion of estimators of the SURE model's parameters in the presence of heteroscedasticity, and the discussion in this chapter is based heavily on his work and results.

8.2 MODEL SPECIFICATION AND ESTIMATION

Consider the SURE model in which the observations are known to be generated from L separate regimes so that the i'th equation,

$$y_i = X_i\beta_i + u_i , \tag{8.1}$$

is expressible in the form

$$\begin{bmatrix} y_i^{(1)} \\ y_i^{(2)} \\ \vdots \\ y_i^{(L)} \end{bmatrix} = \begin{bmatrix} X_i^{(1)} \\ X_i^{(2)} \\ \vdots \\ X_i^{(L)} \end{bmatrix} \beta_i + \begin{bmatrix} u_i^{(1)} \\ u_i^{(2)} \\ \vdots \\ u_i^{(L)} \end{bmatrix} . \tag{8.2}$$

Here $y_i^{(\ell)}$ is a $(T^{(\ell)} \times 1)$ vector of $T^{(\ell)}$ observations on the variable to be explained in the i'th equation during the ℓ'th regime, $X_i^{(\ell)}$ is a $(T^{(\ell)} \times K_i)$ matrix of $T^{(\ell)}$ observations on the K_i regressors, and $u_i^{(\ell)}$ is a $(T^{(\ell)} \times 1)$ vector of disturbances with

$$E(u_i^{(\ell)}) = 0$$

$$E(u_i^{(\ell)}u_i^{(\ell^*)\prime}) = \begin{cases} \sigma_{ii}^{(\ell)} I_{T^{(\ell)}} & \text{if } \ell = \ell^* \\ 0 & \text{if } \ell \neq \ell^* \end{cases} \tag{8.3}$$

and $T^{(1)} + T^{(2)} + \cdots + T^{(L)} = T$.

For any choice of the i'th and j'th equations, the cross moments for the disturbances are given by

$$E(u_i^{(\ell)}u_j^{(\ell^*)\prime}) = \begin{cases} \sigma_{ij}^{(\ell)} I_{T^{(\ell)}} & \text{if } \ell = \ell^* \\ 0 & \text{if } \ell \neq \ell^* \end{cases} \tag{8.4}$$

from which we have

$$E(u_i u_j') = \begin{bmatrix} \sigma_{ij}^{(1)} I_{T(1)} & 0 & \cdots & 0 \\ 0 & \sigma_{ij}^{(2)} I_{T(2)} & \cdots & 0 \\ \vdots & \vdots & & \vdots \\ 0 & 0 & \cdots & \sigma_{ij}^{(L)} I_{T(L)} \end{bmatrix} = \Psi_{ij} . \tag{8.5}$$

Thus the SURE model with heteroscedastic disturbances is defined by

$$y = X\beta + u$$

$$E(u) = 0, \quad E(uu') = \Psi = \begin{bmatrix} \Psi_{11} & \cdots & \Psi_{1M} \\ \vdots & & \vdots \\ \Psi_{M1} & \cdots & \Psi_{MM} \end{bmatrix} . \tag{8.6}$$

Notice that such a specification increases the number of distinct unknown disturbance variance and covariance parameters from $\frac{1}{2}M(M+1)$ to $\frac{1}{2}LM(M+1)$.

The OLS estimator of β in (8.6) is

$$b_0 = (X'X)^{-1}X'y , \tag{8.7}$$

which discards the heteroscedasticity of the disturbances as well as the correlations across the equations. Let us call b_0 the DHDC estimator. The FGLS estimator which discards the heteroscedasticity but recognizes the correlated nature of the disturbances across the equations of the model (and will be labelled the DHRC estimator) is

$$b_F = [X'(S^{-1} \otimes I_T)X]^{-1}X'(S^{-1} \otimes I_T)y , \tag{8.8}$$

where $S = ((s_{ij}))$ is any consistent estimator of $\Sigma = ((\sigma_{ij}))$ under the assumption that $\sigma_{ij}^{(1)}, \sigma_{ij}^{(2)}, \ldots, \sigma_{ij}^{(L)}$ are all equal to σ_{ij}. Choosing S as $\tilde{S}$ and $\hat{S}$ leads to the SUUR and SURR estimators of β respectively.

Treating the i'th equation, given by (8.2), of the model as a SURE model in its own right, one can apply the FGLS estimator to each of the M equations separately, to allow for the different regimes. This provides the following estimator, which can be termed the RHDC estimator as it recognizes the presence of heteroscedasticity but completely discards the correlated nature of the disturbances across the equations:

$$b_F^{\dagger} = (X'\hat{\Psi}_D^{-1}X)^{-1}X'\hat{\Psi}_D^{-1}y\ , \tag{8.9}$$

where $\hat{\Psi}_D$ is any consistent estimator of Ψ_D (the block diagonal matrix formed from Ψ), defined as

$$\Psi_D = \begin{bmatrix} \Psi_{11} & 0 & \cdots & 0 \\ 0 & \Psi_{22} & \cdots & 0 \\ \vdots & \vdots & & \vdots \\ 0 & 0 & \cdots & \Psi_{MM} \end{bmatrix} . \tag{8.10}$$

A consistent estimator of the ℓ'th diagonal element of Ψ_{ii} is given by

$$\hat{\psi}_{ii}^{(\ell)} = \frac{1}{T^{(\ell)}} y_i^{(\ell)'} \bar{P}_{X_i^{(\ell)}} y_i^{(\ell)} . \tag{8.11}$$

Taking the divisor in (8.11) as $(T^{(\ell)} - K_i)$ instead of $T^{(\ell)}$, and using this estimator in (8.9), yields an estimator for β which is very similar to that considered by Taylor (1977).

When $T^{(1)} = T^{(2)} = \cdots = T^{(L)} = T_*$ (say), several other estimators for the elements of Ψ_{ii} are available, such as the minimum norm quadratic unbiased estimators developed by Rao (1970), the average of the squared residuals estimators discussed by Rao (1973), and the almost unbiased estimators proposed by Horn, Horn and Duncan (1975) (see also Horn and Horn, 1975, for an interesting comparison). When used in (8.9), these estimators for the variances provide other RHDC estimators of β.

The estimation of β can probably be carried out in an efficient manner by an RHRC procedure that takes into account both the heteroscedasticity and the correlated nature of the disturbances. For example, if we apply the FGLS estimator to (8.6), we get

$$\overset{*}{b}_F = (X'\hat{\Psi}^{-1}X)^{-1}X'\hat{\Psi}^{-1}y \tag{8.12}$$

where $\hat{\Psi}$ is any consistent estimator of Ψ. For instance, we may take

$$\hat{\psi}_{ij}^{(\ell)} = \frac{1}{T^{(\ell)}} y_i^{(\ell)'} \bar{P}_{X_i^{(\ell)}} \bar{P}_{X_j^{(\ell)}} y_j^{(\ell)} \tag{8.13}$$

$(i, j = 1, 2, \ldots, M; \; \ell = 1, 2, \ldots, L)$.

Assuming normality of the disturbances, and following Magnus (1978), the maximum likelihood procedure can be used to obtain consistent and asymptotically efficient estimators of the model's parameters. However, the associated estimating equations are typically nonlinear and do not enable us to write down this estimator of β in closed form. Numerical optimization procedures must be adopted to obtain the maximum likelihood estimator of β in this context.

8.3 ESTIMATOR PROPERTIES

In order to investigate the properties of the DHDC, DHRC, RHDC and RHRC estimators, the conventional conditions (such as the finiteness of the elements of $\frac{1}{T}X_i'X_j$ when T grows large) needed when studying the distributional properties of estimators are assumed to be satisfied. Now it can be seen easily that the asymptotic distribution of $\sqrt{T}$ times the estimation error of each of the four estimators b_0, b_F, $b_F^{\dagger}$ and $\overset{*}{b}_F$ is multivariate Normal with mean vector 0. If we look at the variance covariance matrices of their asymptotic distributions, it may be observed that the estimator $\overset{*}{b}_F$ is asymptotically the most efficient of those being considered. However, natural questions to ask are how far this property applies in finite samples, and how the other estimators perform in finite-sample situations. To investigate these matters, asymptotic approximations to the sampling distributions of the estimators, as well as the bias vectors and mean squared error matrices can be obtained by using the techniques introduced in Chapter 3, but the results turn out to be difficult to interpret and do not permit us to draw any clear conclusions. With this in mind, in addition to the usual limitations of asymptotic approximations, we do not discuss the details of these derivations here. Instead, we consider some exact results assuming that the disturbances of the SURE model are normally distributed. It can be shown quite easily that all of the four estimators being considered are unbiased, provided that their mean vectors exist. To examine their variance covariance matrices more closely, we take a two equation model and restrict our attention to the estimators of β_1, the coefficient vector in the first equation. As noted already, the results which follow should be attributed to Low (1982).

8.3.1 The DHDC Estimator

From (8.7), it is easy to see that the variance covariance matrix of the OLS estimator of β_1 in the two equation SURE model is

$$V(b_{(1)0}) = (X_1'X_1)^{-1}X_1'\Psi_{11}X_1(X_1'X_1)^{-1}$$
$$= (X_1^{(1)'}X_1^{(1)} + X_1^{(2)'}X_1^{(2)})^{-1}(\psi_{11}^{(1)}X_1^{(1)'}X_1^{(1)} + \psi_{11}^{(2)}X_1^{(2)'}X_1^{(2)})$$
$$\cdot (X_1^{(1)'}X_1^{(1)} + X_1^{(2)'}X_1^{(2)})^{-1} . \tag{8.14}$$

Suppose that Γ is a nonsingular matrix such that

$$\frac{1}{\psi_{11}^{(1)}}\Gamma'X_1^{(1)'}X_1^{(1)}\Gamma = I_{K_1}$$
$$\frac{1}{\psi_{11}^{(2)}}\Gamma'X_1^{(2)'}X_1^{(2)}\Gamma = \Lambda \tag{8.15}$$

where Λ is a $(K_1 \times K_1)$ diagonal matrix whose diagonal elements are the characteristic roots of the determinantal equation

$$\frac{1}{\psi_{11}^{(2)}}X_1^{(2)'}X_1^{(2)} - \lambda\frac{1}{\psi_{11}^{(1)}}X_1^{(1)'}X_1^{(1)} = 0 .$$

Using (8.15), we have

$$V(b_{(1)0}) = \Gamma H \Gamma' \tag{8.16}$$

where H is a diagonal matrix with k'th diagonal element

$$h_k = \frac{1 + \left[\frac{\psi_{11}^{(2)}}{\psi_{11}^{(1)}}\right]^2 \lambda_k}{\left[1 + \frac{\psi_{11}^{(2)}}{\psi_{11}^{(1)}}\lambda_k\right]^2} . \tag{8.17}$$

8.3.2 The RHDC Estimator

From (8.9) we have

$$
\begin{aligned}
(b^{\dagger}_{(1)F} - \beta_1)
&= (X_1'\hat{\Psi}_{11}^{-1}X_1)^{-1}X_1'\hat{\Psi}_{11}^{-1}u_1 \\
&= \Big(\frac{1}{\hat{\psi}_{11}^{(1)}}X_1^{(1)'}X_1^{(1)} + \frac{1}{\hat{\psi}_{11}^{(2)}}X_1^{(2)'}X_1^{(2)}\Big)^{-1}\Big(\frac{1}{\hat{\psi}_{11}^{(1)}}X_1^{(1)'}u_1^{(1)} + \frac{1}{\hat{\psi}_{11}^{(2)}}X_1^{(2)'}u_1^{(2)}\Big) \qquad (8.18)
\end{aligned}
$$

where we have taken

$$
\begin{aligned}
\hat{\psi}_{11}^{(\ell)} \; (\ell=1,2) &= \frac{1}{(T^{(\ell)} - K_1)} y_1^{(\ell)'}\bar{P}_{X_1^{(\ell)}} y_1^{(\ell)} \\
&= \frac{1}{(T^{(\ell)} - K_1)} u_1^{(\ell)'}\bar{P}_{X_1^{(\ell)}} u_1^{(\ell)} \; . \qquad (8.19)
\end{aligned}
$$

By virtue of the normality of the disturbances, it is easy to see that $X_1^{(1)'}u_1^{(1)}$, $X_1^{(2)'}u_1^{(2)}$, $\hat{\psi}_{11}^{(1)}$ and $\hat{\psi}_{11}^{(2)}$ are stochastically independent so that the variance covariance matrix of $b^{\dagger}_{(1)F}$ is, from (8.18),

$$
\begin{aligned}
V(b^{\dagger}_{(1)F}) &= E(X_1'\hat{\Psi}_{11}^{-1}X_1)^{-1}X_1'\hat{\Psi}_{11}^{-1}\Psi_{11}\hat{\Psi}_{11}^{-1}X_1(X_1'\hat{\Psi}_{11}^{-1}X_1)^{-1} \\
&= E\Big(\sum_{\ell=1}^{2}\frac{1}{\hat{\psi}_{11}^{(\ell)}}X_1^{(\ell)'}X_1^{(\ell)}\Big)^{-1}\Big(\sum_{\ell=1}^{2}\frac{\psi_{11}^{(\ell)}}{\hat{\psi}_{11}^{(\ell)2}}X_1^{(\ell)'}X_1^{(\ell)}\Big)\Big(\sum_{\ell=1}^{2}\frac{1}{\hat{\psi}_{11}^{(\ell)}}X_1^{(\ell)'}X_1^{(\ell)}\Big)^{-1}
\end{aligned}
$$

which, using (8.15), can be written as

$$
V(b^{\dagger}_{(1)F}) = \Gamma H^{\dagger}\Gamma' \qquad (8.20)
$$

where $H^{\dagger}$ is a diagonal matrix with k'th diagonal element as

$$h_k^{\dagger} = E\left[\frac{1+\left[\dfrac{\psi_{11}^{(2)}\hat{\psi}_{11}^{(1)}}{\psi_{11}^{(1)}\hat{\psi}_{11}^{(2)}}\right]^2 \lambda_k}{\left[1+\dfrac{\psi_{11}^{(2)}\hat{\psi}_{11}^{(1)}}{\psi_{11}^{(1)}\hat{\psi}_{11}^{(2)}}\lambda_k\right]^2}\right]$$

$$= E\left[\frac{1}{1+\dfrac{\psi_{11}^{(2)}\hat{\psi}_{11}^{(1)}}{\psi_{11}^{(1)}\hat{\psi}_{11}^{(2)}}\lambda_k}\right]^2 + \frac{1}{\lambda_k} E\left[1-\frac{1}{1+\dfrac{\psi_{11}^{(2)}\hat{\psi}_{11}^{(1)}}{\psi_{11}^{(1)}\hat{\psi}_{11}^{(2)}}\lambda_k}\right]^2 . \tag{8.21}$$

Now, to evaluate $h_k^{\dagger}$, note that as $(T^{(\ell)} - K_1)\hat{\Psi}_{11}^{(\ell)}/\Psi_{11}^{(\ell)}$ follows a χ^2 distribution with $(T^{(\ell)} - K_1)$ degrees of freedom, and as these statistics are independent for $\ell = 1, 2$, then the ratio

$$f = \frac{\psi_{11}^{(2)}\hat{\psi}_{11}^{(1)}}{\psi_{11}^{(1)}\hat{\psi}_{11}^{(2)}}$$

follows an F distribution with $(T^{(1)} - K_1)$ and $(T^{(2)} - K_1)$ degrees of freedom. Thus we find that the first expectation on the right hand side of (8.21) is equal to

$$\frac{\Gamma\left(\dfrac{T^{(1)}+T^{(2)}}{2} - K_1\right)}{\Gamma\left(\dfrac{T^{(1)}-K_1}{2}\right)\Gamma\left(\dfrac{T^{(2)}-K_1}{2}\right)} \left[\frac{T^{(1)}-K_1}{T^{(2)}-K_2}\right]^{(T^{(1)}-K_1)/2}$$

$$\cdot \int_0^{\infty} \frac{f^{(T^{(1)}-K_1)/2-1}}{(1+f\lambda_k)^2\left(1+\dfrac{T^{(1)}-K_1}{T^{(2)}-K_1}f\right)^{\frac{T^{(1)}+T^{(2)}}{2}-K_1}}\, df .$$

Recalling that

$$T^{(1)} + T^{(2)} = T$$

and applying the transformation

$$t = \frac{1}{1 + f\lambda_k} ,$$

this expression is equal to

$$\frac{\Gamma\left(\frac{T^{(1)} + T^{(2)}}{2} - K_1\right)}{\Gamma\left(\frac{T^{(1)} - K_1}{2}\right)\Gamma\left(\frac{T^{(2)} - K_1}{2}\right)} \left[\frac{T^{(1)} - K_1}{T^{(2)} - K_1} \lambda_k\right]^{(T^{(1)}-K_1)/2}$$

$$\cdot \int_0^1 \frac{t^{(T^{(2)}-K_1)/2-1} (1-t)^{(T^{(1)}-K_1)/2-1}}{\left[1 - \left(1 - \frac{T^{(2)} - K_1}{T^{(1)} - K_1}\lambda_k\right)t\right]^{\frac{T}{2}-K_1}} dt$$

which involves a hypergeometric integral and can be straightforwardly evaluated to yield the following expression:

$$\frac{\Gamma\left(\frac{T}{2} - K_1\right)\Gamma\left(\frac{T^{(2)} - K_1}{2} + 2\right)}{\Gamma\left(\frac{T}{2} - K_1 + 2\right)\Gamma\left(\frac{T^{(2)} - K_1}{2}\right)} \, {}_2F_1\left(\frac{T^{(1)} - K_1}{2}, 2; \frac{T}{2} - K_1 + 2; 1 - \frac{T^{(2)} - K_1}{T^{(1)} - K_1}\lambda_k\right) , \tag{8.22}$$

provided that

$$(T^{(2)} - K_1) > 4 ,$$

$$0 < \left(\frac{T^{(2)} - K_1}{T^{(1)} - K_1}\lambda_k\right) < 2 . \tag{8.23}$$

The hypergeometric function is defined as

$$
{}_2F_1(a_1, a_2; a_3; x) = \sum_{\alpha=0}^{\infty} \frac{\Gamma(a_1 + \alpha)\Gamma(a_2 + \alpha)\Gamma(a_3)}{\Gamma(a_1)\Gamma(a_2)\Gamma(a_3 + \alpha)} \cdot \frac{x^{\alpha}}{\alpha!} \tag{8.24}
$$

provided that $a_3 > 0$ and $|x| < 1$.

The second term on the right hand side of (8.21) can be evaluated in a similar way, and may be shown to be:

$$
\begin{aligned}
&\frac{1}{\lambda_k}\left[1 + \frac{\Gamma\left(\frac{T}{2} - K_1\right)\Gamma\left(\frac{T^{(2)} - K_1}{2} + 2\right)}{\Gamma\left(\frac{T}{2} - K_1 + 2\right)\Gamma\left(\frac{T^{(2)} - K_1}{2}\right)}\left\{{}_2F_1\left(\frac{T^{(1)} - K_1}{2},\ 2;\ \frac{T}{2} - K_1 + 2;\right.\right.\right. \\
&\left. 1 - \frac{T^{(2)} - K_1}{T^{(1)} - K_1}\lambda_k\right) - 2\left[\frac{\frac{T}{2} - K_1 + 1}{\frac{T^{(2)} - K_1}{2} + 1}\right]{}_2F_1\left(\frac{T^{(1)} - K_1}{2},\ 1;\ \frac{T}{2} - K_1 + 1;\right. \\
&\left.\left.\left. 1 - \frac{T^{(2)} - K_1}{T^{(1)} - K_1}\lambda_k\right)\right\}\right]
\end{aligned} \tag{8.25}
$$

provided that (8.23) is satisfied.

Defining

$$
\begin{aligned}
\phi_j = {} & \frac{\Gamma\left(\frac{T}{2} - K_1\right)\Gamma\left(\frac{T^{(2)} - K_1}{2} + j + 1\right)}{\Gamma\left(\frac{T^{(1)} - K_1}{2}\right)\Gamma\left(\frac{T^{(2)} - K_2}{2}\right)} \sum_{\alpha=0}^{\infty} \frac{\Gamma\left(\frac{T^{(1)} - K_1}{2} + \alpha\right)}{\Gamma\left(\frac{T}{2} - K_1 + \alpha + j + 1\right)} (j\alpha + 1) \\
& \cdot \left[1 - \frac{T^{(2)} - K_1}{T^{(1)} - K_1}\lambda_k\right]^{\alpha} \qquad (j = 0,\ 1)
\end{aligned} \tag{8.26}
$$

and combining (8.22) and (8.25), we obtain from (8.21) the following expression:

$$h_k^{\dagger} = \frac{1}{\lambda_k}[(1 + \lambda_k)\phi_1 - 2\phi_0 + 1] \ , \tag{8.27}$$

provided that the conditions (8.23) are satisfied. This expression for $h_k^{\dagger}$ is used to form the diagonal elements of $H_k^{\dagger}$, and this gives the variance covariance matrix of the RHDC estimator of β_1, from (8.20).

8.3.3 The DHRC Estimator

Next, we consider the SURR and SUUR versions of the estimator (8.8). The expressions for the variance covariance matrices of these two estimators are quite difficult to derive even in the case of a two equation model. To get some idea of their complexity, let us consider a two equation model with orthogonal regressors. For this case, the variance covariance matrices of the corresponding estimators when heteroscedasticity is not present have been obtained already in Section 4.4. We assume further, for the sake of simplicity in the exposition, that there are merely two regimes each having the same number of observations, viz., T/2.

From (8.8), the SURR estimator of β_1 is

$$\hat{\beta}_{(1)SR} = (X_1'X_1)^{-1}X_1'y_1 - \frac{\hat{s}_{12}}{\hat{s}_{22}}(X_1'X_1)^{-1}X_1'y_2 \tag{8.28}$$

where

$$\frac{\hat{s}_{12}}{\hat{s}_{22}} = \frac{y_1'\bar{P}_{X_1}\bar{P}_{X_2}y_2}{y_2'\bar{P}_{X_2}y_2} = \frac{u_1'\bar{P}_{X_1}\bar{P}_{X_2}u_2}{u_2'\bar{P}_{X_2}u_2} \ . \tag{8.29}$$

It is easy to see that the variance covariance matrix of the estimator (8.28) is

$$V(\hat{\beta}_{(1)SR}) = (X_1'X_1)^{-1}X_1'E\left[\Psi_{11} - \frac{\hat{s}_{12}}{\hat{s}_{22}}(u_1u_2' + u_2u_1') + \left[\frac{\hat{s}_{12}}{\hat{s}_{22}}\right]^2 u_2u_2'\right]X_1(X_1'X_1)^{-1} \ . \tag{8.30}$$

Now, to evaluate the expectation in (8.30) we introduce the following notation:

$$\Psi_{11\cdot 2} = \Psi_{11} - \Psi_{12}\Psi_{22}^{-1}\Psi_{21} ,$$

$$A = \begin{bmatrix} \Psi_{11\cdot 2}^{1/2} & \Psi_{12}\Psi_{22}^{-1/2} \\ 0 & \Psi_{22}^{1/2} \end{bmatrix} , \quad v = A^{-1}u ,$$

$$N_1 = \begin{bmatrix} 0 & \Psi_{11\cdot 2}^{1/2}\bar{P}_{X_1}\bar{P}_{X_2}\Psi_{22}^{1/2} \\ 0 & \Psi_{22}^{-1/2}\Psi_{21}\bar{P}_{X_2}\Psi_{22}^{1/2} \end{bmatrix}, \quad N_2 = \begin{bmatrix} 0 & 0 \\ 0 & \Psi_{22}^{1/2}\bar{P}_{X_2}\Psi_{22}^{1/2} \end{bmatrix} .$$

Observing that

$$\begin{aligned} u_1'\bar{P}_{X_1}\bar{P}_{X_2}u_2 &= v'N_1v \\ u_2'\bar{P}_{X_2}u_2 &= v'N_2v \end{aligned} \tag{8.31}$$

with v following a multivariate Normal distribution with mean vector μ (here zero) and variance covariance matrix I_T, it is easy to see that

$$E\left[\left[\frac{v'N_1v}{v'N_2v}\right]^a vv'\right]$$

$$= \left[(2\pi)^{-\frac{T}{2}}\int_{-\infty}^{\infty}\cdots\int_{-\infty}^{\infty}\left[\frac{v'N_1v}{v'N_2v}\right]^a vv' \exp\left\{-\frac{1}{2}(v-\mu)'(v-\mu)\right\} dv\right]_{\mu=0}$$

$$= \left[(2\pi)^{-\frac{T}{2}}\int_{-\infty}^{\infty}\cdots\int_{-\infty}^{\infty}\left[\frac{v'N_1v}{v'N_2v}\right]^a [(v-\mu)(v-\mu)' + \mu(v-\mu)' + (v-\mu)\mu' + \mu\mu'] \right.$$

$$\left. \cdot \exp\left\{-\frac{1}{2}(v-\mu)'(v-\mu)\right\} dv\right]_{\mu=0}$$

$$= \left[(2\pi)^{-\frac{T}{2}}\int_{-\infty}^{\infty}\cdots\int_{-\infty}^{\infty}\left[\frac{v'N_1v}{v'N_2v}\right]^a \left[\frac{\partial^2}{\partial\mu\,\partial\mu'} + \mu\frac{\partial}{\partial\mu'} + \left(\mu\frac{\partial}{\partial\mu'}\right)' + \mu\mu' + I_T\right] \right.$$

$$\left. \cdot \exp\left\{-\frac{1}{2}(v-\mu)'(v-\mu)\right\} dv\right]_{\mu=0} \tag{8.32}$$

$$= \left[\left[\frac{\partial^2}{\partial\mu\,\partial\mu'} + \mu\frac{\partial}{\partial\mu} + \left(\mu\frac{\partial}{\partial\mu}\right)' + \mu\mu' + I_T \right] E\left[\frac{v'N_1v}{v'N_2v}\right]^a \right]_{\mu=0}$$

$$= \left[\left[\frac{\partial^2}{\partial\mu\,\partial\mu'} + I_T \right] E\left[\frac{v'N_1v}{v'N_2v}\right]^a \right]_{\mu=0}$$

where a is any positive scalar and $\partial/\partial\mu$ denotes the vector operator of partial derivatives with respect to the elements of μ.

Now, the moment generating function of $v'N_1v$ and $v'N_2v$ is

$$g(\theta_1, \theta_2) = E[\exp\{\theta_1 v'N_1v + \theta_2 v'N_2v\}]$$

$$= (2\pi)^{-\frac{T}{2}} \int_{-\infty}^{\infty} \cdots \int_{-\infty}^{\infty} \exp\left\{ \theta_1 v'N_1v + \theta_2 v'N_2v - \frac{1}{2}(v-\mu)'(v-\mu) \right\} dv$$

$$= (2\pi)^{-\frac{T}{2}} \exp\left\{-\frac{1}{2}\mu'(I_T - B^{-1})\mu\right\}$$

$$\cdot \int_{-\infty}^{\infty} \cdots \int_{-\infty}^{\infty} \exp\left\{-\frac{1}{2}(v - B^{-1}\mu)'B(v - B^{-1}\mu)\right\} dv$$

$$= |B|^{-1/2} \exp\left\{-\frac{1}{2}\mu'(I_T - B^{-1})\mu\right\} \tag{8.33}$$

where θ_1 and θ_2 are such that $B = [I_T - \theta_1(N_1 + N_1') - 2\theta_2 N_2]$ is positive definite, and (8.33) is obtained by "completing the square."

Further, we observe that

$$E\left[\frac{v'N_1v}{v'N_2v}\right]^a = E[(v'N_1v)^a (v'N_2v)^{-a}]$$

$$= E\left[\left(\frac{\partial^a}{\partial\theta_1^a} \exp\{\theta_1 v'N_1v\}\right)_{\theta_1=0} \frac{1}{\Gamma(a)} \int_0^{\infty} \theta_2^{a-1} \exp\{-\theta_2 v'N_2v\}\, d\theta_2\right]$$

$$= E\left[\left(\frac{\partial^a}{\partial\theta_1^a} \exp\{\theta_1 v'N_1v\}\right)_{\theta_1=0} \frac{1}{\Gamma(a)} \int_{-\infty}^{0} (-\theta_2)^{a-1} \exp\{\theta_2 v'N_2v\}\, d\theta_2\right]$$

$$= \frac{1}{\Gamma(a)} E\left[\int_{-\infty}^{0} (-\theta_2)^{a-1}\left(\frac{\partial^a}{\partial\theta_1^a} \exp\{\theta_1 v'N_1 v + \theta_2 v'N_2 v\}\right)_{\theta_1=0} d\theta_2\right]$$

$$= \frac{1}{\Gamma(a)} \int_{-\infty}^{0} (-\theta_2)^{a-1}\left(\frac{\partial^a}{\partial\theta_1^a} E[\exp\{\theta_1 v'N_1 v + \theta_2 v'N_2 v\}]\right)_{\theta_1=0} d\theta_2$$

$$= \frac{1}{\Gamma(a)} \int_{-\infty}^{0} (-\theta_2)^{a-1}\left(\frac{\partial^a}{\partial\theta_1^a} g(\theta_1, \theta_2)\right)_{\theta_1=0} d\theta_2 \qquad (8.34)$$

provided that the expectation of $(v'N_1v/v'N_2v)^a$ exists.

Now, from (8.32) and (8.34), we have

$$E\left[\frac{v'N_1 v}{v'N_2 v} vv'\right]$$

$$= \left[\left[\frac{\partial^2}{\partial\mu\partial\mu'} + I_T\right] E\left(\frac{v'N_1 v}{v'N_2 v}\right)\right]_{\mu=0}$$

$$= \left[\left[\frac{\partial^2}{\partial\mu\partial\mu'} + I_T\right] \int_{-\infty}^{0} \left(\frac{\partial}{\partial\theta_1} g(\theta_1, \theta_2)\right)_{\theta_1=0} d\theta_2\right]_{\mu=0}$$

$$= \int_{-\infty}^{0} \left[\left[\frac{\partial^2}{\partial\mu\partial\mu'} + I_T\right] \left(\frac{\partial}{\partial\theta_1} g(\theta_1, \theta_2)\right)_{\theta_1=0}\right]_{\mu=0} d\theta_2$$

$$= \int_{-\infty}^{0} \left[\frac{\partial^2}{\partial\mu\partial\mu'} \left(\frac{\partial}{\partial\theta_1} g(\theta_1, \theta_2)\right)_{\theta_1=0}\right]_{\mu=0} d\theta_2 + \int_{-\infty}^{0} \left[\left(\frac{\partial}{\partial\theta_1} g(\theta_1, \theta_2)\right)_{\theta_1=0}\right]_{\mu=0} I_T \, d\theta_2 . \qquad (8.35)$$

However, using results from Sawa (1978; p. 171),

$$\left(\frac{\partial}{\partial\theta_1} g(\theta_1, \theta_2)\right)_{\theta_1=0}$$

$$= \left(\frac{1}{2} |B|^{-\frac{1}{2}} [\mathrm{tr}(B^{-1}(N_1 + N_1')) + \mu' B^{-1}(N_1 + N_1') B^{-1}\mu] \exp\left\{-\frac{1}{2}\mu'(I_T - B^{-1})\mu\right\}\right)_{\theta_1=0}$$

$$= \frac{1}{2}|I_T - 2\theta_2 N_2|^{-\frac{1}{2}}\Big[\mathrm{tr}\,((I_T - 2\theta_2 N_2)^{-1}(N_1 + N_1'))$$

$$+ \mu'(I_T - 2\theta_2 N_2)^{-1}(N_1 + N_1')(I_T - 2\theta_2 N_2)^{-1}\mu\Big] \exp\Big\{-\frac{1}{2}\mu'[I_T - (I_T - 2\theta_2 N_2)^{-1}]\mu \quad ,$$

so that

$$\left[\left(\frac{\partial}{\partial\theta_1} g(\theta_1, \theta_2)\right)_{\theta_1=0}\right]_{\mu=0} = \frac{1}{2}|I_T - 2\theta_2 N_2|^{-\frac{1}{2}} \mathrm{tr}\Big((I_T - 2\theta_2 N_2)^{-1}(N_1 + N_1')\Big) \tag{8.36}$$

$$\left[\frac{\partial^2}{\partial\mu\partial\mu'}\left(\frac{\partial}{\partial\theta_1} g(\theta_1, \theta_2)\right)_{\theta_1=0}\right]_{\mu=0} = \frac{1}{2}|I_T - 2\theta_2 N_2|^{-\frac{1}{2}}\Big[\mathrm{tr}\Big((I_T - 2\theta_2 N_2)^{-1}(N_1 + N_1')\Big)$$

$$\cdot \{I_T - (I_T - 2\theta_2 N_2)^{-1}\} + 2(I_T - 2\theta_2 N_2)^{-1}(N_1 + N_1')(I_T - 2\theta_2 N_2)^{-1}\Big] . \tag{8.37}$$

Substituting (8.36) and (8.37) in (8.35), we find

$$G_1 \equiv E\left[\frac{v'N_1 v}{v'N_2 v} vv'\right]$$

$$= \frac{1}{2}\int_{-\infty}^{0} |I_T - 2\theta_2 N_2|^{-\frac{1}{2}} \overset{*}{B}_1 \, d\theta_2 \tag{8.38}$$

where

$$\overset{*}{B}_1 = \{\mathrm{tr}\,((I_T - 2\theta_2 N_2)^{-1}(N_1 + N_2))\}(I_T - 2\theta_2 N_2)^{-1}$$

$$+ 2(I_T - 2\theta_2 N_2)^{-1}(N_1 + N_1')(I_T - 2\theta_2 N_2)^{-1} .$$

Similarly, from (8.32), (8.34), (8.36) and (8.37),

$$G_2 \equiv E\left[\left(\frac{v'N_1 v}{v'N_2 v}\right)^2 vv'\right]$$

$$= -\frac{1}{4}\int_{-\infty}^{0} \theta_2 |I_T - 2\theta_2 N_2|^{-\frac{1}{2}} \overset{*}{B}_2 \, d\theta_2 \tag{8.39}$$

where

$$\overset{*}{B}_2 = [\{\mathrm{tr}\,((I_T - 2\theta_2 N_2)^{-1}(N_1 + N_1'))\}^2 + 2\,\mathrm{tr}\,\{(I_T - 2\theta_2 N_2)^{-1}(N_1 + N_1')\}^2](I_T - 2\theta_2 N_2)^{-1}$$
$$+ 4\{\mathrm{tr}\,((I_T - 2\theta_2 N_2)^{-1}(N_1 + N_1'))\}(I_T - 2\theta_2 N_2)^{-1}(N_1 + N_1')(I_T - 2\theta_2 N_2)^{-1}$$
$$+ 8(I_T - 2\theta_2 N_2)^{-1}(N_1 + N_1')(I_T - 2\theta_2 N_2)^{-1}(N_1 + N_1')(I_T - 2\theta_2 N_2)^{-1} .$$

Closed form expressions for the quantities G_1 and G_2 were obtained by Low (1982; Appendix D) but they turn out to be extremely complicated and are not capable of providing any clear conclusions.

If we partition

$$\underset{(k=1,2)}{G_k} = \begin{bmatrix} G_{k(1,1)} & G_{k(1,2)} \\ G_{k(2,1)} & G_{k(2,2)} \end{bmatrix}$$

with all submatrices of order $\left(\frac{T}{2} \times \frac{T}{2}\right)$, it is easy to verify that

$$E\left[\frac{\hat{s}_{12}}{\hat{s}_{22}} u_1 u_2'\right] = E\left[\frac{\hat{s}_{12}}{\hat{s}_{22}} (I_{\frac{T}{2}} \quad 0) uu' \begin{pmatrix} 0 \\ I_{\frac{T}{2}} \end{pmatrix}\right]$$
$$= (I_{\frac{T}{2}} \quad 0)\, E\left[\frac{\hat{s}_{12}}{\hat{s}_{22}} uu'\right] \begin{pmatrix} 0 \\ I_{\frac{T}{2}} \end{pmatrix}$$
$$= (I_{\frac{T}{2}} \quad 0)\, A E\left[\frac{v'N_1 v}{v'N_2 v} vv'\right] A' \begin{pmatrix} 0 \\ I_{\frac{T}{2}} \end{pmatrix}$$
$$= (\Psi_{11\cdot 2}^{1/2} \quad \Psi_{12}\Psi_{22}^{-1/2})\, G_1 \begin{pmatrix} 0 \\ \Psi_{22}^{1/2} \end{pmatrix}$$
$$= \Psi_{11\cdot 2}^{1/2} G_{1(1,2)} \Psi_{22}^{1/2} + \Psi_{12}\Psi_{22}^{-1/2} G_{1(2,2)} \Psi_{22}^{1/2} . \qquad (8.40)$$

Similarly, we have

$$E\left[\left(\frac{\hat{s}_{12}}{\hat{s}_{22}}\right)^2 u_2u_2'\right] = (0 \quad I_{\frac{T}{2}})\, AE\left[\left(\frac{v'N_1v}{v'N_2v}\right)^2 vv'\right] A' \begin{pmatrix} 0 \\ I_{\frac{T}{2}} \end{pmatrix}$$

$$= (0 \quad \Psi_{22}^{1/2})\, G_2 \begin{pmatrix} 0 \\ \Psi_{22}^{1/2} \end{pmatrix}$$

$$= \Psi_{22}^{1/2} G_{2(2,2)} \Psi_{22}^{1/2} . \tag{8.41}$$

Using (8.40) and (8.41), we find from (8.30) the following expression

$$V(\hat{\beta}_{(1)SR}) = (X_1'X_1)^{-1} X_1' H_{SR} X_1 (X_1'X_1)^{-1} \tag{8.42}$$

where

$$H_{SR} = \Psi_{11} - (\Psi_{11\cdot 2}^{1/2} G_{1(1,2)} + \Psi_{12}\Psi_{22}^{-1/2} G_{1(2,2)}\Psi_{22}^{1/2}$$
$$- \Psi_{22}^{1/2}(G_{1(1,2)}'\Psi_{11\cdot 2}^{1/2} + G_{1(2,2)}'\Psi_{22}^{-1/2}\Psi_{21}) + \Psi_{22}^{1/2} G_{2(2,2)}\Psi_{22}^{1/2} . \tag{8.43}$$

Low (1982; Chapter 3) also obtained the expression for the variance covariance matrix of the SUUR estimator of β_1:

$$V(\tilde{\beta}_{(1)SU}) = (X_1'X_1)^{-1} X_1' H_{SU} X_1 (X_1'X_1)^{-1} \tag{8.44}$$

where H_{SU} is the same as H_{SR} except that $\bar{P}_{X_1}$ and $\bar{P}_{X_2}$ are replaced by $(I_T - P_{X_1} - P_{X_2})$.

Finally we note that there are no available results relating to the finite sample properties of the estimators b_F^* given in (8.12).

8.4 EFFICIENCY COMPARISONS

The complicated nature of the exact expressions for the variance covariance matrices of these estimators motivated Low (1982; Chapter 3) to carry out a numerical evaluation. For this purpose, he considered a two equation model with each equation containing merely one regressor and no intercept term, and he selected two sets of ten observations on each of the two regressors such that the observations on the first regressor in the first equation were trended in the first set, while the second set did not contain any trended observations. Each regime comprised five observations. Three values (0.1, 0.5 and 0.9) were chosen for the correlation coefficient between the disturbances. Taking the ratio $\delta = (\Psi_{11}^{(2)}/\Psi_{11}^{(1)})$ as a measure of the degree of heteroscedasticity in the model, three values of this measure were considered, viz., $\delta = 2, 5, 10$. The investigation considered the OLS estimator (8.7) from the DHDC category, the FGLS estimator (8.9) from the RHDC category, and the SURR estimator from the DHRC category. Numerical methods were employed to evaluate the integrals involved in the variance expressions.

Basing his comparison on the exact expressions for the variances of the estimators, Low's numerical investigations revealed that the OLS estimator was more efficient than the other two estimators when the correlation coefficient between the disturbances in the two equations was as low as 0.1 and δ was 2. However, the efficiency of the OLS estimator declined substantially when this correlation was relatively high and the heteroscedasticity became more pronounced. The relative performances of the FGLS and SURR estimators depended upon the degrees of correlation and heteroscedasticity, as expected. No marked difference was noticed in the performance of the estimators when the data observations were trended. With nontrended observations there was a slight tendency for the FGLS estimator to perform better than the others.

Low (1982; Chapter 5) also conducted a Monte Carlo experiment in order to analyze the performances of various estimators. From the DHDC category, the OLS estimator was chosen while three estimators were chosen from the RHDC category, viz., the FGLS estimator (8.9) and two Stein-rule estimators—one with and the other without a Lindley-like mean correction. Similarly, the SUUR estimator and the SURR estimator along with its Stein-rule version were chosen from the DHRC category. Finally, there were two estimators from the RHRC category, viz., the FGLS estimator (8.12) and a preliminary test (PT) estimator. This PT estimator was based on the outcome of a test for the heteroscedasticity of the disturbances in each equation. Suppose H_{i0}: $\Psi_{ii}^{(1)} = \Psi_{ii}^{(2)}$ denotes the null hypothesis of homoscedasticity in the i'th equation. This hypothesis can be examined easily by a conventional F-test. If the null hypotheses for all of the equations are accepted, implying the absence of heteroscedasticity, the PT estimator is taken to be any estimator from the DHRC category, say the SURR estimator. On the other hand, if all of the null hypotheses are rejected, indicating the presence of heteroscedasticity in all of the equations, the PT estimator is just the FGLS estimator (8.12). When some of the hypotheses are accepted while the remaining are rejected, then the variance covariance matrix Ψ is structured accordingly. This specific structure explicitly incorporates the absence of heteroscedasticity in those equations for which the null hypothesis is accepted, and the presence of heteroscedasticity in the remaining equations for which the null hypothesis is rejected. Assuming such a specific structure of the variance covariance matrix for the disturbances in the model, the FGLS estimator is obtained and is taken as the PT estimator in this case.

Low's Monte Carlo study comprised three experiments dealing with both two equation and three equation models. In all cases, each equation included an intercept term and two regressors. The experiment was replicated five hundred times in the first and second experiments, and one thousand times in the case of the third experiment. Sample sizes of 20 and 40 were considered. Several interesting findings emerged from this study. It was found that no substantial loss in efficiency occurred if the estimator did not account for the heteroscedasticity and the correlated nature of the disturbances across the equations, provided that the heteroscedasticity and these correlations were of low degree. The efficiency loss became substantial when the disturbances were highly correlated and highly heteroscedastic. Neglecting heteroscedasticity may result in significant efficiency losses but unnecessarily accounting for it does not result in serious losses in efficiency. Thus, if there is uncertainty about the presence of heteroscedasticity, an estimation procedure that nevertheless allows for it may be recommended. This emphasizes the importance of a preliminary test for homoscedasticity. According to Low's results, although the associated PT estimator strikes a compromise between the DHRC and RHRC estimators, it does not turn out to be the best choice in both the situations of homoscedastic as well as heteroscedastic disturbances. The best choice continues to be the DHRC estimator in the case of homoscedastic disturbances, and the RHRC estimator in the case of heteroscedastic disturbances. In spite of this, not much loss in efficiency is incurred if a PT estimator is employed. Low observed that this small loss may be worthwhile, while hedging against a possible significant loss arising from the use of an inappropriate estimator. Such a statement, however, may not apply for models containing more than two equations.

Comparing biased and unbiased estimators, Low found that biased estimators generally outperform the unbiased estimators with respect to the criterion of precision of estimation and robustness of heteroscedasticity. The gain was found to be substantial when the absolute values of the coefficients are small. The use of a Lindley-like mean correction in conjunction with Stein-rule estimators was found to improve their performance. When the number of equations in the model was increased (from two to three), no clear pattern regarding the ranking of estimators emerged. Discriminating between the SURR and SUUR estimators, it was observed that the use of unrestricted residuals leads to efficient estimation of coefficients when the disturbances are strongly correlated, despite the presence of heteroscedasticity. Finally, it was found that the values of the regressors have little effect on the ranking of the various estimators.

8.5 RESEARCH SUGGESTIONS

The presence of heteroscedasticity in the disturbances in a SURE model poses many interesting issues, and suggests the need for an analytical study of the properties of the various estimators which might be employed. This may lead to the construction of estimators which are efficient in finite samples. Iterative and shrinkage estimators have yet to be fully explored in this context, and also the theory of minimum norm quadratic estimation can probably be fruitfully used here to obtain efficient estimators. It would be worthwhile to develop tests for detecting disturbance heteroscedasticity and contemporaneous correlations simultaneously, and from these to derive appropriate estimators (see, e.g., Harvey and Phillips, 1981). Procedures for testing hypotheses about the SURE model's coefficients and constructing confidence regions are yet to be developed in the context of heteroscedastic disturbances. We have considered a particular type of heteroscedasticity in the SURE model, but there are other forms of heteroscedasticity that may realistically be formulated. Judge, Griffiths, Hill and Lee (1980; Chapter 4) consider some possibilities in a single-equation context, and extending these to the SURE model framework and considering the associated estimation of parameters and tests of hypotheses are areas for further research.

EXERCISES

8.1 Verify that the expectation

$$E\left[\frac{\psi_{11}^{(1)}\hat{\psi}_{11}^{(2)}}{\psi_{11}^{(1)}\hat{\psi}_{11}^{(2)} + \lambda_k\psi_{11}^{(2)}\hat{\psi}_{11}^{(1)}}\right]^2$$

in (8.21) is equal to the expression (8.22) under the conditions given by (8.23).

8.2 Assuming that the disturbances in a heteroscedastic SURE model as specified by (8.6) are normally distributed, obtain the variance covariance matrix, to order $O(T^{-2})$, of the estimator $b_F^{\dagger}$ given by (8.9) when $L = 2$ and $T^{(1)} = T^{(2)}$. [Hint: See Srivastava (1970).]

8.3 Consider a two equation SURE model with heteroscedastic disturbances:

$$y_1 = \beta_1 x_1 + u_1$$

$$y_2 = \beta_2 x_2 + u_2$$

where

$$E(u_i u_j') = \text{Diag}\,(\sigma_{ij}^{(1)}, \sigma_{ij}^{(2)}, \ldots, \sigma_{ij}^{(T)})$$

$$\sigma_{ij}^{(t)} = \gamma_1 + \gamma_2 x_{it} x_{jt}$$

$$(i, j = 1, 2; \quad t = 1, 2, \ldots, T) .$$

Suggest a suitable estimator for β_1 and β_2 and state its asymptotic properties. Consider the case when γ_1 is known to be zero a priori.

8.4 Interpreting an FGLS estimator of β in a SURE model as the OLS estimator in a multivariate regression model subject to a set of exclusion restrictions on the coefficients, and assuming that the disturbances are heteroscedastic, discuss the consistent estimation of the variance covariance matrix of the FGLS estimator. [Hint: See Duncan (1983; Section 4).]

8.5 In a SURE model with heteroscedastic disturbances, the variance covariance matrix of the disturbance vector u is $\sigma^2 \Psi$ where σ^2 is an unknown scalar and Ψ is a fully known and nonsingular matrix of order $(MT \times MT)$. If b_G denotes the GLS estimator of β and if σ^2 is estimated by

$$\hat{\sigma}^2 = \frac{1}{\alpha}(y - Xb_G)'\Psi^{-1}(y - Xb_G) ,$$

obtain the expressions for the bias and mean squared error of $\hat{\sigma}^2$ assuming α to be any fixed scalar. Also find the value of α for which (i) $\hat{\sigma}^2$ is unbiased; (ii) $\hat{\sigma}^2$ has smallest mean squared error.

REFERENCES

Duncan, G. M. (1983). "Estimation and inference for heteroscedastic systems of equations," International Economic Review 24, 559-566.

Harvey, A. C., and G. D. A. Phillips (1981). "Testing for heteroscedasticity in simultaneous equation models," Journal of Econometrics 15, 311-340.

Horn, S. D., and R. A. Horn (1975). "Comparison of estimators of heteroscedastic variances in linear models," Journal of the American Statistical Association 70, 872-879.

Horn, S. D., R. A. Horn and D. B. Duncan (1975). "Estimating heteroscedastic variances in linear models," Journal of the American Statistical Association 70, 380-385.

Judge, G. G., W. E. Griffiths, R. C. Hill and T. S. Lee (1980). The Theory and Practice of Econometrics (Wiley, New York).

Low, C. K. (1982). "Seemingly unrelated regressions with heteroskedastic disturbances: some finite sample results," Ph.D. Thesis, Department of Econometrics and Operations Research, Monash University, Melbourne.

Magnus, J. R. (1978). "Maximum likelihood estimation of the GLS model with unknown parameters in the disturbance covariance matrix," Journal of Econometrics 7, 281-312.

Rao, C. R. (1970). "Estimation of heteroscedastic variances in linear models," Journal of the American Statistical Association 65, 161-172.

Rao, J. N. K. (1973). "On the estimation of heteroscedastic variances," Biometrics 29, 11-24.

Sawa, T. (1978). "The exact moments of the least squares estimator for the autoregressive model," Journal of Econometrics 8, 159-172.

Srivastava, V. K. (1970). "The efficiency of estimating seemingly unrelated regression equations," Annals of the Institute of Statistical Mathematics 22, 483-493.

Taylor, W. E. (1977). "Small sample properties of a class of two stage Aitken estimators," Econometrica 45, 497-508.

9

Constrained Error Covariance Structures

9.1 INTRODUCTION

In the last chapter we relaxed the assumption of homoscedasticity of the disturbances in the SURE model, and considered the case of heteroscedastic disturbances. There are, however, other interesting and useful alternative specifications relating to the disturbances such that their variance covariance matrices have special patterns. For example, the disturbances may be known to have positive/negative correlations across the equations of the model, and/or it may be that their variances can be arranged so as to follow a strict ranking with respect to magnitude. Patterned structures arising from error components are other such examples. Such specifications often bring a significant reduction in the number of parameters characterizing the covariance structure of the disturbances and provide a suitable framework for developing efficient estimation procedures. An illustration of this point is the subject matter of the present chapter, where we base our discussion primarily on the work of Avery (1977), Baltagi (1980), Srivastava and Mishra (1986) and Zellner (1979).

9.2 VARIANCE INEQUALITIES AND POSITIVITY OF CORRELATIONS

9.2.1 Model Specification

Consider the SURE model

$$y = X\beta + u$$
$$E(u) = 0, \quad E(uu') = (\Sigma \otimes I_T). \tag{9.1}$$

Sometimes the variance covariance matrix of the disturbances in the SURE model is known to be constrained in such a manner that the variance of the disturbance term in an equation is larger (or smaller) than the variance of the disturbance term in the preceding equation, and/or all of the correlation coefficients between the disturbances across the equations are either positive or negative. Without any loss of generality, let us follow the analysis of Zellner (1979) and assume that

$$\sigma_{11} < \sigma_{22} < \cdots < \sigma_{MM}$$
$$0 < \frac{\sigma_{ij}}{(\sigma_{ii}\sigma_{jj})^{\frac{1}{2}}} < 1 \qquad (i \neq j;\ i, j = 1, 2, \ldots, M). \tag{9.2}$$

This specification is achieved by expressing the disturbance vectors $u_1, u_2, \ldots, u_M$ as follows:

$$u_i = \epsilon_1 + \epsilon_2 + \cdots + \epsilon_i \qquad (i = 1, 2, \ldots, M) \tag{9.3}$$

where $\epsilon_1, \epsilon_2, \ldots, \epsilon_M$ are $(T \times 1)$ vectors such that

$$E(\epsilon_i) = 0$$
$$E(\epsilon_i\epsilon_j') = \begin{cases} \theta_i I_T & \text{if } i = j \\ 0 & \text{if } i \neq j \end{cases} \tag{9.4}$$

with θ_i an unknown scalar quantity. From (9.3) and (9.4), we observe that

$$\sigma_{ij} = \theta_1 + \theta_2 + \cdots + \theta_i \qquad (i \leq j) \tag{9.5}$$

from which it is easy to see that the inequalities (9.2) are satisfied.

Note that the introduction of the structure (9.3) reduces the number of unknown parameters characterizing Σ. In the general case, the matrix Σ

involves $M(M+1)/2$ distinct σ_{ij}'s while, according to (9.5), it is completely specified by M quantities, the θ_i's. Further, it may be observed that some minor modifications to the specification (9.2) provide alternative sets of constraints. For example, if we consider the case of a two equation SURE model, then the inequalities (9.2) are

$$\sigma_{11} < \sigma_{22} \tag{9.6}$$

$$0 < \frac{\sigma_{12}}{(\sigma_{11}\sigma_{22})^{\frac{1}{2}}} < 1 . \tag{9.7}$$

If we wish to replace (9.7) by

$$-1 < \frac{\sigma_{12}}{(\sigma_{11}\sigma_{22})^{\frac{1}{2}}} < 0 ,$$

and continue to retain (9.6), we may, as Zellner (1979; p. 681) notes, set

$$u_1 = \epsilon_1 ; \; u_2 = -\epsilon_1 + \epsilon_2 .$$

If we wish to drop inequality (9.6) and retain merely the constraint (9.7), we may write

$$u_1 = \epsilon_1 + \epsilon_2 ; \; u_2 = \epsilon_1 + \epsilon_3 .$$

Using (9.3) we can express the model (9.1) as

$$\begin{bmatrix} y_1 \\ y_2 - y_1 \\ y_3 - y_2 \\ \vdots \\ y_{M-1} - y_{M-2} \\ y_M - y_{M-1} \end{bmatrix} = \begin{bmatrix} X_1 & 0 & 0 & \cdots & 0 & 0 \\ -X_1 & X_2 & 0 & \cdots & 0 & 0 \\ 0 & -X_2 & X_3 & \cdots & 0 & 0 \\ \vdots & \vdots & \vdots & & \vdots & \vdots \\ 0 & 0 & 0 & \cdots & X_{M-1} & 0 \\ 0 & 0 & 0 & \cdots & -X_{M-1} & X_M \end{bmatrix} \begin{bmatrix} \beta_1 \\ \beta_2 \\ \beta_3 \\ \vdots \\ \beta_{M-1} \\ \beta_M \end{bmatrix} + \begin{bmatrix} \epsilon_1 \\ \epsilon_2 \\ \epsilon_3 \\ \vdots \\ \epsilon_{M-1} \\ \epsilon_M \end{bmatrix}$$

or

$$w = W\beta + \epsilon . \tag{9.8}$$

From (9.4) we have

$$E(\epsilon) = 0 , \qquad E(\epsilon\epsilon') = (\Theta \otimes I_T) \tag{9.9}$$

where Θ is an $(M \times M)$ diagonal matrix with diagonal elements $\theta_1, \theta_2, \ldots, \theta_M$ (see Zellner, 1979; p. 686).

9.2.2 Estimation Procedures

The application of OLS to models (9.1) and (9.8) respectively provides the estimators

$$b_0 = (X'X)^{-1}X'y \tag{9.10}$$

$$b_0^{\dagger} = (W'W)^{-1}W'w . \tag{9.11}$$

Notice that the estimator (9.10) ignores the constraints (9.2) completely while the estimator (9.11) takes care of them through a restructuring of the model. However, this is not the case with the GLS method, the application of which to the models (9.1) and (9.8) yields identical estimators:

$$\begin{aligned} b_G &= [X'(\Sigma^{-1} \otimes I_T)X]^{-1}X'(\Sigma^{-1} \otimes I_T)y \\ &= [W'(\Theta^{-1} \otimes I_T)W]^{-1}W'(\Theta^{-1} \otimes I_T)w . \end{aligned} \tag{9.12}$$

The SUUR and SURR estimators arising from the first line of (9.12) are given by

$$\begin{aligned} \tilde{\beta}_{SU} &= [X'(\tilde{S}^{-1} \otimes I_T)X]^{-1}X'(\tilde{S}^{-1} \otimes I_T)y \\ \hat{\beta}_{SR} &= [X'(\hat{S}^{-1} \otimes I_T)X]^{-1}X'(\hat{S}^{-1} \otimes I_T)y \end{aligned} \tag{9.13}$$

but they do not incorporate the inequality constraints (9.2). Following Zellner (1979; p. 686) an alternative FGLS estimator based on the second line of (9.12) is

$$b_F^{\dagger} = [W'(\hat{\Theta}^{-1} \otimes I_T)W]^{-1}W'(\hat{\Theta}^{-1} \otimes I_T)w , \tag{9.14}$$

where $\hat{\Theta}$ is a diagonal matrix with diagonal elements

$$\hat{\theta}_1 = \frac{1}{T} y_1'[I_T - X_1(X_1'X_1)^{-1}X_1']y_1$$
$$\hat{\theta}_i = \frac{1}{T} (y_i - y_{i-1})'[I_T - W_i(W_i'W_i)^{-1}W_i'](y_i - y_{i-1}) \tag{9.15}$$
$$(i = 2, 3, \ldots, M)$$

with $W_i = (-X_{i-1} \quad X_i)$.

9.2.3 Efficiency Comparisons

It is easy to see that both of the OLS estimators (9.10) and (9.11) are unbiased with variance covariance matrices given by

$$V(b_0) = (X'X)^{-1}X'(\Sigma \otimes I_T)X(X'X)^{-1} \tag{9.16}$$

$$V(b_0^\dagger) = (W'W)^{-1}W'(\Theta \otimes I_T)W(W'W)^{-1} . \tag{9.17}$$

All three FGLS estimators, as given by (9.13) and (9.14), are consistent and their variance covariance matrices, to order $O(T^{-1})$, are identical and equal to

$$\Omega = [X'(\Sigma^{-1} \otimes I_T)X]^{-1} = [W'(\Theta^{-1} \otimes I_T)W]^{-1} . \tag{9.18}$$

It follows that each of these feasible versions of the GLS estimator is more efficient, to order $O(T^{-1})$, than the two OLS estimators. To discuss the approximate gain in efficiency from using an FGLS estimator, we now report the analysis of Srivastava and Mishra (1986). Assume orthogonal regressors:

$$X_i'X_j = 0 \quad (i, j = 1, 2, \ldots, M ; \quad i \neq j) . \tag{9.19}$$

It is then easy to see from (9.16) and (9.17) that the variance covariance matrices of the OLS estimators of β_i (the coefficient vector in the i'th equation of the SURE model) are given by

$$V(b_{(i)0}) = \sigma_{ii}(X_i'X_i)^{-1}$$
$$= (\theta_1 + \theta_2 + \cdots + \theta_i)(X_i'X_i)^{-1} \tag{9.20}$$

$$V(b^{\dagger}_{(i)0}) = \begin{cases} \frac{1}{4}(\theta_i + \theta_{i+1})(X_i'X_i)^{-1} & \text{if } i = 1, 2, \ldots, (M-1) \\ \theta_M(X_M'X_M)^{-1} & \text{if } i = M . \end{cases} \tag{9.21}$$

A comparison of (9.20) and (9.21) may reveal the effect of incorporating the constraints on the variance covariance matrix of the disturbances, on the efficiency of OLS estimation despite the fact that both of the estimators ignore the correlated nature of the disturbances across the equations. Further, we see that the application of OLS to estimate β_i ($i \neq M$) in the modified model (9.8) provides a more efficient estimator than the one obtained from the original model (9.1) when

$$(\theta_i + \theta_{i+1}) < 4(\theta_1 + \theta_2 + \cdots + \theta_i) , \tag{9.22}$$

which will definitely hold if θ_{i+1} is smaller than $3\theta_i$. However, the estimation of β_M by OLS from the model (9.8) rather than from the model (9.1) is clearly a superior strategy.

Observing from (9.18) that the i'th diagonal block of the matrix Ω is

$$\Omega_i = \begin{cases} \left(\frac{\theta_i\theta_{i+1}}{\theta_i + \theta_{i+1}}\right)(X_i'X_i)^{-1} & \text{if } i = 1, 2, \ldots, (M-1) \\ \theta_M(X_M'X_M)^{-1} & \text{if } i = M , \end{cases}$$

and comparing this with (9.20) and (9.21), we find

$$V(b_{(i)0}) - \Omega_i = \begin{cases} \left[\theta_1 + \theta_2 + \cdots + \theta_{i-1} + \frac{\theta_i^2}{\theta_i + \theta_{i+1}}\right](X_i'X_i)^{-1} & \text{if } i = 1, 2, \ldots, (M-1) \\ (\theta_1 + \theta_2 + \cdots + \theta_{M-1})(X_M'X_M)^{-1} & \text{if } i = M , \end{cases} \tag{9.23}$$

$$V(b^{\dagger}_{(i)0}) - \Omega_i = \begin{cases} \frac{(\theta_i - \theta_{i+1})^2}{4(\theta_i + \theta_{i+1})}(X_i'X_i)^{-1} & \text{if } i = 1, 2, \ldots, (M-1) \\ 0 & \text{if } i = M , \end{cases} \tag{9.24}$$

which gives some indication of the gain in asymptotic efficiency arising from FGLS estimation (see Srivastava and Mishra, 1986).

To make further comparisons among the estimators defined in (9.13) and (9.14), we assume normality of the disturbances and use as the performance criterion the variance covariance matrix to order $O(T^{-2})$. From equations (3.32) and (3.56) of Section 3.3, it may be recalled that both the SUUR and SURR estimators for β_i have identical variance covariance matrices, to the order of our approximation:

$$V_i = \theta_i(1 - \phi_i)\left(1 + \frac{M-1}{T}\right)(X_i'X_i)^{-1} \qquad (i \neq M) \tag{9.25}$$

where $\phi_i = \theta_i/(\theta_i + \theta_{i+1})$; $i = 1, 2, \ldots, (M-1)$.

For the FGLS estimator (9.14), we observe that

$$b^{\dagger}_{(i)F} = \begin{cases} b_{11} - \hat{\phi}_1 b_{12} & \text{if } i = 1 \\ b_{ii} - b_{ii-1} - \hat{\phi}_i(b_{ii+1} - b_{ii-1}) & \text{if } i = 2, 3, \ldots, (M-1) \\ b_{MM} - b_{MM-1} & \text{if } i = M \end{cases} \tag{9.26}$$

where $b_{ij} = (X_i'X_i)^{-1}X_i'y_j$ and $\hat{\phi}_i = \hat{\theta}_i/(\hat{\theta}_i + \hat{\theta}_{i+1})$.

Notice that the FGLS estimator of β_M coincides with the OLS estimator obtained from the modified model (9.8) so that no gain in efficiency is achieved. Next, it may be observed that the expression (9.26) for the FGLS estimators continues to hold when the system is partially orthogonal in the sense that

$$X_i'X_{i+1} = 0 \qquad (i = 1, 2, \ldots, (M-1)) \tag{9.27}$$

which is a somewhat less restrictive case than (9.19). In other words, the expressions for the FGLS estimators remain valid when, instead of pair-wise orthogonality of the regressor matrices in the SURE model, the M equations of the model are so arranged that the regressor matrix in any particular equation is orthogonal to the regressor matrix in the preceding, if any, as well as the succeeding, if any, equations.

Now, let us restrict our attention to the estimators of $\beta_1, \beta_2, \ldots, \beta_{M-1}$. From (9.26) and (9.27), we can write

$$(b^{\dagger}_{(i)F} - \beta_i) = (X_i'X_i)^{-1}X_i'[(1 - \hat{\phi}_i)\epsilon_i - \hat{\phi}_i\epsilon_{i+1}] \qquad (i \neq M) \quad . \tag{9.28}$$

By virtue of the assumed normality of the disturbances, the $\hat{\phi}_i$'s and $X_i'\epsilon_j$'s are independently distributed. As the expression on the right hand side of (9.28) is an odd function of the disturbances, the estimator $b^{\dagger}_{(i)F}$ is unbiased. Its variance covariance matrix is

$$V(b^{\dagger}_{(i)\,F}) = [\theta_i E(1 - \hat{\phi}_i)^2 + \theta_{i+1} E(\hat{\phi}^2_i)](X'_i X_i)^{-1}$$

$$= [\theta_i - 2\theta_i E(\hat{\phi}_i) + (\theta_i + \theta_{i+1})E(\hat{\phi}^2_i)](X'_i X_i)^{-1}$$

$$= \theta_i[1 - 2E(\hat{\phi}_i) + \frac{1}{\phi_i}E(\hat{\phi}^2_i)](X'_i X_i)^{-1} \; . \tag{9.29}$$

To simplify this expression let us first evaluate $E(\hat{\phi}_i)$ and $E(\hat{\phi}^2_i)$ to order $O(T^{-1})$. For this purpose it is easy to see from (9.1), (9.3), (9.15) and (9.19) that (Srivastava and Mishra, 1986)

$$\hat{\theta}_i = \frac{1}{T}\epsilon'_i(I_T - P_{x_i} - P_{x_{i-1}})\epsilon_i$$

where P_{x_0} should be treated as a null matrix.

Observing that $e_i = (\hat{\theta}_i - \theta_i)$ is of order $O_p(T^{-\frac{1}{2}})$, we can write

$$\hat{\phi}_i = \frac{\theta_i + e_i}{\theta_i + \theta_{i+1} + e_i + e_{i+1}}$$

$$= \phi_i\left(1 + \frac{e_i}{\theta_i}\right)\left[1 + \frac{\phi_i}{\theta_i}(e_i + e_{i+1})\right]^{-1} \; . \tag{9.30}$$

Expanding and retaining terms to order $O_p(T^{-1})$, we find

$$\hat{\phi}_i = \phi_i + \frac{\phi_i}{\theta_i}[(1 - \phi_i)e_i - \phi_i e_{i+1}] - \left(\frac{\phi_i}{\theta_i}\right)^2 (e_i + e_{i+1})[(1 - \phi_i)e_i - \phi_i e_{i+1}] \; . \tag{9.31}$$

Observing that

$$E(e_i) = -\left(\frac{K_i + K_{i-1}}{T}\right)\theta_i$$

$$E(e^2_i) = \frac{2}{T}\left(1 - \frac{K_i + K_{i-1}}{T}\right)\theta^2_i + \theta^2_i\left(\frac{K_i + K_{i-1}}{T}\right)^2 \; , \tag{9.32}$$

where K_0 is definitionally taken as 0, we have, to order $O(T^{-1})$:

$$E(\hat{\phi}_i) = \phi_i + \frac{\phi_i}{\theta_i}[(1 - \phi_i)E(e_i) - \phi_i E(e_{i+1})]$$

$$- \left(\frac{\phi_i}{\theta_i}\right)^2 [(1 - \phi_i)E(e_i^2) - \phi_i E(e_{i+1}^2) + (1 - 2\phi_i)E(e_i)E(e_{i+1})]$$

$$= \phi_i + \frac{1}{T}\phi_i(1 - \phi_i)[K_{i+1} - K_{i-1} + 2(1 - 2\phi_i)] \ . \tag{9.33}$$

Similarly, we have

$$E(\hat{\phi}_i^2) = \phi_i^2 + \frac{2}{T}\phi_i^2(1 - \phi_i)[K_{i+1} - K_{i-1} + 2(2 - 3\phi_i)] \tag{9.34}$$

to order $O(T^{-1})$.

Substituting (9.33) and (9.34) in (9.29), we get the expression for the variance covariance matrix to order $O(T^{-2})$:

$$V(b^{\dagger}_{(i)F}) = \theta_i(1 - \phi_i)\left[1 + \frac{4}{T}\phi_i(1 - \phi_i)\right](X_i'X_i)^{-1} \ . \tag{9.35}$$

We see from (9.25) and (9.35) that the FGLS estimator (9.26) of β_i dominates the SUUR/SURR estimator with respect to the variance covariance matrix to order $O(T^{-2})$ as the performance criterion when

$$\phi_i(1 - \phi_i) < \left(\frac{M - 1}{4}\right) , \tag{9.36}$$

which definitely holds as long as the SURE model contains more than four equations, as ϕ_i lies between 0 and 1 (see Srivastava and Mishra, 1986).

If we wish to obtain the exact expression for the variance covariance matrix of the FGLS estimator (9.26), we simply need to evaluate the first two moments of $\hat{\phi}_i$ exactly. For this purpose let us write

$$v_i = \frac{\hat{\theta}_i}{\theta_i} = \frac{1}{\theta_i T}\epsilon_i'[I_T - P_{x_i} - P_{x_{i-1}}]\epsilon_i \qquad (P_{x_0} \equiv 0) \tag{9.37}$$

so that the v_i's are independently distributed with v_i following a χ^2 distribution with $2m_i$ degrees of freedom, where

$$m_i = \begin{cases} \frac{1}{2}(T - K_1) & \text{if } i = 1 \\ \frac{1}{2}(T - K_i - K_{i-1}) & \text{if } i = 2, 3, \ldots, M \ . \end{cases}$$

Now, consider the α'th moment of $\hat{\phi}_i$:

$$E(\hat{\phi}_i^{\alpha}) = \int_0^{\infty}\int_0^{\infty}\left[\frac{v_i}{v_i + \left(\frac{1-\phi_i}{\phi_i}\right)v_{i+1}}\right]^{\alpha} \frac{v_i^{m_i-1}\exp\left\{-\frac{v_i}{2}\right\}}{2^{m_i}\Gamma(m_i)} \frac{v_{i+1}^{m_{i+1}-1}\exp\left\{-\frac{v_{i+1}}{2}\right\}}{2^{m_{i+1}}\Gamma(m_{i+1})} dv_i\, dv_{i+1}$$

$$= \frac{1}{2^{m_i+m_{i+1}}\Gamma(m_i)\Gamma(m_{i+1})}\int_0^{\infty}\int_0^{\infty}\frac{v_i^{m_i+\alpha-1} v_{i+1}^{m_{i+1}-1}\exp\left\{-\frac{1}{2}(v_i+v_{i+1})\right\}}{\left[v_i + \left(\frac{1-\phi_i}{\phi_i}\right)v_{i+1}\right]^{\alpha}} dv_i\, dv_{i+1} . \tag{9.38}$$

Applying the transformation

$$v_i = v_i$$

$$v_{i+1} = \left(\frac{t}{1-t}\right)v_i \qquad (0 < t < 1)$$

and then integrating with respect to v_i, we find that the α'th moment of $\hat{\phi}_i$ is equal to an expression containing a standard hypergeometric integral which can be evaluated straightforwardly. The result obtained in this way can be used in (9.29) to get the exact variance covariance matrix of the FGLS estimator (see Srivastava and Mishra, 1986). However, it is clear that the resulting expression is quite involved and does not permit us to draw any clear conclusions.

9.3 ERROR COMPONENTS STRUCTURE

Sometimes it may be possible to decompose the disturbance term into a number of stochastically independent and additive components. This kind of specification provides an interesting framework for enlarging the available data base through a combination of cross-section and time-series observations. To illustrate this point, let us consider the following version of the SURE model:

$$\begin{bmatrix} y_1 \\ y_2 \\ \vdots \\ y_M \end{bmatrix} = \begin{bmatrix} X_1 & 0 & \cdots & 0 \\ 0 & X_2 & \cdots & 0 \\ \vdots & \vdots & & \vdots \\ 0 & 0 & \cdots & X_M \end{bmatrix} \begin{bmatrix} \beta_1 \\ \beta_2 \\ \vdots \\ \beta_M \end{bmatrix} + \begin{bmatrix} u_1 \\ u_2 \\ \vdots \\ u_M \end{bmatrix}$$

or

$$y = X\beta + u \tag{9.39}$$

where y_i is an $(NT \times 1)$ vector of T observations on N cross-sectional units on the variable to be explained by the i'th equation of the model, X_i is an $(NT \times K_i)$ matrix of observations on K_i regressors and β_i is the vector of coefficients. Here, we consider the work of Avery (1977) and Baltagi (1980), and assume that the disturbance term has three additive components associated with the cross-section, the time period and the interaction of cross-section and time period, so that the disturbance vector u_i can be written

$$u_i = (I_N \otimes e_T)\mu_i + (e_N \otimes I_T)\lambda_i + \epsilon_i , \tag{9.40}$$

where e_T and e_N are $(T \times 1)$ and $(N \times 1)$ vectors respectively with all elements equal to unity. The $(N \times 1)$ vector μ_i, $(T \times 1)$ vector λ_i and $(NT \times 1)$ vector ϵ_i are assumed to be stochastic with null mean vectors and variance covariance matrix

$$E\begin{bmatrix} \mu_i \\ \lambda_i \\ \epsilon_i \end{bmatrix} [\mu_i' \quad \lambda_i' \quad \epsilon_i'] = \begin{bmatrix} \sigma^2_{\mu ij} I_N & 0 & 0 \\ 0 & \sigma^2_{\lambda ij} I_T & 0 \\ 0 & 0 & \sigma^2_{\epsilon ij} I_{NT} \end{bmatrix},$$

where $\sigma^2_{\mu ij}$, $\sigma^2_{\lambda ij}$ and $\sigma^2_{\epsilon ij}$ are positive scalar quantities. The fourth-order moments of the elements of the vectors μ_i, λ_i and ϵ_j are assumed to be finite.

Defining

$$E(uu') = \Psi = \begin{bmatrix} \Psi_{11} & \cdots & \Psi_{1M} \\ \vdots & & \vdots \\ \Psi_{M1} & \cdots & \Psi_{MM} \end{bmatrix} \tag{9.41}$$

we have

$$\Psi_{ij} = \sigma^2_{\mu ij}(I_N \otimes e_T e'_T) + \sigma^2_{\lambda ij}(e_N e'_N \otimes I_T) + \sigma^2_{\epsilon ij} I_{NT}$$

$$= \theta^2_{ij(1)} J_1 + \theta^2_{ij(2)} J_2 + \theta^2_{ij(3)} J_3 + \theta^2_{ij(4)} J_4 \tag{9.42}$$

where

$$\theta^2_{ij(1)} = \sigma^2_{\epsilon ij}, \quad J_1 = I_{NT} - \frac{1}{T}(I_N \otimes e_T e'_T) - \frac{1}{N}(e_N e'_N \otimes I_T) + \frac{1}{NT} e_{NT} e'_{NT},$$

$$\theta^2_{ij(2)} = \sigma^2_{\epsilon ij} + T\sigma^2_{\mu ij}, \quad J_2 = \frac{1}{T}(I_N \otimes e_T e'_T) - \frac{1}{NT} e_{NT} e'_{NT},$$

$$\theta^2_{ij(3)} = \sigma^2_{\epsilon ij} + N\sigma^2_{\lambda ij}, \quad J_3 = \frac{1}{N}(e_N e'_N \otimes I_T) - \frac{1}{NT} e_{NT} e'_{NT},$$

$$\theta^2_{ij(4)} = \sigma^2_{\epsilon ij} + T\sigma^2_{\mu ij} + N\sigma^2_{\lambda ij}, \quad J_4 = \frac{1}{NT} e_{NT} e'_{NT}.$$

Notice (Avery, 1977; p. 204) that $\theta^2_{ij(1)}$, $\theta^2_{ij(2)}$, $\theta^2_{ij(3)}$ and $\theta^2_{ij(4)}$ are the four distinct latent roots of Ψ_{ij}, with multiplicities $(N - 1)(T - 1)$, $(N - 1)$, $(T - 1)$ and 1 respectively. If Σ_μ, Σ_λ and Σ_ϵ denote the $(M \times M)$ matrices with their (i, j)'th elements as $\sigma^2_{\mu ij}$, $\sigma^2_{\lambda ij}$ and $\sigma^2_{\epsilon ij}$ respectively, we have

$$\Theta_1 = \Sigma_\epsilon, \qquad \Theta_2 = \Sigma_\epsilon + T\Sigma_\mu,$$
$$\Theta_3 = \Sigma_\epsilon + N\Sigma_\lambda, \qquad \Theta_4 = \Sigma_\epsilon + T\Sigma_\mu + N\Sigma_\lambda, \tag{9.43}$$

where

$$\Theta_k = \begin{bmatrix} \theta^2_{11(k)} & \cdots & \theta^2_{1M(k)} \\ \vdots & & \vdots \\ \theta^2_{M1(k)} & \cdots & \theta^2_{MM(k)} \end{bmatrix} \quad (k = 1, 2, 3, 4). \tag{9.44}$$

Using (9.42) and (9.43), we can write

$$\Psi = (\Theta_1 \otimes J_1) + (\Theta_2 \otimes J_2) + (\Theta_3 \otimes J_3) + (\Theta_4 \otimes J_4). \tag{9.45}$$

Assuming that the Θ_k matrices are non-singular and observing that J_1, J_2, J_3 and J_4 are symmetric and idempotent, such that (Baltagi, 1980; p. 1548):

$$J_k J_{k*} = 0 \qquad \text{if } k \neq k* \qquad (k, k* = 1, 2, 3, 4) ,$$

$$J_1 + J_2 + J_3 + J_4 = I_{NT} ,$$

it can be seen that

$$\Psi^{-1} = (\Theta_1^{-1} \otimes J_1) + (\Theta_2^{-1} \otimes J_2) + (\Theta_3^{-1} \otimes J_3) + (\Theta_4^{-1} \otimes J_4) \tag{9.46}$$

because

$$\begin{aligned} \Psi\Psi^{-1} &= \sum_{k=1}^{4} (\Theta_k \Theta_k^{-1} \otimes J_k^2) + \sum_{\substack{k=1 \\ k \neq k*}}^{4} \sum_{k*=1}^{4} (\Theta_k \Theta_k^{-1} \otimes J_k J_{k*}) \\ &= \sum_{k=1}^{4} (I_M \otimes J_k) \\ &= (I_M \otimes \sum_{k=1}^{4} J_k) \\ &= I_{MNT} . \end{aligned}$$

Regarding the regressor matrices, Baltagi (1980) assumed that when N and T tend to infinity in every way, (1/NT) times the matrices $X_i'X_j$ and $X_i'J_1X_j$ approach positive definite matrices with finite elements, and that each column vector of the matrices $X_1, X_2, \ldots, X_M$ adds up to zero, i.e.,

$$e_{NT}'X_i = 0 \qquad (i = 1, 2, \ldots, M) , \tag{9.47}$$

so that

$$\begin{aligned} X'(\Theta_4^{-1} \otimes J_4)X &= \frac{1}{NT} X'(\Theta_4^{-1} \otimes e_{NT}e_{NT}')X \\ &= 0 . \end{aligned} \tag{9.48}$$

The specification (9.47) essentially amounts to a SURE model in which the i'th equation is

$$y_i = e_{NT}\alpha_i + X_i\beta_i + u_i \qquad (i = 1, 2, \ldots, M)$$

where α_i is a scalar quantity representing the intercept term.

Now, if we apply the GLS estimator equation by equation in the model (9.39), we obtain

$$b = (X'\Psi_0^{-1}X)^{-1}X'\Psi_0^{-1}y \ , \tag{9.49}$$

where Ψ_0 is a block diagonal matrix obtained from Ψ by setting its off-diagonal blocks to null matrices. That is,

$$\Psi_0 = \begin{bmatrix} \Psi_{11} & \cdots & 0 \\ \vdots & & \vdots \\ 0 & \cdots & \Psi_{MM} \end{bmatrix} .$$

If b_i denotes the i'th subvector of b, partitioned conformably with β, so that it is the GLS estimator of β_i in the i'th equation of the model (9.39), it is easy to see that

$$b_i = \Big(\sum_{k=1}^{3} \frac{1}{\theta_{ii(k)}^2} X_i' J_k X_i\Big)^{-1} \Big(\sum_{k=1}^{3} \frac{1}{\theta_{ii(k)}^2} X_i' J_k y_i\Big) \ (i = 1, 2, \ldots, M) \tag{9.50}$$

where use has been made of (9.47) and (9.49).

On the other hand, if we apply the GLS estimator to the entire system in (9.39), we obtain (Baltagi, 1980; pp. 1548-1549)

$$\begin{aligned} b_g &= (X'\Psi^{-1}X)^{-1}X'\Psi^{-1}y \\ &= \Big[X'\Big(\sum_{k=1}^{4} (\Theta_k^{-1} \otimes J_k)\Big)X\Big]^{-1} X'\Big(\sum_{k=1}^{4} (\Theta_k^{-1} \otimes J_k)\Big)y \\ &= \Big[\sum_{k=1}^{3} X'(\Theta_k^{-1} \otimes J_k)X\Big]^{-1} \sum_{k=1}^{3} X'(\Theta_k^{-1} \otimes J_k)y \ , \end{aligned} \tag{9.51}$$

where we have used (9.46) and (9.48). Thus the estimator b_G is a weighted combination of three estimators (viz., between cross-section, between time periods and within time periods) of β, each estimator being weighted by the inverse of its variance covariance matrix.

It is obvious from (9.49) and (9.51) that when the disturbances across the equations are uncorrelated, implying that

$$\Psi_{ij} = 0 \quad (i \neq j;\ i, j = 1, 2, \ldots, M) \ ,$$

the estimators b and b_G coincide. This result is compatible with the equivalence of the OLS and GLS estimators of β in the conventional SURE model (i.e., without an error component structure for the disturbances). Another interesting case in which the OLS and GLS estimators are equivalent in the context of the conventional SURE model is when all of the equations have identical regressors. Such a result, of course, does not hold here. To see this, suppose that $X_1 = X_2 = \cdots = X_M = Z$ (say). Then the estimator (9.51) reduces to (Baltagi, 1980; p. 1549)

$$b_G = \left[\sum_{k=1}^{3} (\Theta_k^{-1} \otimes Z'J_k Z)\right]^{-1} \left[\sum_{k=1}^{3} (\Theta_k^{-1} \otimes Z'J_k y)\right] ,$$

the i'th subvector of which clearly is not equal to the estimator b_i, given by (9.50) with X_i replaced by Z, unless the matrices Θ_1, Θ_2 and Θ_3 are diagonal (see Avery, 1977; p. 206). The possibility of Θ_2 and Θ_3 being null matrices is obviously excluded, in which case the model (9.39) collapses to the conventional SURE model.

Next, let us examine the asymptotic properties of the estimators. For this purpose, we first observe that

$$\lim_{T\to\infty} \Theta_2^{-1} = \lim_{N\to\infty} \Theta_3^{-1} = 0$$

so that

$$\lim_{\substack{N\to\infty \\ T\to\infty}} b_G = \lim_{\substack{N\to\infty \\ T\to\infty}} b_{GD} \tag{9.52}$$

where

$$b_{GD} = [X'(\Theta_1^{-1} \otimes J_1)X]^{-1} X'(\Theta_1^{-1} \otimes J_1)y \tag{9.53}$$

is the GLS estimator of β in a SURE model with dummy variables (see Baltagi, 1980). Consequently, the asymptotic distributions of b_G and b_{GD} are identical. Further, we observe that the asymptotic distributions of $\sqrt{NT}(b_G - \beta)$ and $\sqrt{NT}(b_{GD} - \beta)$ are multivariate Normal with mean vector 0 and variance covariance matrix

$$\left[\lim_{\substack{N\to\infty \\ T\to\infty}} \frac{1}{NT} X'(\Theta_1^{-1} \otimes J_1)X\right]^{-1} . \tag{9.54}$$

Similarly, the asymptotic distribution of $\sqrt{NT}(b - \beta)$ can be obtained, and it is easy to see that both the estimators b_G and b_{GD} are asymptotically more efficient than the estimator b.

In view of the limited usefulness of GLS estimators, we now consider their feasible versions. For this purpose, we introduce the matrices

$$\underset{(NT\times M)}{U} = (u_1, u_2, \ldots, u_M) ,$$

$$\underset{(M\times M)}{W_1} = \frac{1}{(N-1)(T-1)} U'J_1U, \quad \underset{(M\times M)}{W_2} = \frac{1}{(N-1)} U'J_2U, \quad \underset{(M\times M)}{W_3} = \frac{1}{(T-1)} U'J_3U . \tag{9.55}$$

It is easy to see that W_k is an unbiased as well as consistent estimator of Θ_k (k = 1, 2, 3). The matrices W_1, W_2 and W_3 will be referred to as analysis of variance estimators of the variance components. Now in order to get their feasible counterparts, Avery (1977) suggested the use of restricted residuals. That is, obtain the residual vectors $\hat{u}_i = \bar{P}_{X_i} y_i$ (i = 1, 2, ..., M) by the application of the OLS estimator to the model (9.39) and then form the matrix $\hat{U}$ by replacing $u_1, u_2, \ldots, u_M$ in U by $\hat{u}_1, \hat{u}_2, \ldots, \hat{u}_M$ respectively. Using $\hat{U}$ in place of U in W_1, W_2 and W_3, we get $\hat{W}_1$, $\hat{W}_2$ and $\hat{W}_3$ which serve as estimators for Θ_1, Θ_2 and Θ_3 respectively. These, in turn, imply the following FGLS estimator from (9.51):

$$b_F = \left[\sum_{k=1}^{3} X'(\hat{W}_k^{-1} \otimes J_k)X\right]^{-1}\left[\sum_{k=1}^{3} X'(\hat{W}_k^{-1} \otimes J_k)y\right] . \tag{9.56}$$

Baltagi (1980) made another suggestion based on restricted residual vectors

$$\hat{u}_{i*} = [I_{NT} - X_i'(X_i'J_1X_i)^{-1}X_i'J_1]y_i \qquad (i = 1, 2, \ldots, M)$$

obtained by the application of the OLS estimator with dummy variables. Utilizing these $\hat{u}_{i*}$'s in place of the u_i's in U involved in the expressions for W_1, W_2 and W_3, we get the estimators $\hat{W}_{1*}$, $\hat{W}_{2*}$ and $\hat{W}_{3*}$ for Θ_1, Θ_2 and Θ_3 respectively. Substituting them in (9.51), we obtain the following FGLS estimator:

$$b_{F*} = \left[\sum_{k=1}^{3} X'(\hat{W}_{k*}^{-1} \otimes J_k)X\right]^{-1}\left[\sum_{k=1}^{3} X'(\hat{W}_{k*}^{-1} \otimes J_k)y\right] . \tag{9.57}$$

Baltagi (1980) stated that the asymptotic distributions of $\hat{W}_{1*}$, $\hat{W}_{2*}$ and $\hat{W}_{3*}$ are, while the asymptotic distributions of $\hat{W}_1$, $\hat{W}_2$ and $\hat{W}_3$ are not, the same as those of W_1, W_2 and W_3, and consequently the estimator b_{F*} is asymptotically as efficient as b_G while the estimator b_F is not. Prucha (1984) examined this issue and established that both of the estimators b_F and b_{F*} are asymptotically as efficient as b_G. In fact, he showed that any FGLS estimator based on the estimators $\hat{\Theta}_1$, $\hat{\Theta}_2$ and $\hat{\Theta}_3$ will have the same asymptotic properties provided that (plim $\hat{\Theta}_1$) = Θ_1 and the two matrices (plim $\hat{\Theta}_2$) and (plim $\hat{\Theta}_3$) are finite and positive definite (see Prucha, 1984; pp. 206-207, for the proof). The maximum likelihood method is another procedure (assuming normality of the disturbances) that can be used to obtain estimates of Θ_1, Θ_2, Θ_3 and β. However, the estimating equations are typically nonlinear and do not lead to a closed-form expression for the estimators, but iterative procedures can be used to obtain numerical estimates.

Finally, it may be remarked that sometimes the role of one component in the structure of the disturbances may be quite unimportant in comparison with the remaining two components, and then the former component can be ignored without causing any significant losses. Suppose that the time period component is negligible so that instead of (9.40), u_i is expressible as a sum of two independent components:

$$u_i = (I_N \otimes e_T)\mu_i + \epsilon_i , \tag{9.58}$$

which characterizes a two component model. The above discussion of the three component model can be straightforwardly specialized for this model. Notice that now under the specification (9.58) we have

$$E(u_i u_j') = (\Theta_1 \otimes J_5) + (\Theta_2 \otimes J_6) \tag{9.59}$$

where $J_5 = \left[I_{NT} - \frac{1}{T}(I_N \otimes e_T e_T')\right]$ and $J_6 = \frac{1}{T}(I_N \otimes e_T e_T')$ (see Avery and Watts, 1976). The maximum likelihood estimators in the present context can be derived without much difficulty as is shown by Verbon (1980, 1980a).

When using cross-section data, Verbon (1980a) pointed out it may be more appropriate to assume that the disturbances are heteroscedastic. Accordingly, he considered a SURE model with two error components specified by (9.58) and postulated the following heteroscedastic error process:

$$E(u_i u_j') = \sigma^2_{\mu ij}(I_N \otimes e_T)F(I_N \otimes e_T') + \sigma^2_{\epsilon ij}(F \otimes I_N) \tag{9.60}$$

where

$$\underset{(N\times N)}{F} = \begin{bmatrix} f(Z_1'\alpha) & 0 & \cdots & 0 \\ 0 & f(Z_2'\alpha) & \cdots & 0 \\ \vdots & \vdots & & \vdots \\ 0 & 0 & \cdots & f(Z_N'\alpha) \end{bmatrix} . \tag{9.61}$$

Here $Z_1, Z_2, \ldots, Z_N$ are $(p \times 1)$ vectors of cross-sectional characteristics but independent of the time period, α is a $(p \times 1)$ vector of constants and $f(\cdot)$ is any function that does not depend upon any particular time period or cross-section or equation of the model. Further, the value of $f(\cdot)$ at $\alpha = 0$ is unity and all of the first-order derivatives of the function with respect to the elements of α do not vanish at $\alpha = 0$. The null hypothesis of homoscedasticity is characterized by $H_0: \alpha = 0$. Assuming normality of the disturbances and using the Lagrange multiplier test proposed by Breusch and Pagan (1980), Verbon (1980a) presented the test statistic

$$LM = \frac{MT}{2}\left[\sum_{n=1}^{N}\left(\frac{d_n}{MT} - 1\right)(Z_n - \bar{Z})'\right]\left[\sum_{n=1}^{N}(Z_n - \bar{Z})(Z_n - \bar{Z})'\right]^{-1}\left[\sum_{n=1}^{N}\left(\frac{d_n}{MT} - 1\right)(Z_n - \bar{Z})\right] \tag{9.62}$$

where $\bar{Z} = \frac{1}{N}\Sigma_{n=1}^{N} Z_n$ and $d_n = \frac{1}{T}\overset{*}{u}'[(\overset{*}{\Theta}{}_1^{-1} \otimes J_5) + (\overset{*}{\Theta}{}_2^{-1} \otimes J_6)]\overset{*}{u}$, with an asterisk on u, Θ_1 and Θ_2 indicating the corresponding maximum likelihood estimates. The asymptotic distribution of the statistic (9.62) under H_0 is χ^2 with p degrees of freedom.

9.4 RESEARCH SUGGESTIONS

Assuming the availability of some constraints on the disturbances, we have considered two specific disturbance structures and have discussed the estimation of the parameters of the SURE model. An analysis of the performances of these estimators in finite samples and developing possibly more efficient estimation procedures remains to be explored further. Even more remains to be done in the case of SURE models with an error component structure, as in this case nothing is known about the properties of the associated estimators beyond a few basic asymptotic results. In addition, there is a need to formulate other realistic but general specifications of the error component structure. For instance, one may specify that the variance covariance matrices of μ_i, λ_i and ϵ_i are more general than scalar multiples of identity matrices (see, e.g., Magnus, 1982). Sometimes it may be preferable to treat some of the components as being fixed and the remaining as random. For example, when T is small and N is large in a three component model, the estimation of the $\sigma^2_{\lambda ij}$'s may not be reliable, and then it may be more appropriate to treat this component as fixed. Another realistic formulation would be to allow for the incompleteness of cross-section and time series data that may arise, for example, due to nonresponse or due to rotating panels, as is detailed by Biørn (1981). Devising techniques to identify appropriate structures for the disturbances is a potential area for future work.

EXERCISES

9.1 Show that the exact expression for the variance covariance matrix of the FGLS estimator $b^\dagger_{(i)F}$ in (9.26), under orthogonality of the regressors and normality of the disturbances, is

$$V(b^\dagger_{(i)F}) = \theta_i\left[1 - 2G_{i1} + \frac{1}{\phi_i}G_{i2}\right](X_i'X_i)^{-1} \qquad (i = 1,\ 2,\ \ldots,\ (M-1))$$

where

$$\underset{(\alpha=1,\,2)}{G_{i\alpha}} = \frac{\Gamma(m_i+\alpha)\Gamma(m_i+m_{i+1})}{\Gamma(m_i)\Gamma(m_i+m_{i+1}+\alpha)}\begin{cases} {}_2F_1\left(\alpha, m_{i+1};\ m_i+m_{i+1}+\alpha;\ 2-\frac{1}{\phi_i}\right) & \text{if } \phi_i > \frac{1}{3} \\ \left(\frac{\phi_i}{1-\phi_i}\right)^{\alpha} {}_2F_1\left(\alpha, m_{i+1}+\alpha;\ m_i+m_{i+1}+\alpha;\ 2-\frac{1}{1-\phi_i}\right) & \\ \qquad\qquad\text{if } \phi_i < \frac{2}{3}\ . \end{cases}$$

9.2 Suppose that $J_1, J_2, \ldots, J_p$ are p mutually orthogonal, symmetric and idempotent matrices of rank $j_1, j_2, \ldots, j_p$ respectively such that their sum is an identity matrix. If $\Theta_1, \Theta_2, \ldots, \Theta_p$ are square matrices and

$$\Psi_1 = \sum_{i=1}^{p} (\Theta_i \otimes J_i)$$

$$\Psi_2 = \sum_{i=1}^{p} (J_i \otimes \Theta_i)\ ,$$

prove that

(i) the characteristic roots of the matrices Ψ_1 and Ψ_2 are identical and equal to the characteristic roots of $\Theta_1, \Theta_2, \ldots, \Theta_p$ with multiplicities $j_1, j_2, \ldots, j_p$.

(ii) A necessary and sufficient condition for the nonsingularity of Ψ_1 and Ψ_2 is that the matrices $\Theta_1, \Theta_2, \ldots, \Theta_p$ are singular. [Hint: See Magnus (1982; p. 242).]

9.3 If A, B and C are nonsingular matrices and a is a column vector such that a'Aa is non-zero, show that

$$\left[(A^{-1} \otimes B^{-1}) + \frac{1}{a'Aa}(aa') \otimes (C^{-1} - B^{-1})\right]^{-1} = (A \otimes B) + \frac{1}{a'Aa}(Aaa'A) \otimes (C - B) .$$

[Hint: See Magnus (1982; p. 243).]

9.4 In a SURE model with a three component structure for the disturbances, the FGLS estimator of β is obtained by substituting $\hat{\Theta}_1$, $\hat{\Theta}_2$ and $\hat{\Theta}_3$ in place of Θ_1, Θ_2 and Θ_3 in the GLS estimator given by (9.51). If $\hat{\Theta}_1$ is a consistent estimator of Θ_1 and if the probability limits of $\hat{\Theta}_1$ and $\hat{\Theta}_2$ are finite, prove that the asymptotic distribution of the FGLS estimator is the same as that of the GLS estimator. [Hint: See Prucha (1984).]

9.5 Consider a SURE model having a two component structure of the form (9.58). Assuming that the disturbances are normally distributed, obtain the maximum likelihood estimators of the variances of individual components.

REFERENCES

Avery, R. B. (1977). "Error components and seemingly unrelated regressions," Econometrica 45, 199-209.

Avery, R. B., and H. W. Watts (1976). "The application of an error components model to experimental panel data," in H. W. Watts and A. Rees (eds.), The New Jersey Experiment (Academic Press, New York), Vol. II, Chapter 14.

Baltagi, B. H. (1980). "On seemingly unrelated regressions with error components," Econometrica 48, 1547-1551.

Biørn, E. (1981). "Estimating seemingly unrelated regression models from incomplete cross-section/time-series data," Report 81/33, Statistisk Sentralbyrå, Oslo.

Breusch, T. S., and A. R. Pagan (1980). "The Lagrange multiplier test and its applications to model specification in econometrics," Review of Economic Studies 47, 239-253.

Magnus, J. R. (1982). "Multivariate error components analysis of linear and nonlinear regression models by maximum likelihood," Journal of Econometrics 19, 239-285.

Prucha, I. R. (1984). "On the asymptotic efficiency of feasible Aitken estimators for seemingly unrelated regression models with error components," Econometrica 52, 203-207.

Srivastava, V. K., and G. D. Mishra (1986). "Efficiency of an OGLS estimator in seemingly unrelated regression equations with constrained covariance structures," Journal of Quantitative Economics 2, 221-230.

Verbon, H. A. A. (1980). "Maximum likelihood estimation of a labour demand system. An application of a model of seemingly unrelated regression equations with the regression error composed of two components," Statistica Neerlandica 34, 33-48.

Verbon, H. A. A. (1980a). "Testing for heteroscedasticity in a model of seemingly unrelated regression equations with variance components (SUREVC)," Economics Letters 5, 149-153.

Zellner, A. (1979). "An error-components procedure (ECP) for introducing prior information about covariance matrices and analysis of multivariate regression models," International Economic Review 20, 679-692.

10

Prior Information

10.1 INTRODUCTION

Very often, especially when modelling economic behaviour, a researcher has available a set of information which is relevant to the inference problem under consideration but which is obtained from some source different from the sample of data currently being used. This extraneous or prior information may be based on other data, or it may be derived from non-data sources. For example, when estimating a model using a set of time-series data some relevant results based on an earlier sample, on data from another country, or on a set of cross-sectional data may be available. These results may suggest likely values of certain of the model's parameters. For instance, if the set of equations which is being treated as a SURE model comprises a system of demand relationships, then the derivation of this system as the solution to a constrained utility-maximization problem implies certain cross-equation restrictions on some of the model's parameters. These restrictions arise, of course, because the parameters have explicit economic interpretations, and the form of such restrictions will depend upon the particular form of utility function underlying the demand system (see Byron, 1970).

Prior information about the parameters of a SURE (or other) model may take a variety of forms. Sometimes, as in the demand system example above, this information may suggest linear or non-linear exact restrictions on the parameter space. In other cases the information may be less definitive, and may merely suggest the signs of certain of the parameters. That is, inequality restrictions may be relevant. Finally, uncertainty regarding the extent to which the prior information should be taken into account may suggest the imposition of stochastic constraints on the parameters of the SURE model. The use of prior information to generate restrictions on the parameter space requires careful consideration, and it is not clear that a frequentist or "classical" statistical framework is best suited for this task. Increasingly, Bayesian methods have been used in econometric modelling to provide a flexible way of combining prior and sample information when drawing inferences about the parameters of interest. Much of the literature relating to the use of prior information when estimating SURE models is based on a "classical" viewpoint, and this is reflected in the presentation in this chapter. However, we also discuss the Bayesian analysis of the SURE model, in Section 10.4, in the hope of maintaining some balance in the overall presentation.

The use of prior information, or more explicitly the incorporation of restrictions on the parameter space, when estimating the model's parameters raises some important issues regarding the properties of the estimators involved, both from large-sample and finite-sample viewpoints. It is well known that in general the imposition of false restrictions will adversely affect these properties in terms of bias, and perhaps inconsistency. There may, however, be a trade-off between bias and precision in finite samples. Conversely, ignoring relevant prior information may result in a loss of estimator efficiency in both large and small samples. Coupled with these statistical considerations is the question of model complexity, and the associated issues of computational cost and degrees of freedom. These matters fall into the area of specification analysis, and indicate the need for formal tests of the validity of any restrictions suggested by the prior information. These and other issues are discussed in this chapter.

10.2 RESTRICTIONS ON THE PARAMETERS

It is often the case that either data-based or non-data-based prior information suggests certain restrictions on the parameters of a SURE model. Accordingly, we consider the model

$$\begin{aligned} y &= X\beta + u \\ E(u) &= 0, \quad E(uu') = (\Sigma \otimes I_T) \ , \end{aligned} \tag{10.1}$$

subject to a set of restrictions on the elements of β. The three cases of exact restrictions, inequality restrictions, and stochastic restrictions are considered in turn.

10.2.1 Exact Linear Restrictions

Let us suppose that strong prior information is available in the form of a set of G independent linear restrictions on β:

$$R\beta = r \ , \tag{10.2}$$

where r and R have non-stochastic elements and R has full row rank, G. (This rank condition ensures the independence and internal consistency of the set of restrictions.) If the OLS estimator is applied to (10.1), with account being taken of (10.2), then it is well known that the restricted ordinary least squares (ROLS) estimator of β is

$$b_{RO} = b_0 + (X'X)^{-1}R'[R(X'X)^{-1}R']^{-1}(r - Rb_0) \tag{10.3}$$

where b_0 is the (unrestricted) OLS estimator of β in (10.1). If u is normally distributed, then b_{RO} is also the restricted maximum likelihood estimator of β. If the restrictions (10.2) are valid, then b_{RO} is unbiased. In any case its variance covariance matrix is given by

$$V(b_{RO}) = CX'(\Sigma \otimes I_T)XC \tag{10.4}$$

where

$$C = (X'X)^{-1} - (X'X)^{-1}R'[R(X'X)^{-1}R']^{-1}R(X'X)^{-1} \ . \tag{10.5}$$

It follows from (10.4) that, if the restrictions are valid, b_{RO} is efficient relative to b_0. If $R\beta \neq r$ then b_{RO} is biased but it is a more precise estimator than b_0, and so there are some parts of the parameter space in which b_{RO} dominates b_0 in terms of mean squared error (see Toro-Vizcarrondo and Wallace, 1968).

Now, suppose that an FGLS estimator is employed. In this case, the estimator which takes account of both the general covariance structure associated with (10.1) and the restrictions in (10.2) is

$$b_{RF} = b_F + W_F R'(RW_F R')^{-1}(r - Rb_F) \ , \tag{10.6}$$

where

$$W_F = [X'(S^{-1} \otimes I_T)X]^{-1} \ , \tag{10.7}$$

and S is a consistent estimator of Σ. We shall refer to b_{RF} as the restricted feasible generalized least squares (RFGLS) estimator of β.

Ruble (1968; pp. 274-278) generalised the above framework by relaxing the assumptions that R has full row rank and that X has full column rank. He merely required that the sum of these two ranks be no less than $K^* = \Sigma_{i=1}^{M} K_i$, and presented an alternative expression for the RFGLS estimator which is computationally simpler to implement than is (10.6):

$$b_{RF} = q + Q(Q'D^{-1}Q)^{-1}Q'(d - D^{-1}q) \quad , \tag{10.8}$$

where

$$\begin{aligned} D &= [(X'X)^{-1} * (S \otimes I_T)] \\ d &= X'[I_{MT} * (S \otimes I_T)]y \end{aligned} \tag{10.9}$$

and the operator * denotes the Hadamard product. In (10.8), $Q' = (I, -Q^{*\prime})$ and $q' = (0, q^{*\prime})$, where I is an identity matrix of order $(K^* - rk(R))$, Q^* is the $[rk(R) \times (K^* - rk(R))]$ matrix formed by reducing R to row-echelon form by a sequence of row operations alone, q^* is a column vector of order rk(R), derived from r by performing the same row operations on the augmented matrix (R, r), and q is $(K^* \times 1)$. As far as the choice of S is concerned, Ruble suggested that $S = \hat{S}$ should be used in preference to $S = \tilde{S}$. For the general case in which the restrictions in (10.2) imply constraints among the parameters of different equations of (10.1), Ruble suggested a modified method for constructing S which takes account of this additional information. It should be added that Ruble's analysis can be extended further, and the requirement that $[rk(R) + rk(X)] \geq K^*$ can be relaxed, along the lines suggested by Gallant and Gerig (1980).

Clearly, if S is any consistent estimator of Σ, then the RFGLS estimator (10.6) will be a consistent estimator of β. Moreover, its variance covariance matrix, to order $O(T^{-1})$, may be shown to be

$$\Omega - \Omega R'(R\Omega R')^{-1}R\Omega \quad , \tag{10.10}$$

where $\Omega = [X'(\Sigma^{-1} \otimes I_T)X]^{-1}$. Thus it follows that, to order $O(T^{-1})$, the RFGLS is more "precise" than is the FGLS estimator which ignores the information in the restrictions (10.2). To this same order, b_{RF} is unbiased if the restrictions are valid, and so in this case it is a more efficient estimator than its FGLS counterpart. If the restrictions in (10.2) are false, the bias and precision trade-off implies that there will still be some parts of the parameter space in which the RFGLS estimator dominates the FGLS estimator in terms of mean squared error.

Clearly, the validity of the restrictions in (10.2) is crucial to the properties of the RFGLS (and ROLS) estimators of β. Thus, our next consideration is the testing of the compatibility of the prior and sample information. That is, we wish to test $H_0 : R\beta = r$ against $H_1 : R\beta \neq r$. Assuming that the disturbance vector, u, in (10.1) has a multivariate Normal distribution, we may adopt the likelihood ratio, Wald or Lagrange multiplier testing procedures (see Harvey, 1981; pp. 159-175). Each is based on one or other of the restricted and unrestricted ML estimators of β which, under the normality assumption, are asymptotically equivalent to b_{RF} and b_F respectively.

To obtain the likelihood ratio test of H_0, recall the concentrated log likelihood function, (5.21):

$$\mathscr{L}^* = \text{constant} - \frac{T}{2} \log \left| \frac{1}{T} (Y - \underline{X}\beta)'(Y - \underline{X}\beta) \right| ,$$

where the notation is defined in Section 5.4. It follows immediately that the likelihood ratio test statistic is

$$\text{LRT} = T \log [|\Sigma_R|/|\Sigma_U|] ,$$

where Σ_R and Σ_U are the maximum likelihood estimators of Σ which impose and ignore the restrictions in H_0 respectively (see Harvey, 1981; p. 165). Under this hypothesis LRT is asymptotically χ^2 with G degrees of freedom.

If Σ is known then the Wald and Lagrange multiplier test statistics are identical:

$$W = LM = (Rb_G - r)'[R(X'(\Sigma^{-1} \otimes I_T)X)^{-1}R']^{-1}(Rb_G - r) ,$$

and are asymptotically χ^2 with G degrees of freedom (see Byron, 1970). When Σ is unknown, feasible versions of the Wald and Lagrange multiplier statistics may be obtained by replacing Σ by a consistent estimator, S. The choice of S does not affect the asymptotic distribution of the test statistic, but will have implications for its finite-sample properties.

Clearly, the W and LM test statistics are closely related to

$$z = \left(\frac{\nu}{G}\right) \left[\frac{(Rb_F - r)'[R(X'(S^{-1} \otimes I_T)X)^{-1}R']^{-1}(Rb_F - r)}{(y - Xb_F)'(S^{-1} \otimes I_T)(y - Xb_F)} \right] , \tag{10.11}$$

where $\nu = (MT - K^*)$. Under H_0, Gz also has an asymptotic distribution which is χ^2 with G degrees of freedom. Of more interest are the properties of these tests in finite samples, but these appear to be unexplored. Theil (1971; pp. 402-403) suggested treating z as being F-distributed with G and ν degrees of freedom in finite samples, but a detailed appraisal of this suggestion is needed before z can be used with confidence in this way, given that S is used to estimate Σ in (10.11).

This test statistic or the LRT, LM or W statistics also may be used to test the hypothesis that the RFGLS estimator dominates the FGLS estimator with respect to either the weak or strong asymptotic mean squared error criteria (see McElroy, 1977). A third application of z is in testing for aggregation bias, as was discussed in detail by Zellner (1962). To see this, consider the situation where $X_1, X_2, \ldots, X_M$ each have the same number of columns, say $\bar{K}$, and comprise variables relating to different micro-units. Simple aggregation of these micro-data into macro-data involves no aggregation bias if the following homogeneity hypothesis is true:

$$H_0^*: \beta_1 = \beta_2 = \cdots = \beta_M = \beta_0 \quad \text{(say)} \ .$$

Equivalently, we may write

$$H_0^*: R\beta = 0 \tag{10.12}$$

where R is $[(M-1)\bar{K} \times M\bar{K}]$, of the form

$$R = \begin{bmatrix} I & -I & 0 & \cdots & 0 & 0 \\ 0 & I & -I & \cdots & 0 & 0 \\ \vdots & \vdots & \vdots & & \vdots & \vdots \\ 0 & 0 & 0 & \cdots & I & -I \end{bmatrix} ,$$

and I is of order $\bar{K}$. In this context z reduces to

$$z^* = \left[\frac{M(T-\bar{K})}{\bar{K}(M-1)}\right]\left[\frac{b_F'R'[R(X'(S^{-1} \otimes I_T)X)^{-1}R']^{-1}Rb_F}{(y - Xb_F)'(S^{-1} \otimes I_T)(y - Xb_F)}\right] , \tag{10.13}$$

the distribution of which (under H_0^*) might be approximated by F with $(M-1)\bar{K}$ and $M(T-\bar{K})$ degrees of freedom. The quality of this approximation is questionable in finite samples, but asymptotically we may take $(M-1)\bar{K}z^*$ to be χ^2 with $(M-1)\bar{K}$ degrees of freedom under H_0^*.

Note that under H_0^*, another consistent estimator of Σ may be considered. If (10.12) holds, then the SURE model may be written as

$$\begin{bmatrix} y_1 \\ y_2 \\ \vdots \\ y_M \end{bmatrix} = \begin{bmatrix} X_1 \\ X_2 \\ \vdots \\ X_M \end{bmatrix} \beta_0 + \begin{bmatrix} u_1 \\ u_2 \\ \vdots \\ u_M \end{bmatrix} ,$$

or

$$y = \underset{\sim}{X}\beta_0 + u \quad , \tag{10.14}$$

where $\underset{\sim}{X}$ is $(MT \times \bar{K})$. In this case the OLS residuals associated with (10.14) may be used to construct an S matrix which is consistent for Σ. This choice of S has some intuitive appeal, for if H_0^* is true then more relevant information is incorporated. Of course, if H_0^* is false, there may be adverse statistical consequences when applying z^* with this S. Asymptotically, the use of this S matrix should have little effect on the distribution of z^*.

Returning to the question of aggregation bias, consider the situation when only macro data are available. As we are then forced to estimate only a macro relationship it would be desirable to have a test (based on the macro data) of the hypothesis of no aggregation bias. To illustrate Zellner's (1962; pp. 356-357) suggestion for this case, consider a simple SURE model in which each equation contains a single regressor:

$$y_{ti} = \beta_{i1} + x_{ti}\beta_{i2} + u_{ti} \quad . \tag{10.15}$$

$$(t = 1, 2, \ldots, T; \; i = 1, 2, \ldots, M)$$

Adding up the micro relationships in (10.15) we obtain the following macro relationship:

$$\sum_{i=1}^{M} y_{ti} = \sum_{i=1}^{M} \beta_{i1} + \sum_{i=1}^{M} x_{ti}\beta_{i2} + \sum_{i=1}^{M} u_{ti}$$

or

$$y_{t\cdot} = \beta_{\cdot 1} + x_{t\cdot}\beta_{M2} + \sum_{j=1}^{M-1} f_{tj}x_{t\cdot}(\beta_{j2} - \beta_{M2}) + u_{t\cdot} \; , \tag{10.16}$$

$$(t = 1, 2, \ldots, T)$$

where

$$x_{t\cdot} = \sum_{i=1}^{M} x_{ti} \; ; \quad f_{tj} = \frac{x_{tj}}{x_{t\cdot}} \; . \tag{10.17}$$

If the necessary micro data were available to permit the construction of the f_{tj}'s, and hence the $(f_{tj}x_{t\cdot})$ variables, we could fit (10.16) as a regression and test the hypothesis that all of the (M - 1) coefficients, $(\beta_{12} - \beta_{M2}), (\beta_{22} - \beta_{M2}), \ldots, (\beta_{M-12} - \beta_{M2})$ are zero. Rejection of this hypothesis implies aggregation bias. In the absence of the necessary micro data one may be willing to postulate that each f_{tj} is a (linear) function of observable macro data, substitute this function in (10.16), and again test for aggregation bias by testing the nullity of the appropriate regression slope parameters. This procedure is easily extended to the case where (10.15) involves more than one regressor (Zellner, 1962; p. 357).

If the null hypothesis is not rejected then (10.14) may be written in terms of the aggregated variables:

$$\sum_{i=1}^{M} y_i = \sum_{i=1}^{M} X_i\beta_0 + \sum_{i=1}^{M} u_i$$

or,

$$\bar{y} = \bar{X}\beta_0 + \bar{u} \; , \tag{10.18}$$

and the application of OLS using the macro data yields

$$\bar{b}_0 = (\bar{X}'\bar{X})^{-1}\bar{X}'\bar{y} \tag{10.19}$$

as an estimator of β_0 which is unbiased if the null hypothesis of no aggregation bias is true.

On the other hand, if micro data were available and β_0 were estimated by applying OLS to each equation of the SURE model in turn, the following M estimators would be obtained:

$$\bar{b}_{(i)0} = (X_i'X_i)^{-1}X_i'y_i \qquad (i = 1, 2, \ldots, M) \; . \tag{10.20}$$

It is easily seen that each of the estimators in (10.20) is unbiased for β_0, and so the expected values of the OLS estimators of β_0 based on either the macro or micro data are identical if H_0^* is true, but generally differ if H_0^* is false. Zellner referred to this discrepancy as aggregation bias.

10.2.2 Inequality Restrictions

We now consider situations in which somewhat less restrictive prior information is available, namely information which suggests only that certain inequality constraints hold among the parameters of the SURE model. In particular, consider the model (10.1), with the hypothesis that

$$R\beta \geq r \ , \tag{10.21}$$

where R is a $[G \times K^*]$ known (non-stochastic) matrix and r is a known $(G \times 1)$ vector. It should be noted that (10.21) is quite general. For example, if our prior information suggests restrictions of the type

$$R_1\beta \leq r_1 \ ; \quad \underline{r}_2 \leq R_2\beta \leq \bar{r}_2 \ ,$$

these may be expressed in the form (10.21) by writing:

$$\begin{bmatrix} -R_1 \\ -R_2 \\ R_2 \end{bmatrix} \beta \geq \begin{bmatrix} -r_1 \\ -\bar{r}_2 \\ \underline{r}_2 \end{bmatrix} .$$

The estimation of β in (10.1), subject to (10.21) may be viewed as a quadratic programming problem. With a single equation, taking a least squares (or maximum likelihood) approach, we choose β so as to maximize

$$-(y - X\beta)'(y - X\beta) \ , \tag{10.22}$$

subject to the restrictions in (10.21). This problem was discussed in general terms by Zellner (1961) and by Judge and Takayama (1966). The latter also included an analysis of the SURE model. After a suitable rearrangement of the problem to ensure that the elements of the coefficient vector satisfy the necessary non-negativity condition, the Kuhn-Tucker equivalence theorem and the duality theorem of Dorn (1961) may be used to obtain the inequality restricted ordinary least squares (IROLS) estimator of β. In particular, Judge and Takayama propose a formulation which facilitates the application of Wolfe's (1959) algorithm based on the simplex method (see also Liew, 1976, Klemm and Sposito, 1980, and Schmidt and Thomson, 1982).

As was discussed by Judge and Takayama, this quadratic programming approach easily handles the imposition of inequality restrictions on the parameters of the SURE model by modifying (10.22) to

$$-(y - X\beta)'(S^{-1} \otimes I_T)(y - X\beta) \ , \tag{10.23}$$

where S is a consistent estimator of Σ, and proceeding as already described.

The resulting estimator may be termed the inequality restricted feasible generalized least squares (IRFGLS) estimator of β. An evaluation and comparison of the properties of the IROLS and IRFGLS estimators of β is a difficult task, this difficulty being compounded by the fact that these estimators are merely numerical solutions to a mathematical programming problem, and are not readily expressed in closed-form. However, for some discussion of this problem the reader is referred to Zellner (1961), Judge and Takayama (1966), Schmidt and Thomson (1982), and Judge and Yancey (1981). The general problem of testing the validity of inequality restrictions was addressed by Yancey, Judge and Bock (1981).

10.2.3 Stochastic Linear Restrictions

A somewhat more flexible form of prior information may involve a degree of uncertainty associated with restrictions on the model's parameters. This uncertainty might be expressed quantitatively by the inclusion of a stochastic component whose properties are determined by the extraneous information itself. For example, we may have a set of G stochastic linear restrictions

$$r = R\beta + v \ , \qquad (10.24)$$

where R and r are as in (10.2), and v is a $(G \times 1)$ random disturbance vector, distributed independently of u in (10.1), with mean 0 and a known positive-definite variance covariance matrix Ψ_0. This "mixed estimation" formulation was discussed in the context of single-equation estimation (with an extension to simultaneous systems) by Theil and Goldberger (1961), and has been extended to general structural equations in the context of instrumental variables estimation by Giles (1982). Srivastava (1980) gives a bibliography of this literature. The mixed estimator has been correctly criticized by some Bayesians, as it takes account of stochastic prior information in an <u>ad hoc</u> way. However, as its implementation requires knowledge only of the first two moments of v, rather than a complete prior density, and for the sake of completeness, we shall discuss it briefly here. The Bayesian analysis of the SURE model is discussed in Section 10.4.

Specification of the restrictions as in (10.24) may arise when unbiased estimates of some known linear combinations of the elements of β are available from some extraneous source, or when two-sided inequality restrictions on some known linear combinations of the elements of β are treated in a manner akin to confidence intervals for these linear combinations. Taking account of the restrictions (10.24) and the SURE model (10.1), we have

$$\begin{bmatrix} y \\ r \end{bmatrix} = \begin{bmatrix} X \\ R \end{bmatrix} \beta + \begin{bmatrix} u \\ v \end{bmatrix} \ . \qquad (10.25)$$

Ignoring, for the moment, the general covariance structure of u, we may apply OLS to (10.25) and obtain the mixed ordinary least squares (MOLS) estimator of β:

$$b_{MO} = [X'X + R'R]^{-1}[X'y + R'r] \quad . \tag{10.26}$$

Similarly, applying an FGLS estimator to (10.25) we obtain the mixed feasible generalised least squares (MFGLS) estimator:

$$b_{MF} = [X'(S^{-1} \otimes I_T)X + R'\Psi_0^{-1}R]^{-1}[X'(S^{-1} \otimes I_T)y + R'\Psi_0^{-1}r] \quad . \tag{10.27}$$

A helpful interpretation of b_{MO} and b_{MF} may be obtained by generating the OLS and GLS "estimators" of β from (10.24) as:

$$b_{OP} = (R'R)^{+}R'r$$

$$b_{GP} = (R'\Psi_0^{-1}R)^{+}R'\Psi_0^{-1}r \quad ,$$

where A^{+} denotes the Moore-Penrose generalised inverse of the matrix A. For complete generality, we have allowed here for the case where $rk(R) = G < K^*$, although unlike the situation in (10.2), the stochastic nature of (10.24) implies that this condition may be violated without causing redundancies or inconsistencies among the restrictions. The MOLS and MFGLS estimators of β then may be expressed as weighted averages of the "estimators" of β implied by the sample and extraneous information:

$$b_{MO} = [X'X + R'R]^{-1}[X'Xb_0 + R'Rb_{OP}] \tag{10.28}$$

$$b_{MF} = [X'(S^{-1} \otimes I_T)X + R'\Psi_0^{-1}R]^{-1}[X'(S^{-1} \otimes I_T)Xb_F + R'\Psi_0^{-1}Rb_{GP}] \, . \tag{10.29}$$

If we define the matrices

$$\begin{aligned} \Lambda_{OS} &= [X'X + R'R]^{-1}X'X \\ \Lambda_{FS} &= [X'(S^{-1} \otimes I_T)X + R'\Psi_0^{-1}R]^{-1}X'(S^{-1} \otimes I_T)X \end{aligned} \tag{10.30}$$

as measures of the shares of sample information, and

$$\Lambda_{OP} = [X'X + R'R]^{-1}R'R$$
$$\Lambda_{FP} = [X'(S^{-1} \otimes I_T)X + R'\Psi_0^{-1}R]^{-1}R'\Psi_0^{-1}R \tag{10.31}$$

as measures of the shares of extraneous information, we can write

$$\begin{aligned} b_{MO} &= b_0 + \Lambda_{OP}(b_{OP} - b_0) \\ &= b_{OP} + \Lambda_{OS}(b_0 - b_{OP}), \end{aligned} \tag{10.32}$$

$$\begin{aligned} b_{MF} &= b_F + \Lambda_{FP}(b_{GP} - b_F) \\ &= b_{GP} + \Lambda_{FS}(b_F - b_{GP}) \ . \end{aligned} \tag{10.33}$$

From (10.32) and (10.33) it follows that b_{MO} approaches b_0 and b_{MF} approaches b_F as the sample information tends to dominate the extraneous information. Similarly, if the extraneous information tends to dominate the sample information then b_{MO} and b_{MF} approach b_{OP} and b_{GP} respectively.

Two further features of this estimation framework may be noted. First, if we can express r in the form $r = R\theta$, where θ is a known vector, then $b_{OP} = b_{GP} = \theta$, and

$$\begin{aligned} b_{MO} &= b_0 + \Lambda_{OP}(\theta - b_0) \\ &= \theta + \Lambda_{OS}(b_0 - \theta) \ , \end{aligned} \tag{10.34}$$

$$\begin{aligned} b_{MF} &= b_F + \Lambda_{FP}(\theta - b_F) \\ &= \theta + \Lambda_{FS}(b_F - \theta) \ . \end{aligned} \tag{10.35}$$

Secondly, if r is a null vector, then both b_{OP} and b_{GP} also are null vectors, and from (10.28) and (10.29) we get

$$b_{MO} = \Lambda_{OS}b_0 \ , \tag{10.36}$$

$$b_{MF} = \Lambda_{FS}b_F \ . \tag{10.37}$$

One interesting feature of (10.34)-(10.37) is that in each case the expressions resemble shrinkage estimators of β.

Now, let us consider the properties of the mixed estimators of (10.1). From (10.26), the MOLS estimator of β is unbiased, if (10.24) holds, and in any case its variance covariance matrix is

$$V(b_{MO}) = [X'X + R'R]^{-1}[X'(\Sigma \otimes I_T)X + R'\Psi_0 R][X'X + R'R]^{-1} . \tag{10.38}$$

When $T \to \infty$, so that the sample information dominates the extraneous information, the large-sample asymptotic approximation for $V(b_{MO})$, to order $O(T^{-1})$, is

$$(X'X)^{-1}X'(\Sigma \otimes I_T)X(X'X)^{-1} , \tag{10.39}$$

which is the exact variance covariance matrix of the OLS estimator of β in (10.1). This estimator ignores both the extraneous information in (10.24) and the general covariance structure of the disturbance vector, u. Similarly, we see from (10.27) that the MFGLS estimator of β is consistent, and it may be shown that the large-sample asymptotic approximation of its variance covariance matrix, to order $O(T^{-1})$, is $[X'(\Sigma^{-1} \otimes I_T)X]^{-1}$ which is identical to the large-sample asymptotic approximation, to order $O(T^{-1})$, to the variance covariance matrix of the FGLS estimator of β. This estimator ignores the extraneous information in (10.24), but does take into account the cross-equation correlations reflected in the covariance structure for u in (10.1).

It is clear that a comparison of the MOLS and MFGLS estimators on the basis of their asymptotic properties is of limited interest. In this case the effects of the extraneous information are totally dominated by the sample information, and asymptotically the MFGLS estimator is efficient relative to the MOLS estimator of β. Of much greater interest, of course, are the finite-sample properties of these estimators, especially when different choices of S are considered. Moreover, the mixed estimation approach itself may give rise to new consistent estimators of Σ. For example, S may be constructed on the basis of the residuals obtained by applying the MOLS, MFGLS, or iterative MFGLS estimators to the SURE model.

Finally, let us consider the matter of testing the compatibility of the sample information and the prior information reflected in (10.24). This compatibility is important in terms of determining the properties of various mixed estimators of β, and so a formal test should be adopted. Theil (1963) considered this testing problem in the context of stochastic extraneous information about the coefficients of a single-equation model. In the context of the SURE model, a suitable statistic is of the form:

$$z = \left(\frac{\nu}{G}\right)\left[\frac{(r - Rb_{MF})'[R\{X'(S^{-1} \otimes I_T)X\}^{-1}R' + \Psi_0]^{-1}(r - Rb_{MF})}{(y - Xb_F)'(S^{-1} \otimes I_T)(y - Xb_F)}\right] \tag{10.40}$$

where $\nu = (MT - K^*)$ are the degrees of freedom for the SURE system as a whole. Under the null hypothesis of compatibility between the sample and stochastic extraneous information about β, the asymptotic distribution of Gz may be approximated by a χ^2 distribution with G degrees of freedom. In finite samples, one might treat z itself as being F-distributed with G and ν degrees of freedom, although this presumption involves an approximation of unknown magnitude. The true finite-sample distribution of z will depend on the choice of S, and whatever such choice is made the power of the test obtained by treating z as being F-distributed is unexplored.

10.3 SPECIFICATION ANALYSIS

When estimating the parameters of a SURE model, as in any econometric analysis, we implicitly assume that the model is correctly specified in a variety of ways. For example, we assume that the linear specification adopted for the model is correct; that the specification of which regressors are included in each equation is appropriate; and that the disturbances satisfy the stochastic assumptions which form part of the model's overall specification. If this specification is lacking in any of these regards, then the results which are discussed throughout this book generally will be invalidated. An important area of econometrics is specification analysis, where we consider the sensitivity of various results (such as estimators' properties, the distribution of test statistics, etc.) to violations of the assumed model specification. Some aspects of this type of analysis, in as much as it relates to generalizations of the assumed structure of the disturbances in the model, were discussed in Chapters 7 and 8. In Section 11.5 we deal briefly with non-linear SURE systems. Here, however, we consider two important practical topics in specification analysis. These are closely related to each other, and are best discussed here as each may be viewed in terms of the validity of particular exact linear restrictions on the parameters of the SURE model. In the next sub-section we discuss the misspecification of the model which arises if some superfluous, irrelevant variables are included as regressors in one or more equations of the system. Of particular interest will be the implications of this misspecification for the properties of various estimators of the SURE model. We then discuss the misspecification which arises from omitting relevant regressors from one or more equations of the model. Much of the presentation is based on Srivastava and Srivastava (1983).

10.3.1 Inclusion of Superfluous Variables

To simplify the exposition we shall consider a two equation SURE model:

$$\begin{aligned} y_1 &= X_1\beta_1 + u_1 \\ y_2 &= X_2\beta_2 + u_2 \ , \end{aligned} \tag{10.41}$$

$$E\begin{bmatrix} u_1 \\ u_2 \end{bmatrix} = \begin{bmatrix} 0 \\ 0 \end{bmatrix}, \quad E\begin{bmatrix} u_1 \\ u_2 \end{bmatrix}[u_1' \quad u_2'] = \begin{bmatrix} \sigma_{11}I_T & \sigma_{12}I_T \\ \sigma_{21}I_T & \sigma_{22}I_T \end{bmatrix} = (\Sigma \otimes I_T) .$$

Suppose that the misspecification involves the inclusion of certain irrelevant regressors in the first equation of (10.41), so that the model which is estimated is

$$\begin{aligned} y_1 &= X_1\beta_1 + X_{1+}\beta_{1+} + v_1 \\ y_2 &= X_2\beta_2 + u_2 , \end{aligned} \tag{10.42}$$

where X_{1+} is a $(T \times K_{1+})$ non-stochastic matrix of rank K_{1+}, comprising observations on K_{1+} additional regressors, and β_{1+} is the corresponding coefficient vector. Writing (10.42) more compactly, we have

$$\begin{bmatrix} y_1 \\ y_2 \end{bmatrix} = \begin{bmatrix} X_1 & 0 \\ 0 & X_2 \end{bmatrix}\begin{bmatrix} \beta_1 \\ \beta_2 \end{bmatrix} + \begin{bmatrix} X_{1+} \\ 0 \end{bmatrix}\beta_{1+} + \begin{bmatrix} v_1 \\ v_2 \end{bmatrix}$$

or,

$$y = X\beta + X_+\beta_{1+} + v . \tag{10.43}$$

Applying OLS, the corresponding estimator of β is

$$b_0 = (X'\bar{P}_{x_+}X)^{-1}X'\bar{P}_{x_+}y , \tag{10.44}$$

where $\bar{P}_{x_+} = I - X_+(X_+'X_+)^{-1}X_+'$, so that the OLS estimators of β_1 and β_2 are

$$\begin{aligned} b_{(1)0} &= (X_1'\bar{P}_{x_{1+}}X_1)^{-1}X_1'\bar{P}_{x_{1+}}y_1 \\ &= (X_1'X_1)^{-1}X_1'y_1 - (X_1'X_1')^{-1}X_1'X_{1+}(X_{1+}'\bar{P}_{x_1}X_{1+})^{-1}X_{1+}'\bar{P}_{x_1}y_1 \end{aligned} \tag{10.45}$$

$$b_{(2)0} = (X_2'X_2)^{-1}X_2'y_2 . \tag{10.46}$$

Note that in obtaining (10.45) and (10.46) we have used the result that for non-singular matrices A_1 and A_2, and general conformable matrix A_3,

$$(A_1 - A_3A_2A_3')^{-1} = A_1^{-1} + A_1^{-1}A_3(A_2^{-1} - A_3'A_1^{-1}A_3)^{-1}A_3'A_1^{-1} . \qquad (10.47)$$

From (10.45) and (10.46) we see that this type of specification error in the first equation of the SURE model (10.41) leaves the OLS estimator of the coefficients in the second equation unaffected. However, the OLS estimator of the coefficient vector in the misspecified equation is affected by the inclusion of the superfluous regressors, in general. An exception is if the latter regressors are orthogonal to those which should be included; i.e., if $X_1'X_{1+} = 0$. This result matches the usual single-equation OLS result for this type of misspecification, and the fact that the estimator for the coefficients in the second equation is unaffected reflects the fact that the OLS estimator ignores the inherent "jointness" of the two equations in (10.41). The effects of the misspecification in one equation are not transmitted to the estimator associated with the other equation in the model.

Taking account of (10.41), we see that $b_{(1)0}$ and $b_{(2)0}$ are unbiased estimators of β_1 and β_2 respectively, so to this extent the misspecification does not adversely affect the properties of the OLS estimator, even if $X_1'X_{1+} \neq 0$. It is also easy to show that the variance covariance matrices for these estimators are

$$\begin{aligned} V(b_{(1)0}) &= E(b_{(1)0} - \beta_1)(b_{(1)0} - \beta_1)' \\ &= \sigma_{11}(X_1'X_1^{-1})[I_{K_1} + X_1'X_{1+}(X_{1+}'\bar{P}_{X_1}X_{1+})^{-1}X_{1+}'X_1(X_1'X_1)^{-1}] \end{aligned} \qquad (10.48)$$

$$\begin{aligned} V(b_{(2)0}) &= E(b_{(2)0} - \beta_2)(b_{(2)0} - \beta_2)' \\ &= \sigma_{22}(X_2'X_2)^{-1} . \end{aligned} \qquad (10.49)$$

So, we see from (10.48) that this type of misspecification in the first equation of the SURE model reduces the efficiency of the OLS estimator of that equation's coefficient vector. This follows from the unbiasedness of $b_{(1)0}$ and the positive semi-definiteness of $(X_{1+}'\bar{P}_{X_1}X_{1+})$. As the OLS estimator of β_2 is itself unaffected by this misspecification, so too is its efficiency. The efficiency loss in the OLS estimation of β_1, due to the inclusion of irrelevant regressors in the first equation of the model, may be measured by the ratio of the generalized variances of the estimators in question:

$$e_1 = \frac{\sigma_{11}|(X_1'X_1)^{-1} + (X_1'X_1)^{-1}X_1'X_{1+}(X_{1+}'\bar{P}_{X_1}X_{1+})^{-1}X_{1+}'X_1(X_1'X_1)^{-1}|}{\sigma_{11}|(X_1'X_1)^{-1}|}$$

$$= \frac{|[X_1'X_1 - X_1'X_{1+}(X_{1+}'X_{1+})^{-1}X_{1+}'X_1]^{-1}|}{|(X_1'X_1)^{-1}|}$$

$$= \prod_{i=1}^{K_1^*} \left(\frac{1}{1 - t_i}\right) , \tag{10.50}$$

where (10.47) has been used. Here $t_1, t_2, \ldots, t_{K_1^*}$ are the squared canonical correlations between the set of variables comprising X_1 and that comprising X_{1+}, and $K_1^* = \min(K_1, K_{1+})$. The $(1 - t_i)$'s are also the non-zero roots of $(X_1'X_1)^{-1}X_1'X_{1+}(X_{1+}'X_{1+})^{-1}X_{1+}'X_1$. This result is a direct generalization of that of Fomby (1981) for the single-equation case (see also Giles, 1983).

Notice that all of the t_i's are zero when X_1 and X_{1+} are orthogonal, and unity when each column of X_{1+} can be expressed as a linear combination of the columns of X_1. In between these extremes the value of e_1 depends on the degree of multicollinearity between the sets of relevant and superfluous variables in the first equation of the model. Clearly, $e_1 \geq 1$, reflecting the efficiency loss due to the misspecification of the model, and the value of e_1 depends not only on the degree of multicollinearity noted already, but also on the actual number of non-zero t_i's (see Fomby, 1981). In the special case where just one variable is wrongly included in the first equation of (10.41), so that X_{1+} is now a column vector, we see from (10.50) that

$$e_1 = (X_{1+}'X_{1+})/(X_{1+}'\bar{P}_{X_1}X_{1+})$$

$$= \left(\frac{1}{1 - R_1^2}\right) , \tag{10.51}$$

where R_1^2 is the square of the multiple correlation coefficient between the wrongly included variable and the correctly included variables of the first equation of the model. The efficiency loss due to misspecification increases monotonically with this multiple correlation.

Now, let us consider alternative and more appropriate estimators of the misspecified SURE model. The SURR estimator of the parameters of (10.43) is

$$\begin{bmatrix} \hat{\beta}_{SR} \\ \hat{\beta}_{1+} \end{bmatrix} = \begin{bmatrix} X'(\hat{S}^{-1} \otimes I_T)X & X'(\hat{S}^{-1} \otimes I_T)X_+ \\ X_+'(\hat{S}^{-1} \otimes I_T)X & X_+'(\hat{S}^{-1} \otimes I_T)X_+ \end{bmatrix}^{-1} \begin{bmatrix} X'(\hat{S}^{-1} \otimes I_T)y \\ X_+'(\hat{S}^{-1} \otimes I_T)y \end{bmatrix} , \tag{10.52}$$

where

$$\hat{S} = \begin{bmatrix} \hat{s}_{11} & \hat{s}_{12} \\ \hat{s}_{21} & \hat{s}_{22} \end{bmatrix} ,$$

and

$$\hat{s}_{11} = \frac{1}{T}(y_1 - X_1 b_{(1)0})' \bar{P}_{x_{1+}} (y_1 - X_1 b_{(1)0})$$

$$\hat{s}_{12} = \hat{s}_{21} = \frac{1}{T}(y_1 - X_1 b_{(1)0})' \bar{P}_{x_{1+}} (y_2 - X_2 b_{(2)0})$$

$$\hat{s}_{22} = \frac{1}{T}(y_2 - X_2 b_{(2)0})'(y_2 - X_2 b_{(2)0}) .$$

From (10.52), we have

$$\hat{\beta}_{SR} = \{X'(\hat{S}^{-1} \otimes I_T)X - X'(\hat{S}^{-1} \otimes I_T)X_+[X_+'(\hat{S}^{-1} \otimes I_T)X_+]^{-1}X_+'(\hat{S}^{-1} \otimes I_T)X\}^{-1}$$
$$\cdot \{X'(\hat{S}^{-1} \otimes I_T)y - X'(\hat{S}^{-1} \otimes I_T)X_+[X_+'(\hat{S}^{-1} \otimes I_T)X_+]^{-1}X_+'(\hat{S}^{-1} \otimes I_T)y\} . \tag{10.53}$$

Using (10.41), (10.45) and (10.46), we find

$$\hat{s}_{11} = \frac{1}{T} u_1' \bar{P}_{x_2} u_1$$

$$\hat{s}_{12} = \hat{s}_{21} = \frac{1}{T} u_1' \bar{P}_{x_1} \bar{P}_{x_2} u_2$$

$$\hat{s}_{22} = \frac{1}{T} u_2' \bar{P}_{x_2} u_2 ,$$

so it is easily seen that $\hat{s}_{ij}$ is a consistent estimator of σ_{ij} (i, j = 1, 2).

Similarly, if we consider the SUUR estimator of β, using the unrestricted OLS residuals from the SURE model (10.43), to estimate Σ, we have

$$\tilde{\beta}_{SU} = \{X'(\tilde{S}^{-1} \otimes I_T)X - X'(\tilde{S}^{-1} \otimes I_T)X_+[X_+'(\tilde{S}^{-1} \otimes I_T)X_+]^{-1}X_+'(\tilde{S}^{-1} \otimes I_T)X\}^{-1}$$
$$\cdot \{X'(\tilde{S}^{-1} \otimes I_T)y - X'(\tilde{S}^{-1} \otimes I_T)X_+[X_+'(\tilde{S}^{-1} \otimes I_T)X_+]^{-1}X_+'(\tilde{S}^{-1} \otimes I_T)y\} . \tag{10.54}$$

It may easily be shown that both of the estimators (10.53) and (10.54) have identical asymptotic properties: they are consistent and asymptotically Normal, with asymptotic variance covariance matrix, to order $O(T^{-1})$, given by

$$V = \{X'(\Sigma^{-1} \otimes I_T)X - X'(\Sigma^{-1} \otimes I_T)X_+[X_+'(\Sigma^{-1} \otimes I_T)X_+]^{-1}X_+'(\Sigma^{-1} \otimes I_T)X\}^{-1}$$

$$= \Omega + \Omega X'(\Sigma^{-1} \otimes I_T)X_+[X_+'QX_+]^{-1}X_+'(\Sigma^{-1} \otimes I_T)X\Omega \ , \qquad (10.55)$$

where

$$\Omega = [X'(\Sigma^{-1} \otimes I_T)X]^{-1}$$

$$Q = (\Sigma^{-1} \otimes I_T) - (\Sigma^{-1} \otimes I_T)X\Omega X'(\Sigma^{-1} \otimes I_T) \ .$$

Thus, we see that misspecification incurred by including superfluous variables in the first equation of the SURE model does not affect the consistency of the SURR or SUUR estimators of the parameters in any of the model's equations, but it does affect the asymptotic variance covariance matrices of these estimators: it reduces their asymptotic efficiency, to order $O(T^{-1})$. A measure of this inefficiency may be obtained, as before, by considering the generalized variance ratio:

$$\xi = \frac{|V|}{|\Omega|} = \prod_i \left(\frac{1}{1 - \theta_i}\right) \ , \qquad (10.56)$$

where, as with the t_i's in the expression (10.50) for e_1, the θ_i's may be interpreted as the squared canonical correlation coefficients between the columns of $(\Sigma^{-\frac{1}{2}} \otimes I_T)X$ and $(\Sigma^{-\frac{1}{2}} \otimes I_T)X_+$. So, all of the θ_i's lie in the unit interval, so $\xi \geq 1$, and from (10.56) we see that the greater the number of (and value of) non-zero θ_i's, the greater the loss in asymptotic efficiency which results when either the SURR or SUUR estimator is applied to the SURE model containing superfluous regressors.

The special case in which just one superfluous regressor is included was studied by Srivastava and Srivastava (1983), and here ξ reduces to

$$\xi = \left(\frac{1}{1 - R^{*2}}\right) \ , \qquad (10.57)$$

where

$$R^{*2} = \frac{X'_+(\Sigma^{-1} \otimes I_T)X\Omega X'(\Sigma^{-1} \otimes I_T)X_+}{X'_+(\Sigma^{-1} \otimes I_T)X_+} .$$

Now, focusing attention just on the estimation of β_1 (the coefficient vector for the correctly included variables in the first equation of the SURE model), let us compare the loss in asymptotic efficiency for the OLS and SURR/SUUR estimators when this first equation is misspecified as in (10.42). For the SUUR and SURR estimators, this loss is

$$\xi_1 = \frac{X'_{1+}(I_T - \rho_{12}^2 P_{x_2})X_{1+}}{X'_{1+}(\bar{P}_{x_1} - \rho_{12}^2 P_{x_2})X_{1+}} , \tag{10.58}$$

where $\rho_{12} = \sigma_{12}/(\sigma_{11}\sigma_{22})^{\frac{1}{2}}$. The corresponding relative efficiency measure for the OLS estimator is e_1 in (10.51).

If we consider the case where there is just one superfluous variable in the first equation, and if $R_2^2 = [(X'_{1+}P_{x_2}X_{1+})/(X'_{1+}X_{1+})]$ is the squared multiple correlation coefficient between this variable and the regressors in the second equation, then

$$\xi_1 = \left[\frac{1 - \rho_{12}^2 R_2^2}{1 - R_1^2 - \rho_{12}^2 R_2^2}\right] , \tag{10.59}$$

so that from (10.51),

$$\xi_1 \geq e_1 , \tag{10.60}$$

with equality when at least one of ρ_{12}, R_1 and R_2 is zero. Thus, in terms of asymptotic efficiency, the SURR and SUUR estimators of β_1 are adversely affected to a greater degree than is the corresponding OLS estimator by the type of misspecification under consideration here.

Now, we turn our attention to the estimation of β_2, the vector of coefficients in the equation which is correctly specified, and recall from (10.46) that the OLS estimator of these coefficients is unaffected by the misspecification in the model's first equation. Thus, defining e_2 analogously with (10.50), we have $e_2 = 1$. On the other hand, considering the SURR or SUUR estimation of β_2, the loss in asymptotic efficiency due to the misspecification is

$$\xi_2 = \left[\frac{X'_{1+}\bar{P}_{x_1}X_{1+}}{X'_{1+}(\bar{P}_{x_1} - \rho_{12}^2 P_{x_2})X_{1+}}\right]$$

$$= \left[\frac{1 - R_1^2}{1 - R_1^2 - \rho_{12}^2 R_2^2}\right] \geq e_2 \quad . \tag{10.61}$$

Again, the SURR and SUUR estimators of β_2 are adversely affected to a greater degree (in terms of loss of asymptotic efficiency) than is the OLS estimator when superfluous regressors are included in the first equation of the model.

Finally, we consider the more general situation where both relationships in a two equation SURE model suffer from misspecification of this type. In this case, the model which is estimated is

$$\begin{aligned} y_1 &= X_1\beta_1 + X_{1+}\beta_{1+} + v_1 \\ y_2 &= X_2\beta_2 + X_{2+}\beta_{2+} + v_2 \end{aligned} \tag{10.62}$$

where X_{i+} is a $(T \times K_{i+})$ matrix of rank K_{i+}, corresponding to the superfluous variables $(i = 1, 2)$. Writing (10.62) more compactly, we have

$$\begin{bmatrix} y_1 \\ y_2 \end{bmatrix} = \begin{bmatrix} X_1 & 0 \\ 0 & X_2 \end{bmatrix}\begin{bmatrix} \beta_1 \\ \beta_2 \end{bmatrix} + \begin{bmatrix} X_{1+} & 0 \\ 0 & X_{2+} \end{bmatrix}\begin{bmatrix} \beta_{1+} \\ \beta_{2+} \end{bmatrix} + \begin{bmatrix} v_1 \\ v_2 \end{bmatrix}$$

or,

$$y = X\beta + X_+\beta_+ + v \quad . \tag{10.63}$$

This will be termed the "apparent" or "contaminated" SURE model, while (10.41) defines the correctly specified model.

The OLS estimator of β in (10.63) is

$$b_0 = (X'\bar{P}_{x_+}X)^{-1}X'\bar{P}_{x_+}y \quad . \tag{10.64}$$

Using (10.41), we see that b_0 is unbiased for β and has variance covariance matrix

$$V(b_0) = (X'\bar{P}_{x_+}X)^{-1}X'\bar{P}_{x_+}(\Sigma \otimes I_T)\bar{P}_{x_+}X(X'\bar{P}_{x_+}X)^{-1} \tag{10.65}$$

On the other hand, if β is estimated from (10.41) by OLS, the associated variance covariance matrix is

$$\overset{*}{V}_0 = (X'X)^{-1}X'(\Sigma \otimes I_T)X(X'X)^{-1} , \tag{10.66}$$

and a comparison of (10.65) and (10.66) gives the loss in the efficiency of the OLS estimator for the two equation SURE model due to the incorrect inclusion of certain regressors in both equations. To explore this efficiency loss more closely, consider the case where it is the same single variable which is a superfluous regressor in each equation, so that $X_{1+} = X_{2+} = x_+$ (a column vector), and

$$X_+ = (I_2 \otimes x_+) . \tag{10.67}$$

In this case,

$$\begin{aligned} |(X_1'\bar{P}_{x_+}X_1)^{-1}| &= \left|\left[X_1'X_1 - \frac{1}{x_+'x_+}X_1'x_+x_+'X_1\right]^{-1}\right| \\ &= |(X_1'X_1)^{-1}| \left[1 + \frac{x_+'P_{x_1}x_+}{x_+'\bar{P}_{x_1}x_+}\right] \\ &= |(X_1'X_1)^{-1}| \left(\frac{1}{1 - R_1^2}\right) , \end{aligned}$$

from which it is clear that the loss of efficiency of the OLS estimator of β_1 is

$$e_1 = \frac{\sigma_{11}|(X_1'\bar{P}_{x_+}X_1)^{-1}|}{\sigma_{11}|(X_1'X_1)^{-1}|} = \left(\frac{1}{1 - R_1^2}\right) . \tag{10.68}$$

Similarly, the loss of efficiency of the OLS estimator of β_2 due to this misspecification of the model is

$$e_2 = \left(\frac{1}{1 - R_2^2}\right) . \tag{10.69}$$

Considering the OLS estimation of the complete β vector, the corresponding efficiency loss may be measured as:

$$e = \frac{|V(b_0)|}{|\overset{*}{V}_0|}$$

$$= \frac{|(X'\bar{P}_{x_+}X)^{-1}|^2 |X'\bar{P}_{x_+}(\Sigma \otimes I_T)\bar{P}_{x_+}X|}{|(X'X)^{-1}|^2 |X'(\Sigma \otimes I_T)X|}$$

$$= \frac{1}{(1 - R_1^2)(1 - R_2^2)} \prod_i \left(\frac{1}{1 - \ell_i}\right) , \qquad (10.70)$$

where the ℓ_i's are the characteristic roots (which lie in the unit interval) of the matrix

$$[X'(\Sigma \otimes I_T)X]^{-1} X'(\Sigma \otimes P_{x_+})X = \frac{1}{x_+' x_+} [X'(\Sigma \otimes I_T)X]^{-1} X'(\Sigma \otimes x_+ x_+')X .$$

These results may be interpreted along the lines of our earlier interpretations of (10.50), (10.51), etc.

Now, consider the application of the SURR estimator for β in the misspecified model (10.63):

$$\hat{\beta}_{SR} = \{X'(\hat{S}^{-1} \otimes I_T)X - X'(\hat{S}^{-1} \otimes I_T)X_+[X_+(\hat{S}^{-1} \otimes I_T)X_+]^{-1} X_+'(\hat{S}^{-1} \otimes I_T)X\}^{-1}$$

$$\cdot \{X'(\hat{S}^{-1} \otimes I_T)y - X'(\hat{S}^{-1} \otimes I_T)X_+[X_+'(\hat{S}^{-1} \otimes I_T)X_+]^{-1} X_+'(\hat{S}^{-1} \otimes I_T)y\} . \qquad (10.71)$$

The form of the corresponding SUUR estimator of β is obvious from (10.71), and it is easily established that both of these estimators are consistent, with asymptotic variance covariance matrix, to order $O(T^{-1})$, given by

$$V = \{X'(\Sigma^{-1} \otimes I_T)X - X'(\Sigma^{-1} \otimes I_T)X_+[X_+'(\Sigma^{-1} \otimes I_T)X_+]^{-1} X_+'(\Sigma^{-1} \otimes I_T)X\}^{-1}$$

$$= \Omega + \Omega X'(\Sigma^{-1} \otimes I_T)X_+(X_+'QX_+)^{-1} X_+'(\Sigma^{-1} \otimes I_T)X\Omega . \qquad (10.72)$$

When applied to the correctly specified model, (10.41), the SUUR and SURR estimators have an asymptotic variance covariance matrix, to order

$O(T^{-1})$, equal to Ω. Taking account of (10.72), it is clear that these estimators lose asymptotic efficiency as a result of the model's misspecification. Again considering the special case where $X_{1+} = X_{2+} = x_+$, this loss of efficiency is measured by:

$$\xi = \frac{|V|}{|\Omega|} = \frac{1}{|I - \Omega X'(\Sigma^{-1} \otimes P_{x_+})X|} = \prod_i \left(\frac{1}{1 - \lambda_i}\right) , \tag{10.73}$$

where the λ_i's are the characteristic roots of $\Omega X'(\Sigma^{-1} \otimes P_{x_+})X$. In turn, these roots are the squared canonical correlation coefficients between the sets of variables associated with $(\Sigma^{\frac{1}{2}} \otimes I_T)X$ and $(\Sigma^{\frac{1}{2}} \otimes x_+)$, so they all lie in the unit interval. As in earlier cases, the magnitudes of these correlations determine the loss in asymptotic efficiency for the SUUR and SURR estimators under misspecification of the SURE model through the inclusion of redundant regressors.

To summarize, we have seen that both the OLS and FGLS estimators for the SURE model are affected to some extent by this form of model misspecification. The OLS estimator remains unbiased, but loses efficiency, while the FGLS estimators remain consistent, but lose asymptotic efficiency. Further, the OLS estimators of the parameters in correctly specified equations are unaltered by the presence of redundant variables in other equations, while any superfluous regressors anywhere in the system will affect the FGLS estimators of all of the model's parameters. In this sense, the OLS estimator is more robust than one of the SUUR and SURR estimators to this type of model misspecification.

10.3.2 Omission of Relevant Variables

A common type of model misspecification in econometrics is that involving the omission, from the model, of certain regressors which form part of the appropriate structure, (10.1). This may arise from attempts to simplify the model in order to conserve degrees of freedom, or as a result of data for some variables being unavailable. In discussing the consequences of such misspecification in SURE models we shall simplify the exposition by considering a two equation system with the inappropriate omission of just one regressor from one equation. The analysis may be generalized easily.

Suppose that the correctly specified SURE model is

$$\begin{aligned} y_1 &= X_1\beta_1 + \gamma x + u_1 \\ y_2 &= X_2\beta_2 + u_2 \\ E(u_1) &= E(u_2) = 0, \quad E(u_i u_j') = \sigma_{ij} I_T \quad (i, j = 1, 2) \end{aligned} \tag{10.74}$$

where x is a (T × 1) vector of observations on the single variable to be omitted wrongly when the model is estimated. This framework, and the following analysis, is drawn from Srivastava and Srivastava (1983).

The correct model may be written as

$$\begin{bmatrix} y_1 \\ y_2 \end{bmatrix} = \begin{bmatrix} X_1 & 0 \\ 0 & X_2 \end{bmatrix} \begin{bmatrix} \beta_1 \\ \beta_2 \end{bmatrix} + \gamma \begin{bmatrix} x \\ 0 \end{bmatrix} + \begin{bmatrix} u_1 \\ u_2 \end{bmatrix}$$

or,

$$y = X\beta + \gamma X_{-} + u \ , \tag{10.75}$$

while the misspecified model which is fitted is

$$\begin{bmatrix} y_1 \\ y_2 \end{bmatrix} = \begin{bmatrix} X_1 & 0 \\ 0 & X_2 \end{bmatrix} \begin{bmatrix} \beta_1 \\ \beta_2 \end{bmatrix} + \begin{bmatrix} v_1 \\ u_2 \end{bmatrix}$$

or,

$$y = X\beta + v \ . \tag{10.76}$$

Estimating (10.76) by OLS yields

$$b_0 = (X'X)^{-1}X'y \ , \tag{10.77}$$

so from (10.75),

$$E(b_0 - \beta) = \gamma(X'X)^{-1}X'X_{-} \tag{10.78}$$

$$M(b_0) - E(b_0 - \beta)(b_0 - \beta)'$$

$$= (X'X)^{-1}X'[(\Sigma \otimes I_T) + \gamma^2 X_{-}X_{-}']X(X'X)^{-1} \ . \tag{10.79}$$

From (10.78) and the definition of X_{-} we see that the OLS estimator of β_2 in (10.76) is unbiased, while the corresponding estimator of β_1 is biased, to an extent "proportional" to the coefficient of the omitted regressor. That is, the misspecification affects the estimator of the parameters of the equation in question adversely. From the mean squared error matrix, (10.79), and the unbiasedness of $b_{(2)0}$, we have

$$M(b_{(1)0}) = \sigma_{11}(X_1'X_1)^{-1} + \gamma^2(X_1'X_1)^{-1}X_1'xx'X_1(X_1'X_1)^{-1}$$
$$V(b_{(2)0}) = \sigma_{22}(X_2'X_2)^{-1} \ . \tag{10.80}$$

On the other hand, the OLS estimator of β based on the correctly specified model, (10.75), is

$$b_0^* = (X'\bar{P}_{\underline{x}}X)^{-1}X'\bar{P}_{\underline{x}}y \ , \tag{10.81}$$

which is unbiased and has variance covariance matrix

$$V(b_0^*) = (X'\bar{P}_{\underline{x}}X)^{-1}X'\bar{P}_{\underline{x}}(\Sigma \otimes I_T)\bar{P}_{\underline{x}}X(X'\bar{P}_{\underline{x}}X)^{-1} \ , \tag{10.82}$$

so that

$$V(b^*_{(1)0}) = \sigma_{11}(X_1'\bar{P}_x X_1)^{-1}$$
$$= \sigma_{11}(X_1'X_1)^{-1} + \frac{\sigma_{11}}{x'x(1 - R_1^2)}(X_1'X_1)X_1'xx'X_1(X_1'X_1)^{-1} \tag{10.83}$$

$$V(b^*_{(2)0}) = \sigma_{22}(X_2'X_2)^{-1} \ , \tag{10.84}$$

where R_1 is the multiple correlation coefficient between the omitted variable and the regressors included in the first equation of the SURE model. From (10.80) and (10.84) we see that the efficiency of the OLS estimator of the coefficients in the correctly specified equation is unaffected by the overall model misspecification. Next, from (10.80) and (10.83), we see that the MSE matrix of the OLS estimator of β_1 in the misspecified model exceeds that of its counterpart in the correct model if

$$\gamma^2 > \frac{\sigma_{11}}{x'x(1 - R_1^2)} \ . \tag{10.85}$$

Considering the generalized variance ratio, allowing for the bias of $b_{(1)0}$, we have

$$e_1 = \frac{|M(b_{(1)0})|}{|V(b^*_{(1)0})|}$$

$$= \left[\frac{1 + \frac{\gamma^2}{\sigma_{11}} x'P_{x_1}x}{1 + \frac{x'P_{x_1}x}{x'x(1 - R_1^2)}}\right] = (1 - R_1^2)\left[1 + \frac{\gamma^2 R_1^2}{\sigma_{11}} x'x\right] , \tag{10.86}$$

which indicates that the loss of efficiency for the OLS estimator of the coefficient vector in the misspecified equation depends upon the disturbance variance, the observations on the omitted variable, and its multiple correlation with the regressors included in that equation.

Two points relating to these results may be noted. First, the OLS estimator of β_1 is unaffected by this specification error if X_1 is orthogonal to x. This corresponds to the usual single-equation model case. Secondly, there can be situations, when

$$\gamma^2 < \frac{\sigma_{11}}{x'x(1 - R_1^2)} , \tag{10.87}$$

such that the bias/variance trade-off results in a gain in efficiency for the OLS estimator of β_1 as a result of misspecifying the model. Note that this gain increases as $R_1 \to 1$, signifying increasing multicollinearity among the regressors of the first equation. So, the common practice of dropping a variable in the presence of strong multicollinearity may have some justification in terms of a MSE improvement for the OLS estimator of the retained regressors' coefficients.

Turning to more appropriate estimators, consider the application of the SURR estimator to the misspecified model, (10.76):

$$\hat{\beta}_{SR} = [X'(\hat{S}^{-1} \otimes I_T)X]^{-1}X'(\hat{S}^{-1} \otimes I_T)y , \tag{10.88}$$

where the elements of $\hat{S}$ are

$$\hat{s}_{ij} = \frac{1}{T} y_i'\bar{P}_{x_i}\bar{P}_{x_j}y_j \quad (i, j = 1, 2) .$$

Applying the SURR estimator to the correct model, (10.75), yields

$$\hat{\beta}^*_{SR} = \{X'(\hat{\Sigma}^{-1} \otimes I_T)X - X'(\hat{\Sigma}^{-1} \otimes I_T)X_-[X'_-(\hat{\Sigma}^{-1} \otimes I_T)X_-]^{-1}X'_-(\hat{\Sigma}^{-1} \otimes I_T)X\}^{-1}$$

$$\cdot \{X'(\hat{\Sigma}^{-1} \otimes I_T)y - X'(\hat{\Sigma}^{-1} \otimes I_T)X_-[X'_-(\hat{\Sigma}^{-1} \otimes I_T)X_-]^{-1}X'_-(\hat{\Sigma}^{-1} \otimes I_T)y\} , \quad (10.89)$$

where the elements of $\hat{\Sigma}$ are

$$\hat{\sigma}_{11} = \frac{1}{T}y'_1[\bar{P}_x - \bar{P}_x X_1(X'_1\bar{P}_x X_1)^{-1}X'_1\bar{P}_x]y_1$$

$$\hat{\sigma}_{12} = \hat{\sigma}_{21} = \frac{1}{T}y'_1[\bar{P}_x - \bar{P}_x X_1(X'_1\bar{P}_x X_1)^{-1}X'_1\bar{P}_x]\bar{P}_{x_2}y_2$$

$$\hat{\sigma}_{22} = \frac{1}{T}y'_2\bar{P}_{x_2}y_2 \; .$$

Using (10.75) it is easily seen that $\hat{\Sigma}$ is consistent for Σ, while $\hat{S}$ is an inconsistent estimator of Σ. Consequently, the SURR estimator of β based on the correctly specified model is consistent while that based on the model from which a variable is omitted is inconsistent. Thus, there is an important cost associated with this type of misspecification of the SURE model, even in large samples.

If we consider the MSE matrices of $\hat{\beta}_{SR}$ and $\hat{\beta}^*_{SR}$ we have, to order $O(T^{-1})$,

$$M(\hat{\beta}_{SR}) = [X'(\Phi^{-1} \otimes I_T)X]^{-1}X'(\Phi^{-1} \otimes I_T)\{(\Sigma \otimes I_T) + \gamma^2 X'_-X_-\}(\Phi^{-1} \otimes I_T)$$

$$\cdot X[X'(\Phi^{-1} \otimes I_T)X]^{-1} \quad (10.90)$$

$$M(\hat{\beta}^*_{SR}) = \{X'(\Sigma^{-1} \otimes I_T)X - X'(\Sigma^{-1} \otimes I_T)X_-[X'_-(\Sigma^{-1} \otimes I_T)X_-]^{-1}X_-(\Sigma^{-1} \otimes I_T)X\}^{-1}$$

$$= \Omega + \frac{1}{X'_-QX_-}\Omega X'(\Sigma^{-1} \otimes I_T)X_-X'_-(\Sigma^{-1} \otimes I_T)X\Omega , \quad (10.91)$$

where

$$\Omega = [X'(\Sigma^{-1} \otimes I_T)X]^{-1}$$

$$Q = (\Sigma^{-1} \otimes I_T) - (\Sigma^{-1} \otimes I_T)X\Omega X'(\Sigma^{-1} \otimes I_T)$$

$$\Phi^{-1} = \begin{bmatrix} \sigma_{11} + \frac{1}{T}\gamma^2(1 - R_1^2)x'x & \sigma_{12} \\ \sigma_{21} & \sigma_{22} \end{bmatrix}^{-1}$$

$$= \Sigma^{-1} - g(\Sigma^{-1} * J)$$

$$g = \frac{1}{T}\sigma_{11}\gamma^2(1 - \rho_{12}^2)(1 - R_1^2)x'x$$

$$J = \begin{bmatrix} 1 & 1 \\ 1 & \rho_{12}^2 \end{bmatrix},$$

and $*$ denotes the Hadamard product operator.

The forms of (10.90) and (10.91) are such that it is difficult to draw any clear general conclusions regarding the relative efficiencies of $\hat{\beta}_{SR}$ and $\hat{\beta}^*_{SR}$ in this context. However, when X_1 and X_2 are orthogonal, (10.90) and (10.91) yield

$$M(\hat{\beta}_{(1)SR}) = \sigma_{11}(1-\rho_{12}^2)(X_1'X_1)^{-1} + \gamma^2(X_1'X_1)^{-1}X_1'xx'X_1(X_1'X_1)^{-1} \tag{10.92}$$

$$M(\hat{\beta}^*_{(1)SR}) = \sigma_{11}(1-\rho_{12}^2)(X_1'X_1)^{-1} + \frac{\sigma_{11}(1-\rho_{12}^2)}{x'x(1-R_1^2-\rho_{12}^2R_2^2)}(X_1'X_1)^{-1}X_1'xx'X_1(X_1'X_1)^{-1} \tag{10.93}$$

and

$$M(\hat{\beta}_{(2)SR}) = \sigma_{22}(1-\rho_{12}^2)\left[1 + \frac{g^2\rho_{12}^2(1-\rho_{12}^2)}{(1-\rho_{12}^2 g)^2}\right](X_2'X_2)^{-1}$$
$$+ \frac{\sigma_{22}\gamma^2\rho_{12}^2(1-g)^2}{\sigma_{11}(1-\rho_{12}^2 g)^2}(X_2'X_2)^{-1}X_2'xx'X_2(X_2'X_2)^{-1} \tag{10.94}$$

$$M(\hat{\beta}^*_{(2)SR}) = \sigma_{22}(1-\rho_{12}^2)(X_2'X_2)^{-1} + \frac{\sigma_{22}\rho_{12}^2(1-\rho_{12}^2)}{x'x(1-R_1^2-\rho_{12}^2R_2^2)}(X_2'X_2)^{-1}X_2'xx'X_2'(X_2'X_2)^{-1}. \tag{10.95}$$

From (10.92) and (10.93) we see that the SURR estimator of β_1 based on the misspecified model is asymptotically more efficient than its counterpart based on the correctly specified model if

$$\gamma^2 < \frac{\sigma_{11}(1 - \rho_{12}^2)}{x'x(1 - R_1^2 - \rho_{12}^2 R_2^2)} , \tag{10.96}$$

and a measure of this relative efficiency is

$$\begin{aligned} \xi_1 &= \frac{|M(\hat{\beta}_{(1)SR})|}{|M(\hat{\beta}^*_{(1)SR})|} \\ &= \left[1 - \frac{R_1^2}{1 - \rho_{12}^2 R_2^2}\right]\left[1 + \frac{\gamma^2 R_1^2 x'x}{\sigma_{11}(1 - \rho_{12}^2)}\right] . \end{aligned} \tag{10.97}$$

Comparing (10.97) and (10.86) we see that $\xi_1 < e_1$ if

$$\gamma^2\left(\frac{x'x}{\sigma_{11}}\right) < \frac{1}{1 - R_1^2 + \dfrac{1 - R_1^2 - R_2^2}{(1 - \rho_{12}^2)R_2^2}} . \tag{10.98}$$

Similarly, from (10.94) and (10.95),

$$\begin{aligned} \xi_2 &= \frac{|M(\hat{\beta}_{(2)SR})|}{|M(\hat{\beta}^*_{(2)SR})|} \\ &= \left[1 - \frac{\rho_{12}^2 R_2^2}{1 - R_2^2}\right]\left[1 + \frac{g^2\rho_{12}^2(1 - \rho_{12}^2)}{(1 - \rho_{12}^2 g)^2}\right]^{K_2}\left[1 + \frac{\sigma_{22}\gamma^2\rho_{12}^2 R_2^2(1 - g)^2 x'x}{\sigma_{11}(1 - g(2 - g)\rho_{12}^2)}\right] \end{aligned} \tag{10.99}$$

and this relative efficiency may be compared with the value unity, this being the efficiency of $b^*_{(2)0}$ relative to $b_{(2)0}$. Finally, it should be noted that results similar to those established above may be obtained if the SUUR estimator is considered.

A more general situation is that in which both equations of our two equation SURE model involve the type of specification error under discussion. Here we present Rao's (1974) analysis, where it is assumed that the same single variable is wrongly omitted from each equation. The correctly specified SURE model is

$$
\begin{aligned}
y_1 &= X_1\beta_1 + \gamma_1 x + u_1 \\
y_2 &= X_2\beta_2 + \gamma_2 x + u_2 \\
E(u_1) &= E(u_2) = 0, \quad E(u_i u_j') = \sigma_{ij} I \qquad (i,\ j = 1,\ 2) \ ,
\end{aligned}
$$

or

$$
\begin{bmatrix} y_1 \\ y_2 \end{bmatrix} = \begin{bmatrix} X_1 & 0 \\ 0 & X_2 \end{bmatrix} \begin{bmatrix} \beta_1 \\ \beta_2 \end{bmatrix} + \begin{bmatrix} x & 0 \\ 0 & x \end{bmatrix} \begin{bmatrix} \gamma_1 \\ \gamma_2 \end{bmatrix} + \begin{bmatrix} u_1 \\ u_2 \end{bmatrix} ,
$$

or

$$
y = X\beta + (I_2 \otimes x)\gamma + u \ . \tag{10.100}
$$

In contrast, the misspecified model in this case is

$$
y = X\beta + v \ . \tag{10.101}
$$

If we estimate (10.101) by OLS, we obtain

$$
b_0 = (X'X)^{-1}X'y \ , \tag{10.102}
$$

while the corresponding estimator based on the correct specification, (10.100), is

$$
b_0^* = [X'(I_2 \otimes x)X]^{-1}X'(I_2 \otimes x)y \ . \tag{10.103}
$$

Using (10.100), we obtain

$$
\begin{aligned}
E(b_0 - \beta) &= (X'X)^{-1}X'(I_2 \otimes x)\gamma \\
E(b_0^* - \beta) &= 0 \ ,
\end{aligned} \tag{10.104}
$$

which again shows that misspecification by the omission of relevant regressors biases the least squares estimator for all of the coefficients, unless the omitted variable is orthogonal to both X_1 and X_2, and that this bias depends on γ.

Similarly,

$$
\begin{aligned}
M(b_0) &= E(b_0 - \beta)(b_0 - \beta)' \\
&= (X'X)^{-1}X'[(\Sigma \otimes I_T) + (I_2 \otimes x)\gamma\gamma'(I_2 \otimes x')]X(X'X)^{-1}
\end{aligned} \tag{10.105}
$$

$$V(b_0^*) = E(b_0^* - \beta)(b_0^* - \beta)'$$

$$= [X'(I_2 \otimes \bar{P}_x)X]^{-1}X'(I_2 \otimes \bar{P}_x)(\Sigma \otimes I_T)(I_2 \otimes \bar{P}_x)X[X'(I_2 \otimes \bar{P}_x)X]^{-1}$$

$$= [X'(I_2 \otimes \bar{P}_x)X]^{-1}X'(\Sigma \otimes \bar{P}_x)X[X'(I_2 \otimes \bar{P}_x)X]^{-1} \ , \qquad (10.106)$$

from which we obtain, for the individual vectors β_i ($i = 1, 2$):

$$M(b_{(i)0}) = \sigma_{ii}(X_i'X_i)^{-1} + \gamma_i^2(X_i'X_i)^{-1}X_i'xx'X_i(X_i'X_i)^{-1} \qquad (10.107)$$

$$V(b_{(i)0}^*) = \sigma_{ii}(X_i'\bar{P}_xX_i)^{-1}$$

$$= \sigma_{ii}(X_i'X_i)^{-1} + \frac{\sigma_{ii}}{x'x(1 - R_i^2)}(X_i'X_i)^{-1}X_i'xx'X_i(X_i'X_i)^{-1} \ . \qquad (10.108)$$

Thus, the approximate MSE matrix of the OLS estimator of β_i in the misspecified model does not exceed (in the usual matrix sense) that of the OLS estimator of β_i in the correctly specified model if

$$\gamma_i^2 \leq \frac{\sigma_{ii}}{x'x(1 - R_i^2)} \qquad (i = 1, 2) \ . \qquad (10.109)$$

Also, we have

$$e_i = \frac{|M(b_{(i)0})|}{|V(b_{(i)0}^*)|} = (1 - R_i^2)\left(1 + \frac{\gamma_i^2R_i^2}{\sigma_{ii}}x'x\right) \ , \qquad (10.110)$$

as a measure of the relative efficiencies of the OLS estimators of β_i in the misspecified and correctly specified models.

If the SURR estimator is adopted, then the estimators obtained from the misspecified and correctly specified models respectively are given by

$$\hat{\beta}_{SR} = [X'(\hat{S}^{-1} \otimes I_T)X]^{-1}X'(\hat{S}^{-1} \otimes I_T)y$$

$$\hat{\beta}^*_{SR} = [X'(\hat{\Sigma}^{-1} \otimes I_T)X - X'(\hat{\Sigma}^{-1} \otimes x)\left(\hat{\Sigma} \otimes \frac{1}{x'x} I_T\right)(\hat{\Sigma}^{-1} \otimes x')X]^{-1}$$

$$\cdot \left[X'(\hat{\Sigma}^{-1} \otimes I_T)y - X'(\hat{\Sigma}^{-1} \otimes x)\left(\hat{\Sigma} \otimes \frac{1}{x'x} I_T\right)(\hat{\Sigma}^{-1} \otimes x')y\right] \quad (10.111)$$

$$= [X'(\hat{\Sigma}^{-1} \otimes \bar{P}_x)X]^{-1}X'(\hat{\Sigma}^{-1} \otimes \bar{P}_x)y$$

where the elements of $\hat{S}$ and $\hat{\Sigma}$ are given by

$$\hat{s}_{ij} = \frac{1}{T} y_i' \bar{P}_{x_i} \bar{P}_{x_j} y_j$$

$$\hat{\sigma}_{ij} = \frac{1}{T} y_i'[\bar{P}_x - \bar{P}_x X_i (X_i' \bar{P}_x X_i)^{-1} X_i' \bar{P}_x][\bar{P}_x - \bar{P}_x X_j (X_j' \bar{P}_x X_j)^{-1} X_j' \bar{P}_x] y_j \ . \quad (10.112)$$

It is easy to verify that $\hat{s}_{ij}$ is not a consistent estimator of σ_{ij}, while $\hat{\sigma}_{ij}$ is. As a result, the SURR estimator of β from the misspecified model is inconsistent, while it is consistent when estimated from the properly specified model. The mean squared error matrices, to order $O(T^{-1})$, of these two estimators are

$$M(\hat{\beta}_{SR}) = [X'(\Phi^{-1} \otimes I_T)X]^{-1} X'(\Phi^{-1} \otimes I_T)[(\Sigma \otimes I_T) + (\gamma\gamma' \otimes xx')](\Phi^{-1} \otimes I_T)$$

$$\cdot X[X'(\Phi^{-1} \otimes I_T)X]^{-1} \quad (10.113)$$

$$M(\hat{\beta}^*_{SR}) = [X'(\Sigma^{-1} \otimes \bar{P}_x)X]^{-1} \ ,$$

where the matrix Φ is now defined as

$$\Phi = \begin{bmatrix} \phi_{11} & \phi_{12} \\ \phi_{21} & \phi_{22} \end{bmatrix} ,$$

with

$$\phi_{ij} = \sigma_{ij} + \frac{1}{T}\gamma_i\gamma_j(x'x)(1 - R_i^2 - R_j^2 + R_{ij})$$

$$R_{ij} = \begin{cases} \dfrac{x'P_{X_i}P_{X_j}x}{x'x} & \text{if } i \neq j \\[2ex] \dfrac{x'P_{X_i}x}{x'x} = R_i^2 & \text{if } i = j \ . \end{cases}$$

A comparison of the two expressions in (10.113) does not lead to any clear conclusions, in general, about the MSE consequences of such misspecification. To simplify matters we therefore consider a SURE model in which not only are X_1 and X_2 orthogonal, but x is also orthogonal to both X_1 and X_2. In this case

$$\Phi = \Sigma + \left(\frac{1}{T}x'x\right)\gamma\gamma' \ ,$$

so that

$$\begin{aligned} \Sigma^{-1} &= \left[\Phi - \left(\frac{1}{T}x'x\right)\gamma\gamma'\right]^{-1} \\ &= \Phi^{-1} + \frac{\left(\frac{1}{T}x'x\right)}{1 - \left(\frac{1}{T}x'x\right)\gamma'\Phi^{-1}\gamma}\Phi^{-1}\gamma\gamma'\Phi^{-1} \ , \end{aligned}$$

and

$$\begin{aligned} \Phi^{-1}\Sigma\Phi^{-1} &= [\Phi\Sigma^{-1}\Phi]^{-1} \\ &= \left[\Phi + \frac{\left(\frac{1}{T}x'x\right)}{1 - \left(\frac{1}{T}x'x\right)\gamma'\Phi^{-1}\gamma}\gamma\gamma'\right]^{-1} \\ &= \Phi^{-1} - \left(\frac{1}{T}x'x\right)\Phi^{-1}\gamma\gamma'\Phi^{-1} \ . \end{aligned}$$

Using these results, the following expressions are obtained from (10.113):

$$\begin{aligned} M(\hat{\beta}_{SR}) &= [X'(\Phi^{-1} \otimes I_T)X]^{-1}X'(\Phi^{-1} \otimes I_T)(\Sigma \otimes I_T)(\Phi^{-1} \otimes I_T)X[X'(\Phi^{-1} \otimes I_T)X]^{-1} \\ &= [X'(\Phi^{-1} \otimes I_T)X]^{-1}X'(\Phi^{-1}\Sigma\Phi^{-1} \otimes I_T)X[X'(\Phi^{-1} \otimes I_T)X]^{-1} \end{aligned}$$

$$= [X'(\Phi^{-1} \otimes I_T)X]^{-1} - \left(\frac{1}{T}x'x\right)\left[X'(\Phi^{-1} \otimes I_T)X\right]^{-1} X'(\Phi^{-1}\gamma\gamma'\Phi^{-1} \otimes I_T)X$$

$$\cdot [X'(\Phi^{-1} \otimes I_T)X]^{-1} \qquad (10.114)$$

$$M(\hat{\beta}^*_{SR}) = [X'(\Sigma^{-1} \otimes I_T)X]^{-1}$$

$$= \left[X'(\Phi^{-1} \otimes I_T)X + \frac{\left(\frac{1}{T}x'x\right)}{1-\left(\frac{1}{T}x'x\right)\gamma'\Phi^{-1}\gamma} X'(\Phi^{-1}\gamma\gamma'\Phi^{-1} \otimes I_T)X\right]^{-1} .$$

$$(10.115)$$

Extracting the corresponding results for the estimators of β_1, we have

$$M(\hat{\beta}_{(1)SR}) = \frac{1}{\phi^{11}}\left[1 - \left(\frac{1}{T}x'x\right)\frac{(\phi^{11}\gamma_1 + \phi^{12}\gamma_2)^2}{\phi^{11}}\right](X_1'X_1)^{-1} \qquad (10.116)$$

$$M(\hat{\beta}^*_{(1)SR}) = \frac{1}{\phi^{11}}\left[1 + \left(\frac{1}{T}x'x\right)\frac{(\phi^{11}\gamma_1 + \phi^{12}\gamma_2)^2}{\phi^{11}\left(1 - \frac{1}{T}x'x\gamma'\Phi^{-1}\gamma\right)}\right](X_1'X_1)^{-1} , \qquad (10.117)$$

where ϕ^{ij} denotes the (i, j)'th element of Φ^{-1}.

Using the result

$$\Phi^{-1} = \left[\Sigma + \left(\frac{1}{T}x'x\right)\gamma\gamma'\right]^{-1}$$

$$= \Sigma^{-1} - \left[\frac{\left(\frac{1}{T}x'x\right)}{1 + \left(\frac{1}{T}x'x\right)\gamma'\Sigma^{-1}\gamma}\right]\Sigma^{-1}\gamma\gamma'\Sigma^{-1} ,$$

we find that

$$0 < \left(\frac{1}{T}x'x\right)\gamma'\Phi^{-1}\gamma = \frac{\left(\frac{1}{T}x'x\right)\gamma'\Sigma^{-1}\gamma}{1 + \left(\frac{1}{T}x'x\right)\gamma'\Sigma^{-1}\gamma} < 1 , \qquad (10.118)$$

by the positive definiteness of Σ.

Now, making use of (10.118), we see from (10.116) and (10.117) that the SURR estimator of β_1 based on the misspecified model is more efficient, to this order of approximation, than its counterpart based on the correctly specified model. This same result holds if the SUUR estimator is applied. However, it should be emphasized that a very special model structure has been considered in the final part of this discussion, and no claim is being made regarding exact results.

To summarize, we have seen that misspecifying the SURE model by omitting relevant regressors from one or more equations biases the OLS estimates of the parameters in the misspecified equations, but it is possible for such misspecification to reduce the MSE of the OLS estimator in certain parts of the parameter space. In the case of FGLS estimators, we have seen that this type of misspecification anywhere in the system leads to their inconsistency. Thus, omitting relevant regressors has extremely serious consequences for the estimators most likely to be applied to the SURE model.

10.3.3 Some Further Remarks

In practice, and often in an attempt to minimize the adverse statistical consequences of omitting relevant regressors, proxy variables may be used in place of variables on which the necessary observations are unavailable. This substitution of a proxy variable for the "correct" variable involves a type of specification error of its own. To illustrate this briefly, in the context of the SURE model, we again consider a two equation system in which one unobservable variable is to be replaced by a proxy:

$$\begin{aligned} y_1 &= X_1\beta_1 + z\delta + u_1 \\ y_2 &= X_2\beta_2 + u_2 \end{aligned} \tag{10.119}$$

where z is a $(T \times 1)$ vector of true but unknown values associated with the unobservable variable.

Let w be a proxy variable for z, with

$$w = z + \epsilon \ , \tag{10.120}$$

where ϵ is a vector of errors assumed to be independent of the disturbances in (10.119). So, the model actually estimated is

$$\begin{bmatrix} y_1 \\ y_2 \end{bmatrix} = \begin{bmatrix} X_1 & 0 \\ 0 & X_2 \end{bmatrix} \begin{bmatrix} \beta_1 \\ \beta_2 \end{bmatrix} + \begin{bmatrix} w \\ 0 \end{bmatrix} \delta + \begin{bmatrix} u_1 - \delta\epsilon \\ u_2 \end{bmatrix}$$

or,

$$y = X\beta + W\delta + v \ . \tag{10.121}$$

The OLS and FGLS estimators of β and δ in (10.121) are given by

$$b_0 = (X'\bar{P}_w X)^{-1} X'\bar{P}_w y \qquad (10.122)$$

$$d_0 = \frac{w'\bar{P}_{x_1} y_1}{w'\bar{P}_{x_1} w} \qquad (10.123)$$

$$b_F = \{X'(S^{-1} \otimes I_T)X - X'(S^{-1} \otimes I_T)W[W'(S^{-1} \otimes I_T)W]^{-1}W'(S^{-1} \otimes I_T)X\}^{-1}$$

$$\cdot \{X'(S^{-1} \otimes I_T)y - X'(S^{-1} \otimes I_T)W[W'(S^{-1} \otimes I_T)W]^{-1}W'(S^{-1} \otimes I_T)y\}$$

$$= \left\{X'(S^{-1} \otimes I_T)X - \frac{1}{W'(S^{-1} \otimes I_T)W} X'(S^{-1} \otimes I_T)WW'(S^{-1} \otimes I_T)X\right\}^{-1}$$

$$\cdot \left\{ X'(S^{-1} \otimes I_T)y - \frac{W'(S^{-1} \otimes I_T)y}{W'(S^{-1} \otimes I_T)W} X'(S^{-1} \otimes I_T)W \right\} \qquad (10.124)$$

$$d_F = \frac{W'\{(S^{-1} \otimes I_T) - (S^{-1} \otimes I_T)X[X'(S^{-1} \otimes I_T)X]^{-1}X'(S^{-1} \otimes I_T)\}y}{W'\{(S^{-1} \otimes I_T) - (S^{-1} \otimes I_T)X[X'(S^{-1} \otimes I_T)X]^{-1}X'(S^{-1} \otimes I_T)\}W}, \qquad (10.125)$$

where S is a consistent estimator of Σ.

Other formulations of this proxy variable analysis may be obtained by replacing (10.120) by one of the following specifications:

$$w = \lambda z + \epsilon \qquad (10.126)$$

$$w = \lambda z + X_1\alpha + \epsilon \qquad (10.127)$$

$$w = \lambda z + \overset{*}{X}_1\overset{*}{\alpha} + \epsilon , \qquad (10.128)$$

where the columns of $\overset{*}{X}_1$ comprise a proper subset of those of X_1, plus other variables, and λ is a scalar reflecting persistent over-statement ($\lambda > 1$) or under-statement ($\lambda < 1$) of the unobservable variable by its proxy. These formulations, especially (10.127), are in the spirit of single-equation analyses by Zellner (1970) and Goldberger (1972). Two final formulations may be mentioned. First, as is discussed by Grether (1974), indicator (or dummy) variables may be introduced. Secondly, (10.120) may be replaced by

$$W\Delta = z + \epsilon , \qquad (10.129)$$

where W denotes a (T × m) matrix of observations on m variables used to develop a proxy for the single variable z, and Δ is an (m × 1) parameter vector.

The properties of the estimators (10.122)-(10.125), and their counterparts obtained by replacing (10.120) by one of (10.126)-(10.129), are essentially unexplored (although see Kmenta, 1981). These properties could be compared with those of the OLS and FGLS estimators of β which could be obtained, in principle, if z were observable. This would give a measure of any "cost" of using the proxy variable, w. The extent to which the FGLS estimators of β are affected by the use of the proxy, relative to the effect on the OLS estimator, also should be explored. One would conjecture that the OLS estimators for the parameters in the second equation of (10.119) are more robust to the use of a proxy variable in the first equation than are the corresponding FGLS estimators of β_2.

10.4 BAYESIAN ANALYSIS

So far we have adopted a frequentist (rather than Bayesian) stance in presenting results associated with estimation and inference in the context of the SURE model. This situation has arisen, not so much as a result of the authors' philosophical beliefs with regard to matters of statistical inference generally, but because of our attempt to survey the existing literature associated with the SURE model in a thorough manner. As it happens, virtually all of this literature adopts a frequentist stance, rather than taking the Bayesian viewpoint. However, one notable exception is the work of Zellner (1971; pp. 240-246), and because the Bayesian approach offers a flexible and unified way of combining prior and sample information to obtain a complete posterior distribution upon which to base the inferences, it is appropriate to discuss this approach here. The relatively small amount of space devoted to the Bayesian analysis of the SURE model should not be taken by the reader as indicating our dissatisfaction with what it has to offer. On the contrary, we believe that the Bayesian approach to inference is fruitful and powerful, and clearly more research along these lines with the SURE model would be well rewarded. Our presentation is based largely on Zellner (1971; pp. 240-246), which in turn draws on Tiao and Zellner (1964) (see also Zellner, 1979; pp. 687-691). With the additional assumption that the disturbance vector, u, has a multivariate Normal distribution, the likelihood function for β and Σ^{-1} (or equivalently for β and Σ) in the model (10.1) is

$$\ell(\beta, \Sigma^{-1} | y) \;\alpha\; |\Sigma^{-1}|^{\frac{T}{2}} \exp\{-\frac{1}{2}(y - X\beta)'(\Sigma^{-1} \otimes I_T)(y - X\beta)\}$$

$$\alpha\; |\Sigma^{-1}|^{\frac{T}{2}} \exp\{-\frac{1}{2} \operatorname{tr} F\Sigma^{-1}\} \;, \qquad (10.130)$$

where F is an $(M \times M)$ symmetric matrix with (i, j)'th element equal to $(y_i - X_i\beta_i)'(y_j - X_j\beta_j)$.

In the Bayesian approach to inference, uncertain prior information about the unknown parameters is reflected in the formulation of a prior density, here $p(\beta, \Sigma^{-1})$. If our prior information is negligible, or "diffuse," then a totally uninformative prior is formulated. In this context, assuming independence between the prior information regarding the elements of β, and that associated with the elements of Σ^{-1}, and following Jeffrey's (1961) invariance theory (see also Savage, 1961), we may write

$$p(\beta, \Sigma^{-1}) = p(\beta)p(\Sigma^{-1}) \tag{10.131}$$

$$p(\beta) \;\alpha\; \text{constant} \tag{10.132}$$

$$p(\Sigma^{-1}) \;\alpha\; |\Sigma^{-1}|^{-(M+1)/2} , \tag{10.133}$$

where in (10.132) we mean that

$$p(\beta_{ij}) \;\alpha\; \text{constant} \;; \quad -\infty < \beta_{ij} < \infty$$

with β_{ij} as the j'th coefficient in the i'th equation of the SURE model. Together, (10.131)-(10.133) represent total prior ignorance regarding the values of the elements of β and the $M(M + 1)/2$ distinct elements of Σ^{-1}.

Now, appealing to Bayes's theorem, from (10.130)-(10.133), the joint posterior density for all of the unknown parameters is

$$p(\beta, \Sigma^{-1}|y) \;\alpha\; |\Sigma^{-1}|^{(T-M-1)/2} \exp\left\{-\frac{1}{2}(y - X\beta)'(\Sigma^{-1} \otimes I_T)(y - X\beta)\right\}$$

$$\alpha\; |\Sigma^{-1}|^{(T-M-1)/2} \exp\left\{-\frac{1}{2}\text{tr } F\Sigma^{-1}\right\} . \tag{10.134}$$

From (10.134) it is clear that the conditional posterior density for β, given Σ^{-1}, is

$$p(\beta|\Sigma^{-1}, y) \;\alpha\; \exp\left\{-\frac{1}{2}(y - X\beta)'(\Sigma^{-1} \otimes I_T)(y - X\beta)\right\} ,$$

which is multivariate Normal in form, with mean vector

$$E(\beta|\Sigma^{-1}, y) = [X'(\Sigma^{-1} \otimes I_T)X]^{-1}X'(\Sigma^{-1} \otimes I_T)y$$

$$= b_G , \tag{10.135}$$

and variance covariance matrix

$$V(\beta \mid \Sigma^{-1}, y) = [X'(\Sigma^{-1} \otimes I_T)X]^{-1}$$
$$= \Omega \ . \tag{10.136}$$

So, conditional on knowledge of Σ, and with diffuse prior information, the posterior mean of β is just the GLS estimator. Under a quadratic loss structure this mean would be used as the Bayes (or the minimum expected loss) estimator of β. Of course, we know that in general b_G is not feasible as Σ is unknown. If S, a consistent estimator of Σ, is substituted for Σ in (10.135), then conditional on S the posterior mean for β is just the FGLS estimator. Such a substitution highlights the connection between the FGLS and Bayes estimators of β, and has asymptotic appeal (see Zellner, 1971; p. 243). However, in finite samples the substitution of S for Σ may be less appealing. In fact, one of the principal advantages of the Bayesian approach to the estimation of β is that no such substitution is necessary. Instead of basing our inferences on the conditional posterior density for β, the elements of Σ^{-1} may be treated as "nuisance parameters," and eliminated from (10.134) by multivariate integration to yield the marginal posterior density for β. Tiao and Zellner (1964) showed that

$$p(\beta \mid y) \ \alpha \ |F|^{-T/2} \ , \tag{10.137}$$

and they noted that although (10.137) is reminiscent of the kernel of a multivariate Student's-t density, in fact it is not possible to arrange it in the latter form because the regressor matrices in the SURE model typically differ from one equation to another. Thus, it is not clear that further analytical interpretations may be drawn from $p(\beta \mid y)$. However, numerical integration techniques may be used to normalize (10.137) and to obtain the marginal densities and moments for elements of β. These in turn provide Bayes estimates of the parameters of the SURE model. Given that the number of elements in β may be quite large, even the normalization of $p(\beta \mid y)$ may represent a considerable computational problem if conventional numerical techniques are applied. In this regard the method of Monte Carlo integration may be especially helpful (see, e.g., Kloek and Van Dijk, 1978).

Two further points may be noted at this juncture. First, if each equation of the SURE model contains the same set of regressors, so that we have a conventional multivariate regression model, then it is possible to analyze $p(\beta \mid y)$ further without resorting to numerical methods (see Tiao and Zellner, 1964, and Zellner, 1971; pp. 224-233). Secondly, if inferences are to be drawn about the elements of Σ then we may obtain the marginal posterior density of Σ^{-1} as

$$p(\Sigma^{-1} \mid y) = \int p(\beta, \Sigma^{-1} \mid y)\, d\beta \tag{10.138}$$

where $p(\beta, \Sigma^{-1} \mid y)$ is given by (10.134) and the integration in (10.138) is over a space of dimension equal to the number of distinct elements in β. This integration generally will have to be performed numerically, and again the Monte Carlo integration technique may be helpful.

Suppose that proper prior information about the parameters is available. In this case, combining the prior density for β and Σ^{-1} with the likelihood function generally will result in a joint posterior density which is analytically intractable, and which will have to be normalized and marginalized by numerical methods. However, let us consider one special form of prior information. Suppose that $p(\beta, \Sigma^{-1})$ is given by (10.131) and (10.133) in conjunction with

$$p(\beta) \;\alpha\; \exp\left\{-\frac{1}{2}(\beta - \beta_p)'\Psi_p^{-1}(\beta - \beta_p)\right\} \quad . \tag{10.139}$$

That is, the marginal prior density for β is multivariate Normal, with mean vector β_p and variance covariance matrix Ψ_p. Accordingly, the joint posterior density for β and Σ^{-1} is

$$\begin{aligned} p(\beta, \Sigma^{-1}|y) \;\alpha\; & |\Sigma^{-1}|^{(T-M-1)/2} \exp\left\{-\frac{1}{2}(y - X\beta)'(\Sigma^{-1} \otimes I_T)(y - X\beta)\right\} \\ & \cdot \exp\left\{-\frac{1}{2}(\beta - \beta_p)'\Psi_p^{-1}(\beta - \beta_p)\right\} \\ \alpha\; & |\Sigma^{-1}|^{(T-M-1)/2} \exp\left\{-\frac{1}{2}[\mathrm{tr}\, F\Sigma^{-1} + (\beta - \beta_p)'\Psi_p^{-1}(\beta - \beta_p)]\right\} \end{aligned} \tag{10.140}$$

and it may be verified easily that the conditional posterior density of β, given Σ^{-1}, is multivariate Normal with mean vector

$$E(\beta|\Sigma^{-1}, y) = [\Psi_p^{-1} + X'(\Sigma^{-1} \otimes I_T)X]^{-1}[\Psi_p^{-1}\beta_p + X'(\Sigma^{-1} \otimes I_T)y] \tag{10.141}$$

and variance covariance matrix

$$V(\beta|\Sigma^{-1}, y) = [\Psi_p^{-1} + X'(\Sigma^{-1} \otimes I_T)X]^{-1} \quad . \tag{10.142}$$

Noting that (10.141) may be written as

$$E(\beta|\Sigma^{-1}, y) = [\Psi_p^{-1} + X'(\Sigma^{-1} \otimes I_T)X]^{-1}[\Psi_p^{-1}\beta_p + X'(\Sigma^{-1} \otimes I_T)Xb_G] \quad ,$$

we see that the Bayes estimator of β, conditional on Σ^{-1} and using a quadratic loss function, is a matrix weighted average of the prior mean vector for β and the GLS estimator, the weighting matrices being the prior and sample precision matrices. Integrating (10.140) with respect to Σ^{-1} yields the following marginal posterior density for β:

$$p(\beta \mid y) \ \alpha \ |F|^{-T/2} \exp\left\{-\frac{1}{2}(\beta - \beta_p)'\Psi_p^{-1}(\beta - \beta_p)\right\} \ , \qquad (10.143)$$

the further analysis of which is best handled by numerical methods.

From the foregoing discussion it is clear that a Bayesian analysis of the SURE model yields some interesting insights. Asymptotically, the Bayes estimator of β converges to the maximum likelihood estimator, and so in large enough samples the Bayes and FGLS estimators will be indistinguishable. However, in finite samples and given proper prior information about at least some of the model's parameters, the Bayesian approach offers a highly flexible way of taking proper account of extraneous information. Moreover, this framework offers (although possibly at some computational expense) a powerful method of eliminating the nuisance parameters comprising the elements of Σ^{-1}, without resorting to asymptotic justifications. Of course, the Bayes estimators have sampling properties which may or may not be appealing to a frequentist, but the Bayesian philosophy places little weight on the properties of estimators in repeated samples. Optimal performance conditional on the given sample is the primary objective.

To complete our brief treatment of Bayesian inference associated with the SURE model we mention the issue of sensitivity analysis. Of particular interest in applications of Bayesian methods is some information concerning the extent to which the posterior distribution, and hence the inferences drawn, are sensitive to the choice of prior density or of the values assigned to the parameters of that density. Of course, the prior density and its parameters are chosen by the researcher to properly reflect the available prior information or beliefs. These beliefs are subjective and personal, and this is a feature of their analysis which Bayesians find to be flexible and helpful. However, the user of the results may hold different beliefs from those of the researcher, and without wishing to replicate the analysis may be interested in the extent to which the final inferences are sensitive to the researcher's choice of prior. If small perturbations in the prior can induce substantial variations in the posterior density, for a given sample, then this is of some interest. This, and related considerations, have led to the notions of both local and global sensitivity (see Leamer, 1978; pp. 170-187). In local sensitivity analysis we are concerned with the effects of small perturbations in the parameters of a prior density, while in global sensitivity analysis we are concerned with the correspondence between an entire class of prior densities and an entire class of posterior densities.

To illustrate these ideas let us consider a SURE model with equal numbers of regressors in each equation, i.e., $K_1 = K_2 = \cdots = K_M = \bar{K}$ (say).

In addition we assume that the disturbance variance covariance has an intraclass structure:

$$\Sigma = (c - d)I_M + de_M e_M' \tag{10.144}$$

where e_M is an $(M \times 1)$ vector with all elements unity, and c and d are scalars satisfying

$$c > 0, \quad \left[\left(\frac{c}{d} - 1\right) + M\right] > 0 \ ,$$

which ensures that Σ is positive definite. This is the model formulation considered by Polasek (1981), and it is easily seen that

$$\Sigma^{-1} = \left(\frac{1}{c - d}\right)[I_M + \alpha e_M e_M'] \tag{10.145}$$

where

$$\alpha = -\left[\frac{d}{c + (M - 1)d}\right] . \tag{10.146}$$

If we take the noninformative prior density (10.131)-(10.133) for β and Σ^{-1}, it is clear that this prior is parameterized just by the two scalars c and d. Recalling that with this choice of prior the conditional posterior distribution for β given Σ^{-1} is multivariate Normal with mean vector b_G and variance covariance matrix Ω, we see that in this context local sensitivity analysis involves studying changes in b_G and Ω resulting from small variations in the values of c and d. Also, as α in (10.146) defines a class of parameters c and d, one interpretation of global sensitivity analysis in this context is to study the changes in b_G and Ω resulting from small changes in α.

Turning to the moments of the conditional posterior density for β, given Σ^{-1}, we have

$$\begin{aligned} \Omega &= [X'(\Sigma^{-1} \otimes I_T)X]^{-1} \\ &= (c - d)[X'X + \alpha X'(e_M e_M' \otimes I_T)X]^{-1} \\ &= (c - d)[(X'X)^{-1} - \alpha(X'X)^{-1}X'(e_M \otimes I_T)V^{-1}(e_M' \otimes I_T)X] \end{aligned} \tag{10.147}$$

$$b_G = [X'(\Sigma^{-1} \otimes I_T)X]^{-1}X'(\Sigma^{-1} \otimes I_T)y$$

$$= [X'X + \alpha X'(e_M e'_M \otimes I_T)X]^{-1}[X'y + \alpha X'(e_M e'_M \otimes I_T)y]$$

$$= [I_{MT} - \alpha(X'X)^{-1}X'(e_M \otimes I_T)V^{-1}(e_M \otimes I_T)X](b_0 + \alpha b_*) , \quad (10.148)$$

where

$$V = [I_{MT} + \alpha(e_M \otimes I_T)X(X'X)^{-1}X'(e_M \otimes I_T)]$$

$$b_* = (X'X)^{-1}X'(e_M e'_M \otimes I_T)y \quad (10.149)$$

and b_0 is the OLS estimator of β. Note that b_0 may also be interpreted as the Bayes estimator of β if the inherent jointness of the SURE model's equations is ignored when specifying the likelihood function, and a diffuse prior density is used.

Now, local sensitivity measures for b_G are given by

$$\left(\frac{\partial b_G}{\partial c}\right) = \frac{1}{(c-d)^2}\Omega X'\left[\alpha^2\left(M - 2 + 2\left(\frac{c}{d}\right)\right)(e_M e'_M \otimes I_T) - I_{MT}\right](y - Xb_G) \quad (10.150)$$

$$\left(\frac{\partial b_G}{\partial d}\right) = \frac{1}{(c-d)^2}\Omega X'\left[\alpha^2\left(M - 1 + \left(\frac{c}{d}\right)^2\right)(e_M e'_M \otimes I_T) + I_{MT}\right](y - Xb_G) . \quad (10.151)$$

To consider a global sensitivity analysis we need to examine how b_G changes if small changes are made to α. From (10.148), we see that $b_G \to b_0$ as $\alpha \to 0$. On the other hand, if α becomes large then V tends to become singular while $(b_0 + \alpha b_*)$ moves away from b_0 in a vector direction determined by the sign of αb_* (see Polasek, 1981). If we make the further assumption that the regressor matrices $X_1, X_2, \ldots, X_M$ are pairwise orthogonal then

$$b_G = \left(\frac{1}{1+\alpha}\right)b_0 + \left(\frac{\alpha}{1+\alpha}\right)b_* ,$$

so that $b_G \to b_0$ as $\alpha \to 0$, and $b_G \to b_*$ as $\alpha \to \infty$. The local and global sensitivity of Ω may be examined in a similar manner. This, and sensitivity analysis with respect to b_G when Σ is an unrestricted matrix are left as exercises for the reader (see also Leamer, 1984). It should also be noted that the sensitivity of b_G and Ω to variations in X may also be examined in this way (see Mosteller and Tukey, 1977).

10.5 RESEARCH SUGGESTIONS

The discussion in this chapter suggests a number of useful topics for further research. For example, in the context of exact restrictions, the finite-sample properties of the RFGLS estimators of β are, to the best of our knowledge, unexplored. For particular choices of S, such as $\hat{S}$ or $\tilde{S}$, it would be interesting to compare the properties of b_{RF} with those of the OLS, ROLS and FGLS estimators. In this same context there is a need to develop hypothesis tests which are powerful in finite samples (see Jayatissa, 1977). As has been emphasized, the tests that we have considered have asymptotic justification, but their application in the context of samples of finite size must be treated cautiously. Obviously, this is a severe practical limitation on their usefulness. Moreover, the application of such tests, with the choice of estimator for the SURE model's parameters being contingent on their outcome, gives rise to preliminary-test strategies. These strategies in turn have implications for the finite-sample properties of the estimators which are effectively being employed, as is well known. This is an area which deserves more study in the context of SURE models, perhaps along the lines of Farebrother (1978).

In the case of stochastic restrictions on the parameters of the SURE model, the finite-sample properties of mixed estimators of β constructed by basing S on residuals from MOLS, MFGLS or iterative MFGLS estimators need to be explored in detail. Moreover, there is a considerable amount of work which could usefully be undertaken to explore the finite-sample properties of compatibility statistics of the form (10.40) in the context of the SURE model. Further, if the choice of estimator for β is based on the outcome of such a test, such that an FGLS estimator is adopted if compatibility is rejected, but an MFGLS estimator is used if compatibility cannot be rejected, then this also gives rise to a preliminary-test strategy. Judge, Yancey and Bock (1973) considered such strategies in the single-equation context, and some related analysis in the context of the SURE model was given by Judge and Bock (1978; pp. 136-141).

Clearly, existing specification analysis results for the SURE model can be extended in various ways. More general model formulations could be studied, with the emphasis on finite-sample results, and with misspecifications affecting several regressors in several equations. Moreover, the simultaneous inclusion of certain redundant regressors and omission of certain relevant regressors could be considered. This general, and very realistic, specification problem has not been considered in the context of a SURE model, although some analyses of the corresponding single-equation problem were given by Guilkey and Price (1981) and Kadiyala (1985). More general formulations of this aspect of specification analysis could be developed along the lines of Riddell and Buse (1980) and Giles (1983). Similarly, the whole matter of proxy variables requires a more thorough discussion in the context of the SURE model, especially when such variables appear in more than one equation of the model. Also, one important comparison which needs full analysis is that between the following alternative strategies: first, omit the unobservable variables from the model entirely; or secondly, use a proxy for the unobservable variables when estimating the SURE model by FGLS. Which strategy is to be preferred, in terms of the bias and mean squared errors of the estimators involved? The single-equation analyses of Aigner (1974) and Frost (1979) may be helpful in addressing this question.

In the Bayesian analysis of the SURE model, further discussion of alternative prior distributions for the parameters would be of some interest. Our brief discussion of sensitivity analysis dealt with a very special model when a diffuse prior is specified for its parameters. A fruitful avenue for further research in this area would be a consideration of local and global sensitivity analyses when informative prior densities are adopted. Leamer's analysis of the single-equation model might be extended to the SURE model, although the extent to which analytical results may be forthcoming is likely to be limited.

EXERCISES

10.1 Consider a two equation SURE model:

$$y_1 = \alpha e_T + \beta_1 x_1 + u_1$$

$$y_2 = \beta_2 x_2 + u_2$$

in which $\alpha = (\beta_1 + \beta_2)$, and e_T denotes a $(T \times 1)$ vector with all elements unity. Compare the ROLS and RFGLS estimators of the coefficients on the basis of their variance covariance matrices to order $O(T^{-1})$.

10.2 Formulate the SURE model as a multivariate regression model subject to a set of exact restrictions on the coefficients and suggest a suitable test of the validity of these restrictions. [Hint: See Harvey (1981; pp. 337-338).]

10.3 Consider a two equation SURE model in which the regressors are orthogonal across equations and the disturbances are normally distributed with common mean 0, common variance 1 and known correlation coefficient ρ_{12}. Describe a test for $H_0: \beta_1 = \beta_0$ against $H_1: \beta_1 \neq \beta_0$, where β_0 is given. Based on the outcome of the test procedure, define a pre-test estimator for β_1 to be β_0 if H_0 is not rejected and to be the GLS estimator if H_0 is rejected. Sketch the derivation of the expressions for the bias vector and mean squared error matrix of this estimator.

10.4 Derive the likelihood ratio, Wald and Lagrange multiplier statistics for testing $H_0: R\beta = r$ against $H_1: R\beta \neq r$ in the SURE model, and prove that they are asymptotically equivalent. [Hint: See Harvey (1981; pp. 159-175).]

10.5 Consider a SURE model for which the variance covariance matrix of the disturbances is known and for which there are linear restrictions on the coefficients. Compare the variance covariance matrices of the restricted GLS and "mixed" GLS estimators of the coefficients and comment on the consequences of introducing variability into the restrictions. [Hint: See Srivastava and Agnihotri (1980).]

10.6 Suppose that unbiased estimates of G linear combinations of the coefficients across the equations of a SURE model are available from some extraneous source. Assuming that the variance covariance matrix of the disturbances is known, suggest a test of the compatibility of the sample and extraneous information. Compare the efficiencies of the OLS and GLS estimators of β.

10.7 Consider a two equation model with subset regressors and

$$\Sigma = \begin{bmatrix} 1 & \sigma_{12} \\ \sigma_{12} & 1 \end{bmatrix}.$$ If a common explanatory variable is mistakenly included in both of the equations, examine its effect on the efficiencies of the OLS and SUUR estimators of the coefficients.

10.8 The coefficient β_1 in a two equation SURE model

$$y_1 = \beta_1 x_1 + \gamma x + u_1$$
$$y_2 = \beta_2 x_2 + u_2 ,$$

is estimated from the following two models:

Model A	Model B
$y_1 = \beta_1 x_1 + v_1$	$y_1 = \beta_1 x_1^* + v_1^*$
$y_2 = \beta_2 x_2 + u_2$	$y_2 = \beta_2 x_2 + u_2$

where x_1^* denotes the vector of OLS residuals obtained from the regression of x_1 on x. Examine the biases and mean squared errors of the OLS estimators of β_1 in these two models. Compare these estimators with the corresponding SUUR estimators on the basis of mean squared error to order $O(T^{-1})$.

10.9 Suppose that all of the equations of a SURE model involve a specification error related to the deletion of some variables. Show that the estimated variance covariance matrix of the OLS estimator of the coefficient vector in a particular equation of the model provides an under-estimate of its true counterpart.

10.10 In a two equation SURE model

$$y_1 = \beta_1 x + \gamma z + u_1$$
$$y_2 = \beta_2 x + u_2 \quad ,$$

z is replaced by its proxy w such that $E(w - z) = 0$, $E(w - z)(w - z)' = \theta I_T$, $E(w - z)u_i' = 0$ $(i = 1, 2)$ where θ is a known positive scalar. Obtain the OLS and SUUR estimators of β_1, and compare their asymptotic properties.

10.11 Suppose that a set of stochastic linear restrictions

$$r = R\beta + v \; ; \quad v \sim N(0, \Psi_0)$$

is obtained from a data source which is independent of the observations available for the SURE model

$$y = X\beta + u \; ; \quad u \sim N(0, (\Sigma \otimes I_T)) \quad .$$

Assuming the matrices Σ and Ψ_0 to be known and choosing non-informative prior distributions for the elements of β and Σ^{-1}, show that the conditional posterior distribution of β given Σ is multivariate Normal with mean vector b_{MG} and variance covariance matrix Ω_{MG} where

$$b_{MG} = [X'(\Sigma^{-1} \otimes I_T)X + R'\Psi_0^{-1}R]^{-1}[X'(\Sigma^{-1} \otimes I_T)y + R'\Psi_0^{-1}r]$$

$$\Omega_{MG} = [X'(\Sigma^{-1} \otimes I_T)X + R'\Psi_0^{-1}R]^{-1} \quad .$$

Obtain the marginal posterior distribution of β. [Hint: See Tiao and Zellner (1964) and Theil (1971; pp. 670-672).]

10.12 Discuss the global sensitivity of the GLS estimator of β in a SURE model under the assumption that $\Sigma = (I_M + \Theta)$ where Θ is a symmetric positive definite matrix bounded from above. [Hint: See Leamer (1984).]

REFERENCES

Aigner, D. J. (1974). "MSE dominance of least squares with errors-of-observations," Journal of Econometrics 2, 365-372.

Byron, R. P. (1970). "The restricted Aitken estimation of sets of demand relations," Econometrica 38, 816-830.

Dorn, W. S. (1961). "Self-dual quadratic programs," Journal of the Society for Industrial and Applied Mathematics 9, 51-54.

Farebrother, R. W. (1978). "Estimating regression coefficients under conditional specification: comment," Communications in Statistics A7, 193-196.

Fomby, T. B. (1981). "Loss of efficiency in regression analysis due to irrelevent variables: a generalization," Economics Letters 7, 319-322.

Frost, P. A. (1979). "Proxy variables and specification bias," Review of Economics and Statistics 61, 323-325.

Gallant, A. R., and T. M. Gerig (1980). "Computations for constrained linear models," Journal of Econometrics 12, 59-84.

Giles, D. E. A. (1982). "Instrumental variable estimation with linear restrictions," Sankhyā B44, 343-350.

Giles, D. E. A. (1983). "Instrumental variables estimation of misspecified regressions," in Proceedings of the American Statistical Association: Business and Economic Statistics Section (American Statistical Association, Washington), 688-691.

Goldberger, A. S. (1972). "Maximum-likelihood estimation of regressions containing unobservable independent variables," International Economic Review 13, 1-15.

Grether, D. M. (1974). "Correlations with ordinal data," Journal of Econometrics 2, 241-246.

Guilkey, D. K., and J. M. Price (1981). "On comparing restricted least squares estimators," Journal of Econometrics 15, 397-404.

Harvey, A. C. (1981). The Econometric Analysis of Time Series (Philip Allan, Oxford).

Jayatissa, W. A. (1977). "Tests of equality between sets of coefficients in two linear regressions when disturbance variances are unequal," Econometrica 45, 1291-1292.

Jeffreys, H. (1961). Theory of Probability (Clarendon, Oxford).

Judge, G. G., and M. E. Bock (1978). The Statistical Implications of Pre-Test and Stein-Rule Estimators in Econometrics (North-Holland, Amsterdam).

Judge, G. G., and T. Takayama (1966). "Inequality restrictions in regression analysis," Journal of the American Statistical Association 61, 166-181.

Judge, G. G., and T. A. Yancey (1981). "Sampling properties of an inequality restricted estimator," Economics Letters 7, 327-333.

Judge, G. G., T. A. Yancey and M. E. Bock (1973). "Properties of estimators after preliminary tests of significance when stochastic restrictions are used in regression," Journal of Econometrics 1, 29-47.

Kadiyala, K. (1985). "Misspecification: excluding and including variables simultaneously," Australian Economic Papers 24, 206-209.

Klemm, R. J., and V. A. Sposito (1980). "Least squares solutions over interval restrictions," Communications in Statistics B9, 423-425.

Kloek, T., and H. K. Van Dijk (1978). "Bayesian estimates of equation system parameters: an application of integration by Monte Carlo," Econometrica 46, 1-19.

Kmenta, J. (1981). "On the problem of missing measurements in the estimation of economic relationships," in E. G. Charatsis (ed.), Proceedings of the Econometric Society European Meeting, 1979: Selected Econometric Papers in Memory of Stefan Valavanis (North-Holland, Amsterdam), 233-257.

Leamer, E. E. (1978). Specification Searches (Wiley, New York).

Leamer, E. E. (1984). "Global sensitivity results for generalized least squares estimates," Journal of the American Statistical Association 79, 867-870.

Liew, C. K. (1976). "Inequality constrained least squares estimation," Journal of the American Statistical Association 71, 746-751.

McElroy, M. B. (1977). "Weaker MSE criteria and tests for linear restrictions in regression models with non-spherical disturbances," Journal of Econometrics 6, 389-394.

Mosteller, F., and J. W. Tukey (1977). Data Analysis and Regression: A Second Course in Statistics (Addison-Wesley, New York).

Polasek, W. (1981). "Seemingly unrelated regression with intraclass correlation structure," Preprint, Institute for Statistics, University of Vienna.

Rao, P. (1974). "Specification bias in seemingly unrelated regressions," in W. Selekaerts (ed.), Econometrics and Economic Theory: Essays in Honor of Jan Tinbergen (Macmillan, London), 101-113.

Riddell, W. C., and A. Buse (1980). "An alternative approach to specification errors," Australian Economic Papers 19, 211-214.

Ruble, W. (1968). "Improving the computation of simultaneous stochastic linear equations estimates," Agricultural Economics Report No. 116, Michigan State University, East Lansing.

Savage, L. J. (1961). "The subjective basis of statistical practice," mimeo, University of Michigan, Ann Arbor.

Schmidt, P., and M. Thomson (1982). "A note on the computation of inequality constrained least squares estimates," Economics Letters 9, 355-358.

Srivastava, S. K., and V. K. Srivastava (1983). "Estimation of seemingly unrelated regression equation model under specification error," Biometrical Journal 25, 181-191.

Srivastava, V. K. (1980). "Estimation of linear single-equation and simultaneous equation models under stochastic linear constraints: an annotated bibliography," International Statistical Review 48, 79-82.

Srivastava, V. K., and B. S. Agnihotri (1980). "Estimation of regression models under linear restrictions," Biometrical Journal 22, 287-288.

Theil, H. (1963). "On the use of incomplete prior information in regression analysis," Journal of the American Statistical Association 58, 401-414.

Theil, H. (1971). Principles of Econometrics (Wiley, New York).

Theil, H., and A. S. Goldberger (1961). "On pure and mixed statistical estimation in economics," International Economic Review 2, 65-78.

Tiao, G. C., and A. Zellner (1964). "On the Bayesian estimation of multivariate regression," Journal of the Royal Statistical Society B26, 277-285.

Toro-Vizcarrondo, C., and T. D. Wallace (1968). "A test of the mean square error criterion for restrictions in linear regression," Journal of the American Statistical Association 63, 558-572.

Wolfe, P. (1959). "The simplex method for quadratic programming," Econometrica 27, 382-398.

Yancey, T. A., G. G. Judge and M. E. Bock (1981). "Testing multiple equality and inequality hypothesis in economics," Economics Letters 7, 249-255.

Zellner, A. (1961). "Linear regression with inequality constraints on the coefficients," Report No. 6109, International Center for Management Science, University of Wisconsin, Madison.

Zellner, A. (1962). "An efficient method of estimating seemingly unrelated regression equations and tests for aggregation bias," Journal of the American Statistical Association 57, 348-368.

Zellner, A. (1970). "Estimation of regression relationships containing unobservable independent variables," International Economic Review 11, 441-454.

Zellner, A. (1971). An Introduction to Bayesian Inference in Econometrics (Wiley, New York).

Zellner, A. (1979). "An error-components procedure (ECP) for introducing prior information about covariance matrices and analysis of multivariate regression models," International Economic Review 20, 679-692.

11
Some Miscellaneous Topics

11.1 INTRODUCTION

In this concluding chapter we discuss a number of miscellaneous topics associated with the SURE model. These topics have received only a modest amount of attention in the literature, but in an attempt to make the coverage of our overall presentation as complete as possible we are including them here. In some cases the topics may be related to material discussed in previous chapters, and we try to indicate where this is so. In other cases the material stands in relative isolation. The chapter concludes with some general remarks concerning directions for future research in connection with the SURE model. These comments should be read in conjunction with similar such remarks at the close of various sections of earlier chapters in this book.

11.2 THE VARYING COEFFICIENTS MODEL

In common with the usual regression model specification, we have assumed throughout this book that the coefficient parameters of the SURE model are unknown but fixed quantities. However, in certain situations there may be empirical and/or theoretical evidence which suggests that these coefficients vary over time. In this case an allowance for this should be made when the model is formulated and estimated. Specifying the model to have coefficients which vary throughout the sample may be appropriate in the context of omitted relevant regressors; with the use of proxy variables; when observations are aggregated in certain ways; or if the functional form of the model is misspecified in some way. For a general discussion, see Raj and Ullah (1981; Chapter 1), from which much of the following material is drawn.

An important specification analysis issue arises in the case of varying coefficient models. If in fact the coefficients vary throughout the sample period, but we specify and estimate the model on the assumption of fixed coefficients, then the estimators adopted will be asymptotically inefficient and will have sampling distributions different from those which would apply if the assumed constancy of the coefficients were in fact correct. Thus, inappropriate inferences may be drawn. Of course, in addition to these statistical shortcomings, estimating the model on the basis of fixed coefficients will fail to provide information about their variability over time.

If we allow coefficients to vary from observation to observation then we can write the SURE model as:

$$y_{ti} = x_{ti1}\beta_{ti1} + x_{ti2}\beta_{ti2} + \cdots + x_{tiK_i}\beta_{tiK_i} + u_{ti} \quad . \tag{11.1}$$

$$(t = 1, 2, \ldots, T; \; i = 1, 2, \ldots, M)$$

Following Singh and Ullah (1974) if we now express the coefficients as random quantities,

$$\beta_{tik} = \beta_{ik} + \epsilon_{tik} \quad , \tag{11.2}$$

where β_{ik} is a fixed scalar and ϵ_{tik} is a random variable, the SURE model with random coefficients is

$$y_{ti} = \sum_{k=1}^{K_i} x_{tik}\beta_{ik} + v_{ti} \quad (t = 1, 2, \ldots, T; \; i = 1, 2, \ldots, M) \tag{11.3}$$

with

$$v_{ti} = u_{ti} + \sum_{k=1}^{K_i} x_{tik}\epsilon_{tik} \quad . \tag{11.4}$$

The equations in (11.3) may be written compactly as

$$\begin{bmatrix} y_1 \\ y_2 \\ \vdots \\ y_M \end{bmatrix} = \begin{bmatrix} X_1 & 0 & \cdots & 0 \\ 0 & X_2 & \cdots & 0 \\ \vdots & \vdots & & \vdots \\ 0 & 0 & \cdots & X_M \end{bmatrix} \begin{bmatrix} \beta_1 \\ \beta_2 \\ \vdots \\ \beta_M \end{bmatrix} + \begin{bmatrix} v_1 \\ v_2 \\ \vdots \\ v_M \end{bmatrix}$$

or,

$$y = X\beta + v \ . \tag{11.5}$$

Notice that the elements of β are just the means of the random coefficients specified in (11.2). Now, to simplify the exposition we shall assume that $K_i = K$ (say) for all i, so that all equations in the model contain the same number of regressors. Of course, we are not restricting the regressors themselves to be the same in each equation. As well as making our usual assumptions about the first and second-order moments of the u_{ti}'s, we also assume

$$\begin{aligned} E(\epsilon_{tik}) &= 0 \qquad \text{for all } t,\ i,\ k \\ E(u_{ti}\epsilon_{\tau jk}) &= 0 \qquad \text{for all } t,\ \tau,\ i,\ j,\ k \\ E(\epsilon_{tik}\epsilon_{\tau j\ell}) &= \begin{cases} \theta_{ijk} & \text{if } t = \tau \text{ and } k = \ell \\ 0 & \text{otherwise} \end{cases} \end{aligned} \tag{11.6}$$

from which it follows that

$$\begin{aligned} E(v_{ti}) &= 0 \qquad \text{for all } t,\ i \\ E(v_{ti}v_{\tau j}) &= \begin{cases} \sigma_{ij} + \sum_{k=1}^{K} x_{tik}\theta_{ijk} & \text{if } t = \tau \\ 0 & \text{if } t \neq \tau \end{cases} \end{aligned} \tag{11.7}$$

where $\sigma_{ij} = E(u_{ti}u_{tj})$.

Let us introduce some general notation:

$$\underset{(T\times KT)}{X_i^*} = \begin{bmatrix} x_{1i1} & \cdots & x_{1iK} & 0 & \cdots & 0 & \cdots & 0 & \cdots & 0 \\ 0 & \cdots & 0 & x_{2i1} & \cdots & x_{2iK} & \cdots & 0 & \cdots & 0 \\ \vdots & & \vdots & \vdots & & \vdots & & \vdots & & \vdots \\ 0 & \cdots & 0 & 0 & \cdots & 0 & \cdots & x_{Ti1} & \cdots & x_{TiK} \end{bmatrix}$$

$$\underset{(MT\times MKT)}{X^*} = \begin{bmatrix} X_1^* & 0 & \cdots & 0 \\ 0 & X_2^* & \cdots & 0 \\ \vdots & \vdots & & \vdots \\ 0 & 0 & \cdots & X_M^* \end{bmatrix}$$

$$\underset{(1\times KT)}{\epsilon_j'} = (\epsilon_{1i1}, \ldots, \epsilon_{1iK}, \epsilon_{2i1}, \ldots, \epsilon_{2iK}, \ldots, \epsilon_{Ti1}, \ldots, \epsilon_{TiK})$$

$$\underset{(K\times K)}{\Theta_{ij}} = \begin{bmatrix} \theta_{ij1} & 0 & \cdots & 0 \\ 0 & \theta_{ij2} & \cdots & 0 \\ \vdots & \vdots & & \vdots \\ 0 & 0 & \cdots & \theta_{ijK} \end{bmatrix}$$

$$\underset{(MK\times MK)}{\Theta} = \begin{bmatrix} \Theta_{11} & \cdots & \Theta_{1M} \\ \vdots & & \vdots \\ \Theta_{M1} & \cdots & \Theta_{MM} \end{bmatrix}$$

Now, it follows from (11.4) that

$$v_i = u_i + X_i^*\epsilon_i \quad , \tag{11.8}$$

so that from (11.7) we may write

$$
\begin{aligned}
E(v_i) &= 0 \\
E(v_i v_j') &= \sigma_{ij} I_T + X_i^*(\Theta_{ij} \otimes I_T) X_j^{*\prime} = \Phi_{ij} \quad \text{(say)}
\end{aligned}
\tag{11.9}
$$

from which we have (Singh and Ullah, 1974; p. 192):

$$
\begin{aligned}
E(v) &= 0 \\
E(vv') &= (\Sigma \otimes I_T) + X^*(\Theta \otimes I_T) X^{*\prime} = \Phi \quad \text{(say)} \ .
\end{aligned}
\tag{11.10}
$$

Notice that if all of the θ_{ijk}'s are zero then $\Phi = (\Sigma \otimes I_T)$, which corresponds to the variance covariance matrix of the disturbances in the conventional SURE model with fixed coefficients. Now, let us consider various estimators of β in (11.5). First, the OLS estimator, which completely ignores the covariance structure of v, is

$$
b_0 = (X'X)^{-1} X'y \ . \tag{11.11}
$$

Secondly, the SURR estimator obtained under the assumption that the coefficients are fixed is

$$
\hat{\beta}^*_{SR} = \left[X'(\hat{S}^{-1} \otimes I_T) X\right]^{-1} X'(\hat{S}^{-1} \otimes I_T) y \ , \tag{11.12}
$$

where the (i, j)'th element of $\hat{S}$ is

$$
\hat{s}_{ij} = \frac{1}{T} y_i' \bar{P}_{x_i} \bar{P}_{x_j} y_j \ . \tag{11.13}
$$

Now, if the SURR estimator is modified to take account not only of the cross-equation correlations between the disturbances but also the fact that the parameters vary throughout the sample, then we obtain

$$
\hat{\beta}_{SR} = (X' \hat{\Phi}^{-1} X)^{-1} X' \hat{\Phi}^{-1} y \ , \tag{11.14}
$$

where $\hat{\Phi}$ is a consistent estimator of Φ, based partly on restricted residuals, as is discussed below.

Notice that the estimation of Φ amounts to the estimation of the σ_{ij}'s and the θ_{ijk}'s in (11.6) and (11.7). In order to estimate these parameters, consider the "restricted" residual vectors

$$
\hat{v}_i = \bar{P}_{x_i} y_i = \bar{P}_{x_i} v_i \qquad (i = 1, 2, \ldots, M) \ .
$$

It follows that

$$E(\hat{v}_i\hat{v}_j') = \bar{P}_{x_i}E(v_iv_j')\bar{P}_{x_j}$$

$$= \bar{P}_{x_i}[\sigma_{ij}I_T + X_i^*(\Theta_{ij} \otimes I_T)X_i^{*\prime}]\bar{P}_{x_j} \quad ,$$

so that

$$w_{ij} = Z_{ij}\delta_{ij} + \omega_{ij} \tag{11.15}$$

where

$$w_{ij} = (\hat{v}_i * \hat{v}_j)$$

$$Z_{ij} = (\bar{P}_{x_i} * \bar{P}_{x_j})[e_T \quad (X_i * X_j)]$$

$$\delta_{ij}' = (\sigma_{ij}, \theta_{ij1}, \theta_{ij2}, \ldots, \theta_{ijK})$$

$$\omega_{ij} = w_{ij} - E(w_{ij})$$

where $*$ denotes the Hadamard product operator, and e_T is a $(T \times 1)$ vector with all elements unity.

If OLS is used to estimate δ_{ij} in (11.15), we obtain

$$\hat{\delta}_{ij} = (Z_{ij}'Z_{ij})^{-1}Z_{ij}'w_{ij} \quad , \tag{11.16}$$

and the elements of $\hat{\delta}_{ij}$ yield estimates of σ_{ij} and the θ_{ijk}'s which may be used to estimate Φ. Because w_{ij} is defined in terms of restricted OLS residuals the estimates of σ_{ij} used to estimate Σ in (11.10) are those corresponding to $\hat{S}$ in our usual notation. Hence (11.14) is described as a modified SURR estimator (Singh and Ullah, 1974; p. 193).

Other consistent estimators of Φ can, of course, be obtained. For example an FGLS estimator may be applied to (11.15). However, whether the OLS or FGLS estimators are applied to this equation, a practical problem can arise: the estimates of some of the diagonal elements of $\hat{\Phi}$ may be negative for some samples. This implies negative estimated variances, which is clearly inappropriate. Secondly, for particular samples, $\hat{\Phi}$ may not be positive definite. Various ad hoc solutions may be considered. For example, the corresponding element(s) of $X^*(\hat{\Theta} \otimes I_T)X^{*\prime}$ may be set to zero in (11.10). Alternatively, δ_{ij} may be chosen so as to minimize the quantity

$$(w_{ij} - Z_{ij}\delta_{ij})'(w_{ij} - Z_{ij}\delta_{ij}) \tag{11.17}$$

subject to appropriate nonnegativity constraints on the elements of δ_{ij}. This, of course, does not yield a closed-form solution for δ_{ij}. Other approaches include obtaining $\hat{\delta}_{ij}$ as a minimum norm quadratic estimator; using the method of mixed estimation (as long as appropriate prior knowledge about the elements of Φ can be specified) along the lines discussed by Srivastava, Mishra and Chaturvedi (1981); or adopting a Bayesian framework, such as that described by Griffiths, Drynan and Prakash (1979).

Now, considering the estimators of β in (11.15) noted so far, it can be shown that the OLS estimator is unbiased with variance covariance matrix

$$V(b_0) = (X'X)^{-1}X'\Phi X(X'X)^{-1} , \tag{11.18}$$

while the SURR estimators (11.12) and (11.14) are each consistent with asymptotic second-order moment matrices (to order $O(T^{-1})$) given by

$$V(\hat{b}^*_{SR}) = [X'(\Sigma^{-1} \otimes I_T)X]^{-1}X'(\Sigma^{-1} \otimes I_T)\Phi(\Sigma^{-1} \otimes I_T)X[X'(\Sigma^{-1} \otimes I_T)X]^{-1} \tag{11.19}$$

$$V(\hat{b}_{SR}) = (X'\Phi^{-1}X)^{-1} . \tag{11.20}$$

If we define the full-rank matrices

$$G_1 = (X'X)^{-1}X' - (X'\Phi^{-1}X)^{-1}X'\Phi^{-1}$$

$$G_2 = [X'(\Sigma^{-1} \otimes I_T)X]^{-1}X'(\Sigma^{-1} \otimes I_T) - (X'\Phi^{-1}X)^{-1}X'\Phi^{-1}$$

then we see that

$$V(b_0) - V(\hat{\beta}_{SR}) = G_1\Phi G_1' \tag{11.21}$$

$$V(\hat{\beta}^*_{SR}) - V(\hat{\beta}_{SR}) = G_2\Phi G_2' , \tag{11.22}$$

so that $\hat{\beta}_{SR}$ is more efficient, to order $O(T^{-1})$, than b_0 or $\hat{\beta}^*_{SR}$ (see Singh and Ullah, 1974; pp. 193-194 for the above results). It is not clear to what extent this efficiency gain holds in small samples.

Other formulations of the varying coefficients SURE model are also possible, and warrant brief mention. These are motivated by practical considerations and may be classified into two broad categories: models with varying coefficients having deterministic components and models with randomly varying coefficients having both deterministic and stochastic components (see Raj and Ullah, 1981; Chapter 2). A simple specification belonging to the first category arises when β_{tik} in (11.1) is assumed to be a function of certain variables, say policy variables. For example, we might assume that

$$\beta_{tik} = \beta_{ik} + \gamma_{ik}p_{tik} , \tag{11.23}$$

where β_{ik} and γ_{ik} are unknown constants and p_{tik} is the t'th observation on the policy variable which determines the variation in β_{tik} in the sample.

Another interesting formulation arises if we assume that the changes in β_{tik} are systematic and discontinuous in nature. For example, suppose that the sample of T observations can be partitioned into q separate regimes, within each of which it is reasonable to presume that the model's coefficients are constant. That is, we have a general "switching regressions" formulation, the details of which may vary from case to case. For example, the partition of the sample into these regimes may be determined, at least in part, by the values taken by certain variables; the boundaries of the regimes either may be known with certainty or else unknown and determined probabilistically; and continuous rather than discrete switching between the regimes may be specified.

Among the other formulations which may be considered is that which arises when the coefficients are of the form

$$\beta_{tik} = \beta_{t-1\ ik} + \epsilon_{tik} , \tag{11.24}$$

so that we have an adaptive regression model. This specification may be generalized to

$$\beta_{tik} = \sum_{j=1}^{M} \sum_{\ell=1}^{K_i} \beta_{t-1\ j\ell}\lambda_{ikj\ell} + \epsilon_{tik} , \tag{11.25}$$

which, with $\lambda_{ikj\ell}$ a constant, is a Kalman filter specification.

Thus, we see that several SURE model formulations may be generated by relaxing the assumption that the coefficients are constant, and describing different mechanisms by which they may be allowed to vary throughout the sample. For each of these formulations, single equation estimators (which ignore the inherent jointness of the SURE model's equations) are available and may be applied to each of the M equations individually (see Raj and Ullah, 1981). Of course, these estimators will lack efficiency in the context of the SURE framework, but except for the type of coefficient variation discussed by Singh and Ullah (1974), appropriate SURE estimators have not been properly explored. Thus, there remain a number of interesting research problems in this area, not only with regard to the construction and properties of different estimators, but also in connection with the testing of hypotheses associated with the possible constancy of the coefficients. Clearly, much of the existing single equation literature on varying coefficients can, in principle, be extended to the SURE model and this would be of considerable practical interest.

11.3 MISSING OBSERVATIONS

In practical applications, data sets frequently are deficient in a variety of ways. One major deficiency arises when certain observations are unavailable for certain of the regressors in the model (see for example Maddala, 1977; pp. 201-207, and Kmenta, 1981). Unless the sample is condensed to match the minimum number of observations on any of the variables, conventional estimators for the SURE model need to be reconsidered in such circumstances. Intuitively, discarding observations in this way is likely to lead to inefficient estimation, and a consideration of this forms part of the following discussion. The dangers of following the alternative naive strategy of omitting from the model those variables on which data are lacking have been discussed at length in Section 10.3.

To illustrate some features of the missing observations problem in the context of the SURE model we consider a two equation system:

$$\begin{bmatrix} y_1^* \\ y_2 \end{bmatrix} = \begin{bmatrix} X_1^* & 0 \\ 0 & X_2 \end{bmatrix} \begin{bmatrix} \beta_1 \\ \beta_2 \end{bmatrix} + \begin{bmatrix} u_1^* \\ u_2 \end{bmatrix}$$

or

$$y = X\beta + u \ . \tag{11.26}$$

Here, y_1^* and u_1^* are $(T^* \times 1)$ vectors and X_1^* is a $(T^* \times K_1)$ matrix. There are $T^{**} = (T - T^*)$ observations missing for one or more of the regressors in the first equation of the model, while the sample for the variables in the second equation is complete. Partitioning y_2, u_2 and X_2 as

$$\underset{(1)}{y_2 = \begin{bmatrix} y_2^* \\ y_2^{**} \end{bmatrix}} \begin{matrix} (T^*) \\ (T^{**}) \end{matrix}, \quad \underset{(1)}{u_2 = \begin{bmatrix} u_2^* \\ u_2^{**} \end{bmatrix}} \begin{matrix} (T^*) \\ (T^{**}) \end{matrix}, \quad \underset{(K_2)}{X_2 = \begin{bmatrix} X_2^* \\ X_2^{**} \end{bmatrix}} \begin{matrix} (T^*) \\ (T^{**}) \end{matrix}$$

it is assumed that

$$E\begin{bmatrix} u_1^* \\ u_2^* \\ u_2^{**} \end{bmatrix} = 0 \ ,$$

$$E\begin{bmatrix} u_1^* \\ u_2^* \\ u_2^{**} \end{bmatrix} [u_1^{*\prime} \ u_2^{*\prime} \ u_2^{**\prime}] = \begin{bmatrix} \sigma_{11}I_{T*} & \sigma_{12}I_{T*} & 0 \\ \sigma_{21}I_{T*} & \sigma_{22}I_{T*} & 0 \\ 0 & 0 & \sigma_{22}I_{T**} \end{bmatrix}$$

$$= \begin{bmatrix} (\Sigma \otimes I_{T*}) & 0 \\ 0 & \sigma_{22}I_{T*} \end{bmatrix} = \Phi \ (\text{say}) \ .$$

Now, the GLS estimator for β in (11.26) is

$$b_G = (X'\Phi^{-1}X)^{-1}X'\Phi^{-1}y \ , \tag{11.27}$$

with variance covariance matrix

$$V(b_G) = (X'\Phi^{-1}X)^{-1} \ . \tag{11.28}$$

So, as is noted by Kmenta (1981; p. 252), the fact that unequal samples are available for each of the equations in the model in no way affects our ability to construct the GLS estimator for the SURE model's parameters. Obviously, if Σ is replaced by a consistent estimator, S, in Φ then the same statement applies for any FGLS estimator of β. The question of just how S may be constructed in this case is taken up below. However, before turning to this, let us consider the consequences of discarding the additional T** observations available for the variables in the second equation of the model, over and above those available for the variables in the first equation. One motivation for proceeding in this way would be to overcome any ambiguity as to how Σ might be estimated when constructing an FGLS estimator of β. In this case, (11.26) becomes

$$\begin{bmatrix} y_1^* \\ y_2^* \end{bmatrix} = \begin{bmatrix} X_1^* & 0 \\ 0 & X_2^* \end{bmatrix} \begin{bmatrix} \beta_1 \\ \beta_2 \end{bmatrix} + \begin{bmatrix} u_1^* \\ u_2^* \end{bmatrix}$$

or

$$y^* = X^*\beta + u^* \quad , \tag{11.29}$$

and the GLS estimator of β is

$$b_G^* = [X^{*\prime}(\Sigma^{-1} \otimes I_{T^*})X^*]^{-1}X^{*\prime}(\Sigma^{-1} \otimes I_{T^*})y^* \tag{11.30}$$

with variance covariance matrix

$$V(b_G^*) = [X^{*\prime}(\Sigma^{-1} \otimes I_{T^*})X^*]^{-1} \quad . \tag{11.31}$$

Note that b_G^* and b_G in (11.27) are both unbiased estimators of β and that b_G may be expressed as

$$b_G = [X^{*\prime}(\Sigma^{-1} \otimes I_{T^*})X^* + G]^{-1}[X^{*\prime}(\Sigma^{-1} \otimes I_{T^*})y^* + g] \tag{11.32}$$

so that

$$V(b_G) = [X^{*\prime}(\Sigma^{-1} \otimes I_{T^*})X^* + G]^{-1} \tag{11.33}$$

where

$$G = \begin{bmatrix} 0 & 0 \\ 0 & \frac{1}{\sigma_{22}} X_2^{**\prime}X_2^{**} \end{bmatrix} , \quad g = \begin{bmatrix} 0 \\ \frac{1}{\sigma_{22}} X_2^{**\prime}y_2^{**} \end{bmatrix} .$$

It follows from a comparison of (11.31) and (11.33) and the positive semi-definiteness of G that b_G^* is inefficient relative to b_G as an estimator of β. Clearly, this result also holds asymptotically if Σ is replaced by any consistent estimator, S. However, the finite-sample consequences of discarding observations in this way will depend upon the precise construction of S and are unexplored.

So far, we have assumed that the T* observations which are unavailable for the first equation in the SURE model (11.26) apply to all regressors in that model. In practice, a more likely situation is that all T observations will be available only for some of the regressors in that equation while T** will be unavailable for certain regressors in the first equation. In this case the two equation SURE model may be written, in conformable notation, as

$$\begin{bmatrix} y_1^* \\ y_1^{**} \\ y_2^* \\ y_2^{**} \end{bmatrix} = \begin{bmatrix} X_{11}^* & X_{12}^* & 0 & 0 \\ X_{11}^{**} & X_{12}^{**} & 0 & 0 \\ 0 & 0 & X_{21}^* & X_{22}^* \\ 0 & 0 & X_{21}^{**} & X_{22}^{**} \end{bmatrix} \begin{bmatrix} \beta_{11} \\ \beta_{12} \\ \beta_{21} \\ \beta_{22} \end{bmatrix} + \begin{bmatrix} u_1^* \\ u_1^{**} \\ u_2^* \\ u_2^{**} \end{bmatrix} \tag{11.34}$$

where the only unobservable data points are the elements of X_{12}^{**}. The extent to which these missing observations reduce the efficiency of FGLS estimators of β, relative to the potential efficiency which could be achieved in principle if X_{12}^{**} were in fact observable, could be considered along the lines adopted by Kmenta (1981; pp. 236-237) and Giles (1986) for the single-equation model. Kmenta (1981; pp. 252-253) also compared the maximum likelihood estimators of the β_{ij}'s in (11.34) obtained if the elements of X_{12}^{**} are treated as additional unknown parameters and all of the available data are used, with those obtained if the data in y_1^{**} and X_{11}^{**} are discarded. It turns out that these two estimators are identical, implying that the data in y_1^{**} and X_{11}^{**} contain no usable information regarding the parameters of (11.34) and so they may be discarded without any adverse consequences. Notice that for each of the two estimators being compared here, the data in $\{y_2^{**}, X_{21}^{**}, X_{22}^{**}\}$ are fully utilized.

As was discussed in Section 2.4, when the regressors in each equation are identical and equal numbers of observations are available on all variables, the OLS, GLS and FGLS estimators of β are identical to each other. The way in which this result may be affected when certain observations are missing for certain variables has been considered by Wallace and Silver (1984) and by Conniffe (1985) for the case where $X_1^* = X_2^*$, so that the two equation SURE model is

$$\begin{bmatrix} y_1^* \\ y_2^* \\ y_2^{**} \end{bmatrix} = \begin{bmatrix} X_2^* & 0 \\ 0 & X_2^* \\ 0 & X_2^{**} \end{bmatrix} \begin{bmatrix} \beta_1 \\ \beta_2 \end{bmatrix} + \begin{bmatrix} u_1^* \\ u_2^* \\ u_2^{**} \end{bmatrix} \tag{11.35}$$

when T^{**} observations are missing for the dependent variable in the first equation.

Using only the available observations, it is easy to verify that the GLS and OLS estimators do not coincide. Further, let $b_{(1)0}$ and $b_{(2)0}$ denote the OLS estimators of β_1 and β_2 respectively. Then the associated variance covariance matrices are

$$\begin{aligned} V(b_{(1)0}) &= \sigma_{11}(X_2^{*\prime}X_2^*)^{-1} \\ V(b_{(2)0}) &= \sigma_{22}(X_2^{*\prime}X_2^* + X_2^{**\prime}X_2^{**})^{-1} \ . \end{aligned} \tag{11.36}$$

The variance covariance matrices of the GLS estimators of β_1 and β_2 are

$$V(b_{(1)G}) = \sigma_{11}[(1-\rho_{12}^2)(X_2^{*\prime}X_2^*)^{-1} + \rho_{12}^2(X_2^{*\prime}X_2^* + X_2^{**\prime}X_2^{**})^{-1}]$$

$$V(b_{(2)G}) = \sigma_{22}(X_2^{*\prime}X_2^* + X_2^{**\prime}X_2^{**})^{-1} , \tag{11.37}$$

where $\rho_{12} = \sigma_{12}/(\sigma_{11}\sigma_{22})^{\frac{1}{2}}$. From (11.36) and (11.37) it is clear that the observations unavailable for the first equation of the model do not affect the efficiency of the OLS estimator of β_2 relative to its GLS counterpart, but for β_1 we have (Wallace and Silver, 1984)

$$V(b_{(1)0}) - V(b_{(1)G}) = \sigma_{11}\rho_{12}^2[(X_2^{*\prime}X_2^*)^{-1} - (X_2^{*\prime}X_2^* + X_2^{**\prime}X_2^{**})^{-1}]$$

$$= \sigma_{11}\rho_{12}^2(X_2^{*\prime}X_2^*)^{-1}[(X_2^{*\prime}X_2^*)^{-1} + (X_2^{**\prime}X_2^{**})^{-1}]^{-1}(X_2^{*\prime}X_2^*)^{-1} , \tag{11.38}$$

which is positive definite. So, the missing observations do affect the relative efficiencies of the OLS and GLS estimators of the parameters in the equation in question. The gain in efficiency from using the GLS estimator for β_1 reflects the use of the additional available data on the variables for the second equation in the model. This gain increases with the correlation between the disturbances for the two equations, as might be anticipated. Finally, notice that there is no efficiency gain at all if $\rho_{12} = 0$. Recall from Section 2.4 that this is the second case in which, in general, the OLS and GLS estimators coincide.

As is noted by Conniffe (1985), the finite-sample details of the statistical consequences of missing observations in this context depend, in general, upon the pattern of the data. One special interesting case is when the observations in X_1^* recur, in say q repetitions, in the second equation so that $X_2^{*\prime}X_2^* = qX_1^{*\prime}X_1^*$. In this case, the right-hand side of (11.38) becomes

$$\sigma_{11}\rho_{12}^2\left(\frac{q}{q+1}\right)(X_1^{*\prime}X_1^*)^{-1} ,$$

so that the relative efficiency of GLS over OLS in the estimation of β_1 increases with q. Some other finite-sample results relating to this problem are reported by Conniffe.

When certain observations are missing for the variables of one equation in the SURE model (11.26), there are several ways of estimating the elements of Σ, as was discussed in the following way by Schmidt (1977). Let

$\hat{u}_1^*$ and $\hat{u}_2$ be the OLS residual vectors for the first and second equations respectively, where $\hat{u}_2' = (\hat{u}_{2*}', \hat{u}_{2**}')$ with its sub-vectors of order T* and T** respectively. Letting s_{ij} relate to the common T* observations and $\hat{\sigma}_{ij}$ relate to all available data points, define

$$s_{11} = \frac{1}{T^*}\hat{u}_1^{*\prime}\hat{u}_1^*$$

$$\hat{\sigma}_{11} = s_{11} - \left(\frac{T^{**}}{T}\right)\left[\frac{\hat{u}_1^{*\prime}\hat{u}_{2*}}{\hat{u}_{2*}'\hat{u}_{2*}}\right]^2\left(\frac{1}{T^*}\hat{u}_{2*}'\hat{u}_{2*} - \frac{1}{T^{**}}\hat{u}_{2**}'\hat{u}_{2**}\right)$$

$$s_{22} = \frac{1}{T^*}\hat{u}_{2*}'\hat{u}_{2*}$$

$$\hat{\sigma}_{22} = \frac{1}{T}\hat{u}_2'\hat{u}_2$$

$$s_{12} = s_{21} = \frac{1}{T^*}\hat{u}_1^{*\prime}\hat{u}_{2*}$$

$$\alpha = \left[\left(\frac{T^*}{T}\right)\frac{\hat{u}_2'\hat{u}_2}{\hat{u}_{2*}'\hat{u}_{2*}}\right]^{1/2} .$$

A first estimator of Σ takes account only of the T* observations:

$$S_{(1)} = \begin{bmatrix} s_{11} & s_{12} \\ s_{21} & s_{22} \end{bmatrix} , \tag{11.39}$$

while a second choice which takes account of all available observations when estimating σ_{22} but suffers from the deficiency of not necessarily being positive definite is

$$S_{(2)} = \begin{bmatrix} s_{11} & s_{12} \\ s_{21} & \hat{\sigma}_{22} \end{bmatrix} . \tag{11.40}$$

To ensure positive definiteness, $S_{(2)}$ might be modified to a third choice of estimator for Σ:

$$S_{(3)} = \begin{bmatrix} s_{11} & \alpha s_{12} \\ \alpha s_{21} & \hat{\sigma}_{22} \end{bmatrix} . \tag{11.41}$$

Two final choices use all of the available data to estimate all elements of Σ:

$$S_{(4)} = \begin{bmatrix} \hat{\sigma}_{11} & \alpha^2 s_{12} \\ \sigma^2 s_{21} & \hat{\sigma}_{22} \end{bmatrix} \tag{11.42}$$

and $S_{(5)}$, the maximum likelihood estimator based on the assumption that the disturbances are normally distributed. The latter estimator is essentially obtained by maximizing the following part of the likelihood function numerically:

$$-\frac{1}{2}\Big[T^{**} \log \sigma_{22} + T^{*} \log |\Sigma| + (y^{*} - X^{*}\beta)'(\Sigma^{-1} \otimes I_{T^{*}})(y^{*} - X^{*}\beta)$$

$$+ \frac{1}{\sigma_{22}}(y_2^{**} - X_2^{**}\beta_2)'(y_2^{**} - X_2^{**}\beta_2)\Big] \quad .$$

These five choices of S, when substituted for Σ in (11.32), lead to five possible FGLS estimators of β in the SURE model with unbalanced data. Asymptotically, of course, these estimators all have identical properties, but this will not be the case in finite samples. To study this problem Schmidt (1977) performed a Monte Carlo experiment, the general design of which was essentially the same as that considered by Kmenta and Gilbert (1968), and discussed in Section 3.5. If $S_{(2)}$ is not positive definite then the corresponding FGLS estimator has moments which need not be finite. To overcome this difficulty Schmidt replaced the absolute value of s_{12} in $S_{(2)}$ by $(s_{11}\hat{\sigma}_{22} - 0.01)^{1/2}$, and this quantity was given the sign of s_{12} whenever $|S_{(2)}| < 0.01$. The OLS estimator of β was also considered along with the five FGLS estimators.

Schmidt's Monte Carlo study suggested that all of the FGLS estimators are more efficient than the OLS estimator. Increasing T^* increased the efficiencies of the estimators, but this was not always true if T^{**} was increased for some fixed value of T^*. The relative performances of the five FGLS estimators were relatively insensitive to changes in T^{**}, and the differences between the efficiencies of these estimators diminished as T^* increased. One interesting finding was that the estimators of Σ which use all of the observations did not necessarily produce FGLS estimators of β which were superior to those based on Σ estimators which ignore the T^{**} observations for the second equation. Of course, this and the other findings must be treated very cautiously. Certainly, there is a case for further Monte Carlo analysis of these estimators, and it would be interesting to extend the investigation to consider the case where varying numbers of observations are available for different regressors in a given equation and to consider models with more than two equations.

Finally, we mention the Bayesian analysis of Swamy and Mehta (1975), for a two equation SURE model with AR(1) disturbances. Part of their study involved a Monte Carlo experiment in which two data sets were generated, these differing only in the assumed disturbance autocorrelations. In each case, $T^* = T^{**} = 10$. With the first data set (in which one autocorrelation parameter was -0.5 and the other 0.8) taking account of the additional T^{**} observations for the second equation shifted the center of the marginal posterior distribution for β_2, and also lowered its mean squared error. With the second data set (in which both autocorrelation coefficients were 0.8) similar results also occurred for the marginal posterior distribution of β_1. However, the effects of the additional observations were more pronounced with respect to the posterior density for β_2 than with respect to that for β_1.

11.4 GOODNESS-OF-FIT MEASURES

In the estimation of SURE models, as with single equation models, one may wish to report summary statistics reflecting some of the features or quality of the results obtained. Statistics for testing for the serial independence and homoscedasticity of the model's disturbances have been discussed in Chapters 7 and 8 respectively, and these might form part of standard result reporting. In addition, one might wish to indicate the extent to which the fitted SURE model "explains" the variability in the data for the dependent variables. That is, some measure of goodness-of-fit, in the form of a coefficient of determination, might be calculated and reported. Here, we consider how such a measure might be constructed if the model is estimated consistently, say, by means of an FGLS estimator. This issue was addressed by McElroy (1977), who linked a coefficient of determination for the SURE model to a statistic for testing for a linear relation between the dependent variables and the regressors. To derive McElroy's measure, R^2, write the SURE model as:

$$\begin{aligned} y &= X\beta + u \\ &= \bar{X}\bar{\beta} + X^\circ\beta^\circ + u \qquad (11.43) \\ E(u) &= 0\,, \quad E(uu') = (\Sigma \otimes I_T) \end{aligned}$$

where

$$\bar{X} = \begin{bmatrix} \bar{X}_1 & 0 & \cdots & 0 \\ 0 & \bar{X}_2 & \cdots & 0 \\ \vdots & \vdots & & \vdots \\ 0 & 0 & \cdots & \bar{X}_M \end{bmatrix}$$

$$X^\circ = \begin{bmatrix} e_T & 0 & \cdots & 0 \\ 0 & e_T & \cdots & 0 \\ \vdots & \vdots & & \vdots \\ 0 & 0 & \cdots & e_T \end{bmatrix}$$

where e_T is a $(T \times 1)$ column vector with all elements unity and the i'th diagonal block of X is

$$X_i = (\bar{X}_i, e_T) \quad .$$

Formulating the model as in (11.43) distinguishes the intercept term in each equation from the other regressors. Note that the following discussion requires the presence of an intercept in every equation of the model. Now, let S be any consistent estimator of Σ, with $S^{-1} = A'A$, where A is nonsingular. Pre-multiplying (11.43) by $(A \otimes I_T)$ and applying OLS yields an FGLS estimator of β, determined by the choice of S. That is,

$$b_F = \begin{bmatrix} \bar{b}_F \\ b_F^\circ \end{bmatrix} = [X'(S^{-1} \otimes I_T)X]^{-1}X'(S^{-1} \otimes I_T)y$$

with the associated residual vector

$$u_F = (y - Xb_F) = (y - \bar{X}\bar{b}_F - X^\circ b_F^\circ) \quad .$$

Now, given that an intercept appears in each equation, the usual orthogonality between OLS residuals and regressors ensures here that

$$X^{\circ\prime}(S^{-1} \otimes I_T)u_F = 0 \quad ,$$

which can be shown to imply that

$$X^{\circ\prime}u_F = 0 \ . \tag{11.44}$$

So, taking deviations about sample means for the data eliminates the intercepts from each equation, as in the single-equation case. Let

$$\bar{P}_{e_T} = \left(I_T - \frac{1}{T}e_Te_T'\right)$$

so that after a little manipulation the FGLS estimated model may be written in deviation form as (McElroy, 1977; p. 384)

$$(I_M \otimes \bar{P}_{e_T})y = (I_M \otimes \bar{P}_{e_T})\bar{X}\bar{b}_F + u_F \ . \tag{11.45}$$

Pre-multiplying (11.45) by $(A \otimes I_T)$ and decomposing the total variation into its "explained" and "unexplained" components, we get

$$y'(S^{-1} \otimes \bar{P}_{e_T})y = \bar{b}_F'\bar{X}'(S^{-1} \otimes \bar{P}_{e_T})\bar{X}\bar{b}_F + u_F'(S^{-1} \otimes I_T)u_F \ , \tag{11.46}$$

where use has been made of the relationship

$$\bar{b}_F'\bar{X}'(S^{-1} \otimes \bar{P}_{e_T})u_F = 0 \ .$$

A coefficient of determination for the full SURE system follows directly from (11.46):

$$R^2 = \frac{\bar{b}_F'\bar{X}'(S^{-1} \otimes \bar{P}_{e_T})\bar{X}b_F}{y'(S^{-1} \otimes \bar{P}_{e_T})y}$$

$$= 1 - \frac{u_F'(S^{-1} \otimes I_T)u_F}{y'(S^{-1} \otimes \bar{P}_{e_T})y} \ . \tag{11.47}$$

As McElroy (1977; p. 384) notes, this R^2 has several desirable properties. Clearly, from (11.46) and (11.47), $0 \leq R^2 \leq 1$. For a particular choice of estimator for Σ its value is maximized by constructing R^2 from b_F (implicit in u_F). It is related monotonically to an appropriate test of

$H_0: \bar{\beta} = 0$ against $H_1: \bar{\beta} \neq 0$. Note that this hypothesis is a special case of that considered in Section 10.2.1, and that in this case the statistic z given in equation (10.11) collapses to

$$z^\circ = \left[\frac{\nu}{G^\circ}\right] \left[\frac{\bar{b}_F'\bar{X}'(S^{-1} \otimes \bar{P}_{e_T})\bar{X}\bar{b}_F}{u_F'(S^{-1} \otimes I_T)u_F}\right] \qquad (11.48)$$

where $\nu = (MT - \Sigma_{i=1}^{M} K_i)$, as before, and $G^\circ = (\Sigma_{i=1}^{M} K_i - M)$. From (11.46)-(11.48), we get

$$z^\circ = \left[\frac{\nu}{G^\circ}\right] \left[\frac{R^2}{1 - R^2}\right] .$$

All of these results mimic the properties of the usual OLS coefficient of determination for a single-equation model. Finally, it can be shown that R^2 may be interpreted as the squared correlation coefficient between $(A \otimes I_T)y$ and $(A \otimes I_T)Xb_F$. This last result exposes the principal weakness of McElroy's measure. It measures the correlation between the actual and fitted values of the dependent variables in a transformed space, not in terms of the original measurements for the data. This deficiency is equally apparent if we interpret R^2, from (11.46) and (11.47), as that proportion of the sample variation in $(A \otimes I_T)y$ explained by $(A \otimes I_T)X$. Note that these last two interpretations have only asymptotic justification, a point not noted by McElroy, as $S = (A'A)^{-1}$ is an asymptotically justified estimator of Σ. It is not clear that we should be interested in a measure based on such a transformation of the original problem. Presumably a more interesting measure of goodness-of-fit would be one which can be given interpretations such as those above when viewed in terms of the original sample space (see Battese and Griffiths, 1980).

McElroy compared her R^2 with other goodness-of-fit measures which have been proposed for sets of equations, and Buse (1979) extended McElroy's analysis to allow for the possibility that the SURE model's disturbances might be autocorrelated or heteroscedastic. Recently, Bewley (1985) considered McElroy's and other measures of goodness-of-fit when the SURE model incorporates the constraints of an allocation model, as typified by a system of demand equations. As noted earlier, McElroy's R^2 is derived on the assumption that each equation in the system includes an intercept. If this is not the case, then the decomposition in (11.46) which ensures the equivalence of the two formulae for R^2 in (11.47) no longer applies. In this case the definition of R^2 is ambiguous and in particular it may take on values in excess of unity if defined by the first relationship in (11.47), or negative values if defined by the second relationship in that equation.

Finally, note that McElroy's measure is one for the goodness-of-fit of the entire SURE model. If the usual single-equation coefficients of determination are calculated on the basis of FGLS estimates and residuals, the usual decomposition of the OLS sums of squares will not hold in general, and ambiguous measures will be obtained (irrespective of the inclusion of an intercept in the equations). It is not clear what meaning can be attached to single equation goodness-of-fit measures in the context of SURE models. The construction of appropriate such measures remains open for investigation, although an obvious suggestion is to compute the correlation between the actual sample values for the dependent variable and those values predicted using the FGLS parameter estimates, equation by equation (see Judge, Griffiths, Hill and Lee, 1980; pp. 251-254). The effect on R^2 of the choice of S, in finite samples also could be considered. This last issue may be important if R^2 is used to rank competing models which explain the same dependent variables. The usefulness of R^2 in this regard has not been explored, although one would conjecture that choosing between competing SURE models by maximizing R^2 would (under certain conditions) lead to the selection of the correct specification, asymptotically (see Low, 1980).

A simple model-selection rule which may be helpful in some cases for choosing between alternative SURE model specifications which explain the same vector of dependent variables may also be mentioned here. It is well known (e.g., Theil, 1970; pp. 212-214) that under certain conditions, choosing between two OLS regressions with the same dependent variable by minimizing the estimated error variance leads to the selection of the correct specification, on average. If we recall the discussion of the FGLS estimator of β in Chapter 2, this estimator is just OLS applied to the transformed model,

$$(S^{-\frac{1}{2}} \otimes I_T)y = (S^{-\frac{1}{2}} \otimes I_T)X\beta + (S^{-\frac{1}{2}} \otimes I_T)u$$

or,

$$y^* = X^*\beta + u^* \quad ,$$

where S is a consistent estimator of Σ. That is,

$$b_F = (X^{*\prime}X^*)^{-1}X^{*\prime}y^* = b_0^* \quad .$$

Let

$$\begin{aligned} \hat{u}^* &= y^* - X^*b_0^* \\ &= y^* - X^*b_F \\ &= (S^{-\frac{1}{2}} \otimes I_T)u_F \quad , \end{aligned}$$

where u_F is the FGLS residual vector in terms of the original sample space. Bearing in mind that the transformed regressors comprising the columns of X* are stochastic, it follows from Schmidt (1974; p. 122) that choosing between two SURE models, explaining the same y and in which neither regressor matrix is nested within the other, by minimizing

$$\left(\frac{1}{T}\hat{u}^{*\prime}\hat{u}^{*}\right) = \frac{1}{T}u_F'(S^{-1}\otimes I_T)u_F ,$$

will lead to the selection of the correct specification with probability one (see also Giles and Low, 1981). This assumes, of course, that the correctly specified model is one of those being compared.

11.5 DYNAMIC MODELS

In all of the discussion so far we have treated the regressors in each equation of our SURE model,

$$\begin{aligned} y &= X\beta + u \\ E(u) &= 0, \qquad E(uu') = (\Sigma\otimes I_T) = \Psi , \end{aligned} \tag{11.49}$$

as being strictly exogenous. That is, X is a nonstochastic matrix. This is essentially what distinguishes the SURE model from the simultaneous equations model of econometrics. Moreover, we have assumed that the model under discussion comprises a set of static relationships. In practice, however, the columns of X in (11.49) are likely to include observations on lagged values of the elements of y if the model is being estimated from time-series data. That is, one or more lagged values of one or more of the dependent variables in the model may appear as regressors in one or more of the equations. In this case, (11.49) still represents a SURE model, but one which is now dynamic. Dynamic SURE models have received only limited attention in the literature (see Spencer, 1979, and Harvey, 1981; pp. 293-300).

The general dynamic SURE model may be written as

$$\begin{aligned} A(L)y_t &= Bz_t + u_t \qquad (t = 1, 2, \ldots, T) \\ A(L) &= I - A_1L - A_2L^2 - \cdots - A_rL^r \end{aligned} \tag{11.50}$$

where L is the usual lag operator and A_k is an $(M \times M)$ matrix of unknown parameters $(k = 1, 2, \ldots, r)$. In (11.50), y_t and u_t are $(M \times 1)$, B is $(M \times g)$ and z_t is a $(g \times 1)$ vector of the t'th observation on g exogenous regressors.

That is, the elements of z_t are associated with the variables in X other than lagged values of the M dependent variables in the model. Lags of up to r periods are allowed for in (11.50), and the model could be written in even more general form by incorporating lagged values of the variables in z_t as additional regressors.

As far as the dynamic properties of (11.50) are concerned, we merely note that the system is stable if all of the roots of $|A(L)|$ lie outside the unit circle (see Harvey, 1981; pp. 293-296). In this case $A^{-1}(L)$ exists and the dynamic SURE model may be written as

$$y_t = A^{-1}(L)Bz_t + A^{-1}(L)u_t \qquad (t = 1, 2, \ldots, T) \quad . \tag{11.51}$$

The estimation of the dynamic SURE model is perhaps best considered in terms of (11.49), where for expository purposes we shall set r = 1. In this case,

$$X = \begin{bmatrix} X_1 & 0 & \cdots & 0 \\ 0 & X_2 & \cdots & 0 \\ \vdots & \vdots & & \vdots \\ 0 & 0 & \cdots & X_M \end{bmatrix}$$

$$X_i = (Y_{-1(i)} \quad Z_i)$$

$$\beta' = (\beta_1', \beta_2', \ldots, \beta_M') \tag{11.52}$$

$$\beta_i' = (\alpha_i', \gamma_i')$$

where, for the i'th equation of the model, Z_i is a $(T \times g_i)$ matrix of exogenous regressors, and $Y_{-1(i)}$ is a $[T \times (K_i - g_i)]$ matrix of one-period lagged values of some of the model's dependent variables; $i = 1, 2, \ldots, M$. Note that $(K_i - g_i) \leq M$.

Assuming for the moment that the disturbances in (11.49) are serially independent, note that the OLS and GLS estimators of β are biased but consistent if X is of the form (11.52). If each equation of the system includes the same set of explanatory variables (including lagged variables) then these two estimators will be identical, as in the static SURE model. All of the finite-sample results presented throughout this book, and in particular those relating to the SUUR and SURR estimators given in Chapter 4 will no longer apply if the SURE model is dynamic, and only limited Monte Carlo evidence is available in this case (see Spencer, 1979, and our discussion in

Section 3.5 of the study by Kmenta and Gilbert, 1968). Asymptotically, the FGLS, ML and GLS estimators of β will be equivalent in this case. In particular, the SUUR and SURR estimators will be consistent and asymptotically efficient. These asymptotic results will not depend upon whether or not the first observation is taken into account, say in formulating the likelihood function in order to proceed with ML estimation. Of course, the treatment of these initial conditions will affect the finite-sample properties of the various estimators of β.

If the disturbances in the dynamic SURE model are serially correlated then the usual estimators of β will be inconsistent. In this case Spencer (1979) suggested adopting the estimator proposed by Hatanaka (1974) and by Dhrymes and Taylor (1976) for the dynamic simultaneous equations model with autocorrelated errors, bearing in mind that the SURE model may be viewed as a special case of the simultaneous equations model. This two-step estimator is consistent and asymptotically efficient, and Spencer conducted several Monte Carlo experiments to compare its finite-sample properties with those of other estimators, including OLS and the estimator for static SURE models with autocorrelated errors proposed by Parks (1967). These experiments were based on six different two equation dynamic SURE models with first-order scalar autoregressive disturbances. Samples of size 30, 60 and 100 were considered and each experiment was replicated 200 times. The results showed that, in general, even with samples as small as 30 observations the asymptotically efficient estimator performed better than the asymptotically inefficient estimators, such as that proposed by Parks. However, the converse was found in cases where the coefficients of the lagged dependent variables were large and/or the disturbances exhibited only very weak autocorrelation. The finite-sample properties of estimators for the dynamic SURE model, with or without autocorrelated errors, deserve further investigation.

11.6 NON-LINEAR MODELS

In this book we have considered SURE models which are linear in both the variables and parameters, and in which the disturbances are additive. In practice, such a model specification may be unduly restrictive. For example, if the system of equations which we wish to estimate is derived from some underlying microeconomic optimization problem, then non-linearities may be a feature of the model. The derivation of a system of demand equations from the constrained maximization of the consumer's utility function is a case in point, for many commonly used utility functions. Non-linearities in the variables often may be eliminated by an appropriate transformation of the model or renaming of the variables. In this case the model is inherently linear and our previous discussion and results apply, provided that the disturbance term remains additive. Of concern here are models which are inherently non-linear. We retain the assumption that the disturbances are additive and discuss Gallant's (1975) analysis of the model

$$y_{ti} = f_i(x_{ti}\,;\,\theta_i) + u_{ti} \ , \quad (i = 1, 2, \ldots, M;\ t = 1, 2, \ldots, T) \qquad (11.53)$$

where x_{ti} is a $(K_i \times 1)$ vector containing the t'th observation on each of the explanatory variables in the i'th equation; $\theta_i \in \Theta_i$ is a $(p_i \times 1)$ vector of unknown parameters in the i'th equation; and $f_i(\cdot\,;\,\cdot)$ denotes the i'th response function. Thus, the nature of the non-linearities may differ from equation to equation.

Writing the model more compactly, we have

$$\begin{bmatrix} y_1 \\ y_2 \\ \vdots \\ y_M \end{bmatrix} = \begin{bmatrix} f_1(\theta_1) \\ f_2(\theta_2) \\ \vdots \\ f_M(\theta_M) \end{bmatrix} + \begin{bmatrix} u_1 \\ u_2 \\ \vdots \\ u_M \end{bmatrix}$$

or,

$$\begin{aligned} y &= f(\theta) + u \\ E(u) &= 0 \ , \quad E(uu') = (\Sigma \otimes I_T) \ . \end{aligned} \qquad (11.54)$$

We have condensed the notation so that $f_i(\theta_i)$ is a $(T \times 1)$ vector with typical element $f_i(x_{ti}\,;\,\theta_i)$, and θ is the $[(\Sigma_i\, p_i) \times 1]$ vector formed by stacking the elements of the θ_i's. In the following analysis it should be recalled that f is a function of the data as well as of θ. Now, treating each equation of the system separately, choosing θ_i to minimize the quantity

$$\frac{1}{T}[y_i - f_i(\theta_i)]'[y_i - f_i(\theta_i)] \qquad (11.55)$$

over Θ_i produces the non-linear OLS estimators, $\hat{\theta}_i$ $(i = 1, 2, \ldots, M)$. If the u_i's are assumed to be normally distributed then it is easily seen that $\hat{\theta}_i$ is also the maximum likelihood estimator of θ_i. The corresponding estimator of θ, $\hat{\theta}$, is obtained directly by stacking the $\hat{\theta}_i$'s, and the elements of Σ may be estimated by constructing

$$s_{ij} = \frac{1}{T}[y_i - f_i(\hat{\theta}_i)]'[y_j - f_j(\hat{\theta}_j)]$$

and $S = ((s_{ij}))$.

Taking account of the jointness of the equations in (11.54) the FGLS estimator, $\hat{\theta}_F$, is obtained by minimizing

$$\frac{1}{T}[y - f(\theta)]'(S^{-1} \otimes I_T)[y - f(\theta)] \tag{11.56}$$

with respect to θ over the parameter space, $\Theta = (\Theta_1 \times \Theta_2 \times \cdots \times \Theta_M)$. It should be kept in mind that the minimization of (11.56) or (11.55) will generally require the use of numerical approximation methods, and that $\hat{\theta}$ and $\hat{\theta}_F$ cannot be written in closed-form. Of course, this considerably limits the extent to which the finite-sample properties of these estimators can be described, but some general asymptotic results were discussed by Gallant. His results require a number of assumptions, including compactness of the parameter and sample spaces; the existence of continuous first and second derivatives of f_i with respect to the elements of θ_i; and the non-singularity of the matrix

$$\Omega = \begin{bmatrix} \sigma^{11}V_{11} & \cdots & \sigma^{1M}V_{1M} \\ \vdots & & \vdots \\ \sigma^{M1}V_{M1} & \cdots & \sigma^{MM}V_{MM} \end{bmatrix}^{-1} \tag{11.57}$$

where V_{ij} is the $(p_i \times p_j)$ matrix,

$$V_{ij} = \int \left[\frac{\partial f_i(x_i ; \theta_i)}{\partial \theta_i}\right] \left[\frac{\partial f_j(x_j ; \theta_j)}{\partial \theta_j'}\right] d\mu(x)$$

and μ is the limit of the measures of the sequence $\{x_t\}$.

Under these conditions Gallant (1975; pp. 45–48) established the strong consistency of $\hat{\theta}$ and $\hat{\theta}_F$ for θ, and showed that $\sqrt{T}(\hat{\theta} - \theta)$ and $\sqrt{T}(\hat{\theta}_F - \theta)$ converge in distribution to multivariate Normal with null mean vectors and (asymptotic) variance covariance matrices Δ and Ω respectively, where Ω is given by (11.57) and

$$\Delta = \begin{bmatrix} \Delta_{11} & \cdots & \Delta_{1M} \\ \vdots & & \vdots \\ \Delta_{M1} & \cdots & \Delta_{MM} \end{bmatrix} \tag{11.58}$$

with $\Delta_{ij} = \sigma_{ij}V_{ii}^{-1}V_{ij}V_{jj}^{-1}$ (i, j = 1, 2, ..., M).

From (11.57) and (11.58) it can be shown that $\hat{\theta}_F$ is asymptotically efficient relative to $\hat{\theta}$, unless Σ is diagonal and/or $f_i = f_j$ for all i, j, in which case the two estimators have equal asymptotic efficiency. Finally, to assist in the construction of asymptotic confidence intervals or tests, Ω can be estimated consistently by

$$\hat{\Omega} = \left[\frac{1}{T}\hat{F}'(S^{-1} \otimes I_T)\hat{F}\right]^{-1} \tag{11.59}$$

where

$$\hat{F} = \begin{bmatrix} \hat{F}_1 & 0 & \cdots & 0 \\ 0 & \hat{F}_2 & \cdots & 0 \\ \vdots & \vdots & & \vdots \\ 0 & 0 & \cdots & \hat{F}_M \end{bmatrix}$$

$$\hat{F}_i = \begin{bmatrix} \dfrac{\partial f_i(x_{1i}\,;\,\theta_i)}{\partial \theta_i'} \\ \vdots \\ \dfrac{\partial f_i(x_{Ti}\,;\,\theta_i)}{\partial \theta_i'} \end{bmatrix}_{\theta=\hat{\theta}_F} .$$

It should be noted that under the assumption of normally distributed disturbances, so that $\hat{\theta}_F$ is the (non-linear) ML estimator of θ, tests of general (non-linear, within or across equations) restrictions on the parameters could be constructed by appealing to the Wald, Lagrange multiplier or likelihood ratio principles in a manner analogous to that discussed in Section 10.2. Although the asymptotic distributions of the test statistics are the same under each of these approaches, the finite-sample properties of these tests are unexplored in the present context.

As noted already, the estimators $\hat{\theta}$ and $\hat{\theta}_F$ cannot be written in closed-form, and so their finite-sample properties also are unknown in general. A limited amount of Monte Carlo evidence is, however, available. Gallant (1975; p. 41) considered a two equation model in which

$$f_1(x_{t1}; \theta_1) = \theta_{11} + \theta_{12}x_{t11} + \theta_{13} \exp\{\theta_{14}x_{t12}\}$$

$$f_2(x_{t2}; \theta_2) = \theta_{21} + \theta_{22} \exp\{\theta_{23}x_{t21}\}$$

$$(t = 1, 2, \ldots, 50)$$

where one set of parameter values was considered and the experiment was replicated 450 times. For this particular model, Gallant found that the mean, skewness and kurtosis for the sampling distribution of $\hat{\theta}_F$ were essentially the same as those for the distribution of $\hat{\theta}$. However, $\hat{\theta}_F$ exhibited less dispersion than did $\hat{\theta}$. Of course, these results may be specific to this particular experimental design.

11.7 FURTHER RESEARCH

In this book we have attempted to summarize the existing literature relating to estimation and inference in the context of models which comprise a system of seemingly unrelated regression equations. We have tried to draw together this disparate literature in a coherent way, introduced some new results, and made a number of suggestions for possible future research in this general area. These suggestions appear in the last sections of several of the major chapters. However, there are a number of other interesting aspects of our topic which have received only scant attention to date and warrant brief mention.

Wherever we have made a specific distributional assumption regarding the disturbances of the SURE model it has been one of normality. This aspect of the model's specification was not questioned throughout our discussion or considered under the heading of specification analysis. In many applications, this assumption may be questionable, and its violation may have serious consequences for our inferences (see, e.g., Judge, Griffiths, Hill and Lee, 1980; Chapter 7). There is a need to address such matters as testing for normality, transformations to achieve normality, inference under alternative disturbance distributions, and the development of procedures which are robust to distributional misspecification in the context of the SURE model. A general framework for handling some of these matters might be based simply on the assumption of the existence of the disturbances' moments to a specified order (see Ullah, Srivastava and Chandra, 1983). The recent work by Sarkar and Krishnaiah (1984) also relates to this topic.

In certain situations the variance covariance matrix for the disturbances may be singular. This is a common occurrence with allocation models, such as those which arise in demand analysis (see Powell, 1969). In some cases this poses no problem as the deletion of one or more equations of the system may result in a non-singular variance covariance matrix, and permit unique parameter estimates to be obtained (including those in the deleted equation(s)) (see Giles and Hampton, 1986, for a generalization of this situation). A singular variance covariance matrix may also occur in other ways, and may not be so easily handled. Judge, Griffiths, Hill and Lee (1980; pp. 275-281) discuss some such situations and present a method of estimation when the matrix in question involves only known parameters. In the more likely case in which the variance covariance matrix is unknown, Court's (1974) analysis of the three stage least squares estimator is relevant, given the connection between this estimator and conventional SURE estimators. However, further finite-sample analysis of this problem would be in order.

Other frontiers remain, such as prediction procedures, general stochastic regressors, the use of unobservable variables, the efficient use of prior information, and diagnostic checking. In these and other areas there is room for considerably more research effort. An emphasis on results which have relevance in samples of the size actually encountered in practice, rather than relying on asymptotic behaviour; and on general analytical results, rather than those based on limited Monte Carlo experiments, must be given a high priority. Analytical finite-sample results relating to various combinations of some of the problems that we have discussed, such as SURE models with disturbances which are both autocorrelated and heteroscedastic; models which are misspecified with regard to both the regressor variables and the disturbance process, etc., remain as matters for further research which would bring the discussion of estimation and inference in the SURE model closer to the reality of applied econometric analysis.

EXERCISES

11.1 Consider a two equation SURE model with random coefficients in which the first equation contains only one regressor while the second equation has two regressors. Assuming that the regressor in the first equation also occurs in the second equation, obtain the feasible estimators of the means of the random coefficients and examine their asymptotic properties.

11.2 In a SURE model, suppose that some of the coefficients in each equation are random while the others are fixed. Formulate the model and develop an asymptotically efficient estimation procedure.

11.3 Consider the following SURE model for which there are $(T + T^* + T^{**})$ observations and two equations:

$$\begin{bmatrix} y_1 \\ y_1^* \\ y_1^{**} \end{bmatrix} = \begin{bmatrix} X_1 \\ X_1^* \\ X_1^{**} \end{bmatrix} \beta_1 + \begin{bmatrix} u_1 \\ u_1^* \\ u_1^{**} \end{bmatrix}$$

$$\begin{bmatrix} y_2 \\ y_2^* \\ y_2^{**} \end{bmatrix} = \begin{bmatrix} X_2 \\ X_2^* \\ X_2^{**} \end{bmatrix} \beta_2 + \begin{bmatrix} u_2 \\ u_2^* \\ u_2^{**} \end{bmatrix} .$$

Examine the efficiency of the OLS and GLS estimators of the coefficients when y_2^* and X_2^{**} are unavailable.

11.4 Consider the two equation SURE model specified by (11.34) and suppose that the disturbances follow a bivariate Normal distribution with mean vector 0 and known variance covariance matrix Σ with both of the diagonal elements unity and the off-diagonal elements equal to ρ. Assuming that X_{12}^* is unavailable, and treating its elements as unknown parameters, obtain the maximum likelihood estimators of the coefficients and show that they are identical to the corresponding GLS estimators applied to the complete observations only. Compare the efficiencies of these estimators with the GLS estimators that could have been obtained, if X_{12}^* were known. [Hint: See Kmenta (1981; pp. 252-253).]

11.5 In a SURE model

$$y_i = X_i\beta_i + \alpha_i e_T + u_i \qquad (i = 1, 2, \ldots, M)$$

where e_T is a $(T \times 1)$ vector with all elements unity, the disturbances are known to be heteroscedastic. Suggest a suitable measure of goodness-of-fit for this model and indicate its properties. [Hint: See Buse (1979).]

11.6 In a dynamic version of the SURE model, suppose that the disturbances are generated by a first-order autoregressive process. Assuming normality of the disturbances and using the Lagrange multiplier principle, obtain an asymptotically valid test of the hypothesis that the disturbances are serially independent. [Hint: See Harvey (1981; pp. 298-299).]

11.7 Consider the following nonlinear SURE model:

$$y_1 = \beta_1 + \exp\{\beta_1\}x + u_1$$
$$y_2 = \beta_2 + u_2 .$$

Determine the maximum likelihood estimators of β_1 and β_2, and show that they are consistent. Are the estimators still consistent when $x' = (1, 2, \ldots, t, \ldots, T)$?

11.8 In the two equation SURE model

$$y_1 = x_1\beta_1 + u_1$$
$$y_2 = x_2\beta_1^2 + u_2 \quad ,$$

suggest an appropriate method of estimating β_1 assuming that

$$\Sigma = \begin{bmatrix} \sigma^2 & 2\sigma^2 \\ 2\sigma^2 & 8\sigma^2 \end{bmatrix} .$$

11.9 Provide some examples where the variance covariance matrix of the disturbances in a SURE model may be singular. Develop a theory for the estimation of the parameters in such models. [Hint: See Judge, Griffiths, Hill and Lee (1980; pp. 275-280).]

REFERENCES

Battese, G. E., and W. E. Griffiths (1980). "On R^2-statistics for the general linear model with non-scalar covariance matrix," Australian Economic Papers 19, 343-348.

Bewley, R. A. (1985). "Goodness-of-fit for allocation models," Economics Letters 17, 227-229.

Buse, A. (1979). "Goodness-of-fit in the seemingly unrelated regressions model: a generalization," Journal of Econometrics 10, 109-113.

Conniffe, D. (1985). "Estimating regression equations with common explanatory variables but unequal number of observations," Journal of Econometrics 27, 179-196.

Court, R. H. (1974). "Three stage least squares and some extensions where the structural disturbance covariance matrix may be singular," Econometrica 42, 547-558.

Dhrymes, P. J., and J. B. Taylor (1976). "On an efficient two-step estimator for dynamic simultaneous equations models with autoregressive errors," International Economic Review 17, 362-376.

Gallant, A. R. (1975). "Seemingly unrelated nonlinear regressions," Journal of Econometrics 3, 35-50.

Giles, D. E. A. (1986). "Missing measurements and estimator inefficiency in linear regression: a generalization," Journal of Quantitative Economics 2, 87-91.

Giles, D. E. A., and P. Hampton (1986). "A regional consumer demand model for New Zealand," forthcoming in Journal of Regional Science.

Giles, D. E. A., and C. K. Low (1981). "Choosing between alternative structural equations estimated by instrumental variables," Review of Economics and Statistics LXIII, 476-478.

Griffiths, W. E., R. G. Drynan and S. Prakash (1979). "Bayesian estimation of a random coefficient model," Journal of Econometrics 10, 201-220.

Harvey, A. C. (1981). The Econometric Analysis of Time Series (Philip Allan, Oxford).

Hatanaka, M. (1974). "An efficient two-step estimator for the dynamic adjustment model with autoregressive errors," Journal of Econometrics 2, 199-220.

Judge, G. G., W. E. Griffiths, R. C. Hill and T. S. Lee (1980). The Theory and Practice of Econometrics (Wiley, New York).

Kmenta, J. (1981). "On the problem of missing measurements in the estimation of economic relationships," in E. G. Charatsis (ed.), Proceedings of the Econometric Society European Meeting, 1979: Selected Econometric Papers in Memory of Stefan Valavanis (North-Holland, Amsterdam), 233-257.

Kmenta, J., and R. F. Gilbert (1968). "Small sample properties of alternative estimators of seemingly unrelated regressions," Journal of the American Statistical Association 63, 1180-1200.

Low, C. K. (1980). "Selection criteria for simultaneous equations models," Working Paper No. 6/80, Department of Econometrics and Operations Research, Monash University, Melbourne.

Maddala, G. S. (1977). Econometrics (McGraw-Hill, New York).

McElroy, M. B. (1977). "Goodness of fit for seemingly unrelated regressions: Glahn's $R^2_{y.x}$ and Hooper's $\bar{r}^2$," Journal of Econometrics 6, 381-387.

Parks, R. W. (1967). "Efficient estimation of a system of regression equations when disturbances are both serially and contemporaneously correlated," Journal of the American Statistical Association 62, 500-509.

Powell, A. A. (1969). "Aitken estimators as a tool in allocating predetermined aggregates," Journal of the American Statistical Association 64, 913-922.

Raj, B., and A. Ullah (1981). Econometrics: A Varying Coefficients Approach (Croom Helm, London).

Sarkar, S., and P. R. Krishnaiah (1984). "Estimation of parameters under correlated regression equations model," mimeo, University of Pittsburgh, Pittsburgh.

Schmidt, P. (1974). "A note on Theil's minimum standard error criterion when the disturbances are autocorrelated," Review of Economics and Statistics LVI, 122-123.

Schmidt, P. (1977). "Estimation of seemingly unrelated regressions with unequal number of observations," Journal of Econometrics 5, 365-377.

Singh, B., and A. Ullah (1974). "Estimation of seemingly unrelated regressions with random coefficients," Journal of the American Statistical Association 69, 191-195.

Spencer, D. E. (1979). "Estimation of a dynamic system of seemingly unrelated regressions with autoregressive disturbances," Journal of Econometrics 10, 227-241.

Srivastava, V. K., G. D. Mishra and A. Chaturvedi (1981). "Estimation of linear regression model with random coefficients ensuring almost non-negativity of variance estimator," Biometrical Journal 23, 3-8.

Swamy, P. A. V. B., and J. S. Mehta (1975). "On Bayesian estimation of seemingly unrelated regressions when some observations are missing," Journal of Econometrics 3, 157-169.

Theil, H. (1970). Economic Forecasts and Policy (North-Holland, Amsterdam).

Ullah, A., V. K. Srivastava and R. Chandra (1983). "Properties of shrinkage estimators in linear regression when disturbances are not normal," Journal of Econometrics 21, 389-402.

Wallace, T. D., and J. L. Silver (1984). "Yet another note on estimating seemingly unrelated regressions," mimeo, Department of Economics, Duke University, Durham.

Appendix

A. SOME MULTIVARIATE RESULTS

Suppose that u is a $(p \times 1)$ random vector following a multivariate Normal distribution $N(0, \Sigma)$ with mean vector 0 and variance covariance matrix Σ, assumed to be nonsingular. Its probability density function is given by

$$f(u) = (2\pi)^{-p/2} |\Sigma|^{-1/2} \exp\left\{-\frac{1}{2}u'\Sigma^{-1}u\right\},$$

and the characteristic function is

$$\phi(h) = \exp\left\{-\frac{1}{2}h'\Sigma^{-1}h\right\}$$

where h is any $(p \times 1)$ real vector.

Further, suppose S is a $(p \times p)$ symmetric and positive definite random matrix following a Wishart distribution $W_p(n, \Sigma)$ with n degrees of freedom and associated matrix Σ, then its probability density function is given by

$$f(S) = \left[2^{np/2}\pi^{p(p-1)/4}|\Sigma|^{n/2}\prod_{i=1}^{p}\Gamma\left(\frac{n-i+1}{2}\right)\right]^{-1}|S|^{(n-p-1)/2}\exp\left\{-\frac{1}{2}\mathrm{tr}(S\Sigma^{-1})\right\}$$

and the characteristic function is

$$\phi(H) = |I_p - \sqrt{-1}\,H\Sigma|^{-n/2}$$

where H is any $(p \times p)$ symmetric matrix with real elements.

Results A.1: If A and B are nonstochastic matrices, then

(i) $u'Au$ has a χ^2 distribution with degrees of freedom equal to the rank of $A\Sigma$ if and only if $A\Sigma A = A$. In particular, when $\Sigma = I_p$, the quantity $u'Au$ has a χ^2 distribution with degrees of freedom as the rank of A if and only if A is idempotent.

(ii) $u'Au$ and $u'Bu$ are stochastically independent if and only if $A\Sigma B$ is a null matrix.

(iii) $u'Au$ and Bu are stochastically independent if and only if $A\Sigma B'$ is a null matrix.

Proof: See Rao (1973; pp. 186-188).

Results A.2: For the Wishart distribution, we have the following results:

(i) If A is any nonstochastic matrix of order $(p^* \times p)$, the distribution of ASA' is $W_{p^*}(n, A\Sigma A')$.

(ii) $$E(S) = n\Sigma ,$$

$$E(S^{-1}) = \frac{1}{(n-p-1)}\Sigma^{-1} \quad \text{provided that} \quad n > (p+1) .$$

(iii) If S and Σ are conformably partitioned as

$$S = \begin{bmatrix} s & s'_{[1]} \\ s_{[1]} & S_{[1]} \end{bmatrix}\begin{matrix}(q)\\(p-q)\end{matrix} \qquad \Sigma = \begin{bmatrix} \sigma & \sigma'_{[1]} \\ \sigma_{[1]} & \Sigma_{[1]} \end{bmatrix}\begin{matrix}(q)\\(p-q)\end{matrix}$$

$$\qquad (q) \quad (p-q) \qquad\qquad\qquad (q) \quad (p-q)$$

we have

(a) $S_{[1]} \sim W_{(p-q)}(n, \Sigma_{[1]})$.

(b) $(s - s'_{[1]}S_{[1]}^{-1}s_{[1]}) \sim W_q(n-p+q, \sigma - \sigma'_{[1]}\Sigma_{[1]}^{-1}\sigma_{[1]})$.

(c) $(s - s'_{[1]} S^{-1}_{[1]} s_{[1]})$ and $s'_{[1]} S^{-1}_{[1]}$ are stochastically independent.

(d) The conditional distribution of $s'_{[1]} S^{-1}_{[1]}$ given $S_{[1]}$ is

$$N(\sigma'_{[1]} \Sigma^{-1}_{[1]}, (\sigma - \sigma'_{[1]} \Sigma^{-1}_{[1]} \sigma_{[1]}) \otimes S^{-1}_{[1]}) \ .$$

(iv) If ℓ_2 denotes the largest characteristic root of the (2×2) matrix $(2S/\mathrm{tr}\ S)$ with S following $W_2(n, I_2)$, then the probability density function of ℓ_2 is

$$(n - 1)(\ell_2 - 1)[\ell_2(2 - \ell_2)]^{(n-3)/2} \qquad (1 \leq \ell_2 \leq 2) \ .$$

Proof: See Kshirsagar (1972; pp. 64, 72, 108–111) for (i), (ii) and (iii) respectively. For result (iv), we observe from Kshirsagar (1972; p. 441) that the joint probability density function of the two characteristic roots f_1 and f_2 of S is given by

$$g(f_1, f_2) = \pi^{1/2}\left[2^n\Gamma\left(\frac{n}{2}\right)\Gamma\left(\frac{1}{2}(n-1)\right)\right]^{-1}(f_1f_2)^{(n-3)/2}|f_1 - f_2| \exp\left\{-\frac{1}{2}(f_1 + f_2)\right\} .$$

Applying the transformation

$$f_1 = \ell_1 \ell_2$$
$$f_2 = \ell_1(2 - \ell_2)$$

and employing the recurrence relationship

$$\sqrt{\pi}\Gamma(n) = 2^{n-1}\Gamma\left(\frac{n}{2}\right)\Gamma\left(\frac{n+1}{2}\right) ,$$

we obtain the joint probability density function of ℓ_1 and ℓ_2 as

$$g(\ell_1, \ell_2) = (n-1)(\ell_2 - 1)[\ell_2(2 - \ell_2)]^{(n-3)/2}\ell_1^{(n-1)} \exp\{(-\ell_1)/\Gamma(n)\} \ .$$

Integrating out ℓ_1 we obtain the marginal probability density function of ℓ_2.

B. SOME USEFUL EXPECTATIONS

Suppose that $u_1, u_2, \ldots, u_M$ are $(T \times 1)$ random vectors following multivariate Normal distributions with null mean vectors. Further, suppose that

$$E(u_i u_j') = \sigma_{ij} I_T \qquad (i, j = 1, 2, \ldots, M) .$$

Results B1: If A_1 and A_2 are nonstochastic matrices of order $(T \times T)$, we have

$$E[u_i'A_1u_j \cdot u_g u_h'] = \sigma_{ij}\sigma_{gh}(\text{tr } A_1) I_T + \sigma_{ig}\sigma_{jh}A_1 + \sigma_{ih}\sigma_{jg}A_1' , \tag{B.1.1}$$

$$\begin{aligned} E[u_i'A_1u_j \cdot u_k'A_2u_\ell \cdot u_g u_h'] = {} & \sigma_{ij}(\text{tr } A_1)[\sigma_{k\ell}\sigma_{gh}(\text{tr } A_2) I_T + \sigma_{gk}\sigma_{h\ell}A_2 + \sigma_{g\ell}\sigma_{hk}A_2'] \\ & + \sigma_{ig}A_1[\sigma_{k\ell}\sigma_{jh}(\text{tr } A_2) I_T + \sigma_{jk}\sigma_{h\ell}A_2 + \sigma_{j\ell}\sigma_{hk}A_2'] \\ & + \sigma_{ih}[\sigma_{k\ell}\sigma_{jg}(\text{tr } A_2) I_T + \sigma_{gk}\sigma_{j\ell}A_2 + \sigma_{jk}\sigma_{g\ell}A_2']A_1' \\ & + \sigma_{i\ell}[\sigma_{jk}\sigma_{gh}(\text{tr } A_1A_2) I_T + \sigma_{jg}\sigma_{hk}A_1'A_2' + \sigma_{jh}\sigma_{gk}A_2A_1] \\ & + \sigma_{ik}[\sigma_{j\ell}\sigma_{gh}(\text{tr } A_1'A_2) I_T + \sigma_{jg}\sigma_{h\ell}A_1'A_2 + \sigma_{jh}\sigma_{g\ell}A_2'A_1] . \end{aligned} \tag{B.1.2}$$

Proof: These results can be obtained easily from the following recurrence relationship derived by Srivastava and Tiwari (1976), for any m nonstochastic matrices $A_1, A_2, \ldots, A_m$:

$$\begin{aligned} & E\left[\left(\prod_{k=1}^{m} u_{i_k}' A_k u_{j_k}\right) u_g u_h'\right] \\ & = \sigma_{i_1 j_1}(\text{tr } A_1) E\left[\left(\prod_{k=2}^{m} u_{i_k}' A_k u_{j_k}\right) u_g u_h'\right] + \sigma_{i_1 g} A_1 E\left[\left(\prod_{k=2}^{m} u_{i_k}' A_k u_{j_k}\right) u_{j_1} u_h'\right] \\ & \quad + \sigma_{i_1 h} E\left[\left(\prod_{k=2}^{m} u_{i_k}' A_k u_{j_k}\right) u_g u_{j_1}'\right] A_1' + \sum_{n=2}^{m} \sigma_{i_1 j_n} E\left[u_{j_1}' A_1' A_n' u_{i_n} \left(\prod_{\substack{k=2 \\ k \neq n}}^{m} u_{i_k}' A_k u_{j_k}\right) u_g u_h'\right] \\ & \quad + \sum_{n=2}^{m} \sigma_{i_1 i_n} E\left[u_{j_1}' A_1' A_n u_{j_n} \left(\prod_{\substack{k=2 \\ k \neq n}}^{m} u_{i_k}' A_k u_{j_k}\right) u_g u_h'\right] . \end{aligned}$$

Putting $m = 1$, we find that

$$E[u_i'A_1u_j \cdot u_g u_h'] = \sigma_{ij}(\text{tr } A_1) E(u_g u_h') + \sigma_{ig} A_1 E(u_j u_h') + \sigma_{ih} E(u_g u_j') A_1'$$

which leads directly to the first result.

Similarly, setting m = 2, we get

$$
\begin{aligned}
E[u_i'A_1u_j \cdot u_k'A_2u_\ell \cdot u_gu_h'] &= \sigma_{ij}(\mathrm{tr}\,A_1)E(u_k'A_2u_\ell \cdot u_gu_h') + \sigma_{ig}A_1E(u_k'A_2u_\ell \cdot u_ju_h') \\
&\quad + \sigma_{ih}E(u_k'A_2u_\ell \cdot u_gu_j')A_1' + \sigma_{i\ell}E(u_j'A_1'A_2'u_k \cdot u_gu_h') \\
&\quad + \sigma_{ik}E(u_j'A_1'A_2u_\ell \cdot u_gu_h') \\
&= \sigma_{ij}(\mathrm{tr}\,A_1)[\sigma_{k\ell}\sigma_{gh}(\mathrm{tr}\,A_2)I_T + \sigma_{gk}\sigma_{h\ell}A_2 + \sigma_{g\ell}\sigma_{hk}A_2'] \\
&\quad + \sigma_{ig}A_1[\sigma_{k\ell}\sigma_{jh}(\mathrm{tr}\,A_2)I_T + \sigma_{jk}\sigma_{h\ell}A_2 + \sigma_{j\ell}\sigma_{hk}A_2'] \\
&\quad + \sigma_{ih}[\sigma_{k\ell}\sigma_{jg}(\mathrm{tr}\,A_2)I_T + \sigma_{gk}\sigma_{j\ell}A_2 + \sigma_{jk}\sigma_{g\ell}A_2']A_1' \\
&\quad + \sigma_{i\ell}[\sigma_{jk}\sigma_{gh}(\mathrm{tr}\,A_1A_2)I_T + \sigma_{jg}\sigma_{hk}A_1'A_2' + \sigma_{jh}\sigma_{gk}A_2A_1] \\
&\quad + \sigma_{ik}[\sigma_{j\ell}\sigma_{gh}(\mathrm{tr}\,A_1'A_2)I_T + \sigma_{jg}\sigma_{h\ell}A_1'A_2 + \sigma_{jh}\sigma_{g\ell}A_2'A_1],
\end{aligned}
$$

which provides the second result.

Results B.2: Suppose that χ^2 is a scalar random variable following a χ^2 distribution with ν degrees of freedom and w is a $(m \times 1)$ random vector following a multivariate Normal distribution with null mean vector and variance covariance matrix I_m. If χ^2 is independent of the elements of w, we have

$$
E\left[\frac{(w'w)^a}{(\chi^2 + w'w)^a}(ww')\right] = \left[\frac{\Gamma\left(\frac{m+\nu}{2}+1\right)\Gamma\left(\frac{m}{2}+a+1\right)}{\Gamma\left(\frac{m+\nu}{2}+a+1\right)\Gamma\left(\frac{m}{2}+1\right)}\right] I_m
$$

$$
E\left[\frac{(w'w)^a}{(\chi^2 + w'w)^{a+1}}(ww')\right] = \left[\frac{\Gamma\left(\frac{m+\nu}{2}\right)\Gamma\left(\frac{m}{2}+a+1\right)}{2\Gamma\left(\frac{m}{2}+1\right)\Gamma\left(\frac{m+\nu}{2}+a+1\right)}\right] I_m
$$

where a is any positive integer.

Proof: Using Theorem 3 of Judge and Bock (1978; p. 323), we observe that

$$
E\left[\frac{(w'w)^a}{(\chi^2 + w'w)^a}(w'w)\right] = E\left(\frac{\chi_1^2}{\chi^2 + \chi_1^2}\right)^a I_m
$$

where χ_1^2 is a random variable following a χ^2 distribution with $(m + 2)$ degrees of freedom. As χ^2 and χ_1^2 are stochastically independent, the quantity $\chi_1^2/(\chi^2 + \chi_1^2)$ follows a Beta distribution with parameters $\left(\frac{m}{2} + 1\right)$ and $\left(\frac{\nu}{2}\right)$, and hence its a'th moment is

$$\frac{\Gamma\left(\frac{m+\nu}{2} + 1\right)\Gamma\left(\frac{m}{2} + a + 1\right)}{\Gamma\left(\frac{m+\nu}{2} + a + 1\right)\Gamma\left(\frac{m}{2} + 1\right)},$$

which leads to the first result.

Using, additionally, the stochastic independence of $(\chi^2 + \chi_1^2)$ and $\chi_1^2/(\chi^2 + \chi_1^2)$, we have

$$E\left[\frac{(w'w)^a}{(\chi^2 + w'w)^{a+1}} ww'\right] = E\left[\frac{(\chi_1^2)^a}{(\chi^2 + \chi_1^2)^{a+1}}\right] I_m$$

$$= E\left(\frac{\chi_1^2}{\chi^2 + \chi_1^2}\right)^a E\left(\frac{1}{\chi^2 + \chi_1^2}\right) I_m$$

$$= \frac{\Gamma\left(\frac{m+\nu}{2} + 1\right)\Gamma\left(\frac{m}{2} + a + 1\right)}{\Gamma\left(\frac{m+\nu}{2} + a + 1\right)\Gamma\left(\frac{m}{2} + 1\right)} \frac{1}{(m + \nu)} I_m$$

$$= \frac{\Gamma\left(\frac{m+\nu}{2}\right)\Gamma\left(\frac{m}{2} + a + 1\right)}{2\Gamma\left(\frac{m+\nu}{2} + a + 1\right)\Gamma\left(\frac{m}{2} + 1\right)} I_m,$$

which provides the second result.

REFERENCES

Judge, G. G., and M. E. Bock (1978). The Statistical Implications of Pre-Test and Stein-Rule Estimators in Econometrics (North-Holland, Amsterdam).

Kshirsagar, A. M. (1972). Multivariate Analysis (Marcel Dekker, New York).

Rao, C. R. (1973). Linear Statistical Inference and its Applications (Wiley, New York).

Srivastava, V. K., and R. Tiwari (1976). "Evaluation of expectations of products of stochastic matrices," Scandinavian Journal of Statistics 3, 135-138.

Index

Milton Keynes UK
Ingram Content Group UK Ltd.
UKHW020015071024
449327UK00031B/2792